I0796950

BORNEO
Capital
Important City
Small city or town
Airport
Other place of interest
Parks and reserves
metres
feet
SULU SEA
CELEBES SEA
MAKASSAR STRAIT
JAVA SEA
MALAYSIA
BRUNEI
INDONESIA
SARAWAK
SABAH
EAST KALIMANTAN
WEST KALIMANTAN
CENTRAL KALIMANTAN
SOUTH KALIMANTAN
Kota Kinabalu
Bandar Seri Begawan
Kuching
Nusantara
Kinabalu Park
Crocker Range National Park
Sepilok Wildlife Reserve
Kinabatangan Wildlife Sanctuary
Tabin Wildlife Reserve
Danum Valley
Maliau Basin
Ulu Temburong National Park
Lambir Hills National Park
Niah Caves National Park
Similajau National Park
Gunung Mulu National Park
Kayan Mentarang National Park
Kubah National Park
Bako National Park
Gunung Gading National Park
Batang Ai National Park
Betung Karimun National Park
Kutai National Park
Bukit Baka National Park
Gunung Palung National Park
Tanjung Puting National Park
Layang-Layang Island
Pulau Banggi Island
Labuan Island
Kudat
Kota Belud
Tuaran
Papar
Tambunan
Tenom
Sandakan
Gomantong Caves
Lahad Datu
Semporna
Tawau
Mabul Island
Kapalai Island
Sipadan Island
Miri
Niah Caves
Mulu Caves
Ba'Kelalan
Bario
Bintulu
Tarakan
Sibu
Kapit
Damai
Matang
Sangkulirang
Pontianak
Samarinda
Balikpapan
Ketapang
Palangkaraya
Sampit
Sukamara
Banjarmasin
Rejang
Kapuas
Melawi
Mahakam
Barito
Sampit
Katingan
Padas
Kinabatangan

A Field Guide to the
Reptiles
of
Borneo

A Field Guide to the Reptiles of Borneo

Indraneil Das

PRINCETON UNIVERSITY PRESS
PRINCETON AND OXFORD

This work is dedicated to the memory of Dr Lim Boo Liat, pioneering zoologist and herpetologist.

Published in 2026 in the United States and Canada by
Princeton University Press
41 William Street, Princeton, New Jersey 08540
press.princeton.edu

First published in the United Kingdom in 2025 by
John Beaufoy Publishing Ltd, 11 Blenheim Court,
316 Woodstock Road, Oxford OX2 7NS, U.K.
www.johnbeaufoy.com

This edition of *A Field Guide to the Reptiles of Borneo* is published by arrangement with John Beaufoy Publishing Ltd.

ISBN 978-0-691-28122-3

Library of Congress Control Number - 2025933883

Photo captions
Title page: *Boiga cynodon* © Indraneil Das
Cover images: (front): *Chrysopelea paradisi* © Chien C. Lee; (back): *Crocodylus porosus*, *Cyclemys enigmatica*, *Gekko gecko* © Indraneil Das; (spine): *Gonocephalus bornensis © Indraneil Das*

Editor and indexer: Krystyna Mayer
Designer: Nigel Partridge
Project Manager: Rosemary Wilkinson
Cover designer: Wanda España

Printed and bound in Malaysia by
Times Offset (M) Sdn. Bhd.

10 9 8 7 6 5 4 3 2 1

Photo Credits:
Main descriptions: photos are denoted by a page number, followed by t (top), m (middle), b (bottom), l (left) or r (right), if relevant.

Amin Jais/Forest Department Sarawak: 152; Christopher Austin: 145 (t); Hamadnurrifat Bin Mohd Azam: 280; Joseph Charles: 250; Shavez Cheema: 228 (t); Jin Chen: 253, 254 (t); Sue Clark: 244; Chua Ee Kiam: 81; Indraneil Das: 18, 21, 23, 24, 25, 26, 27, 28, 29, 30, 32, 33, 34, 35, 37, 38, 39, 40, 41, 42, 43, 44, 48, 49, 50, 51, 52, 53, 55, 57, 58, 59, 60, 61, 62, 63, 64, 65, 66, 67, 68, 69, 70, 71, 75, 76, 77, 78, 83, 85, 86, 87, 90, 91, 92, 93, 94, 95, 96, 98, 99, 100, 101, 102, 103, 104, 106, 107, 108, 109, 110, 111, 113, 115, 116, 117, 118, 119, 120, 121, 122, 124, 125, 126, 127, 129, 130, 131, 132, 133, 134, 135, 136, 138, 139, 140, 142, 143, 144, 147, 148, 149, 150, 153; 154, 155, 156, 162, 164, 166, 167, 170, 173, 174, 176, 177, 179, 180, 181, 182, 184, 185, 187, 188, 189, 190, 191, 193, 194, 195, 196, 199, 200, 201, 202, 203, 204, 206, 207, 208, 209, 210, 211, 212, 215, 217, 218, 219, 221, 223, 227 (b), 228 (b), 229, 230, 232, 233, 234, 235, 236, 238, 239, 241, 242, 245, 246, 247, 248, 249, 251, 252, 255, 257, 258, 259, 262, 263, 265, 266, 267, 268, 269, 271, 272, 275, 278, 281, 282, 284, 285, 286, 287, 288, 289, 290, 292, 293, 296, 298, 299, 300, 301, 302, 303, 304, 307, 317, 319, 321, 326, 329, 330, 331, 332, 333, 335, 336, 337, 340, 341, 345, 346, 348, 349, 350, 351, 352, 353, 354, 355, 356, 357, 359, 360, 362; Patrick David: 277; Esther Dondorp / RMNH: 160, 178; Riki Subargia Empunak: 322, 334; Xavier Fenoy: 294; John G. Frazier: 36; Ibuki Fukuyama / Magnolia Press: 151; Maren Gaulke: 256; Ruzaini bin Ghazali/USNM: 84; Claire Goiran: 327; Philip Griffin: 146; Robert Inger: 243, 261; James Jolokia: 20; Benjamin Karin: 186; Desita Dyah D. A. Kusumaningrum: 270; Thornton Lawson: 74; Björn Lardner: 72, 73, 79, 342 (b); 263; Chien C. Lee: 56, 60, 168, 291, 342 (t), 344; Kelvin Kok Ping Lim: 198, 216, 224 (t), 225, 227 (t), 306, 310; 316, 318, 324; Norman Lim: 273; Aaron Lobo: 311, 312, 313, 314, 315, 320, 323, 325; Ulrich Manthey: 47, 240, 276; Joshua Matta, FMNH: 89; 141, 161, 163, 165, 171, 183; Andy Paul: 260; Pui Yong Min: 105, 343; Neil Rowntree: 364; Benjamin Schweinhart / Tremarctos Photography: 264; Showichi Sengoku: 123; Jeet Sukumaran: 297; Tan Heok Hui: 145 (b), 338; Artur Tomaszek 112; Nicholas Vidal / MNHN: 172; Gernot Vogel: 205, 237, 254 (b).

Contents

Introduction

Borneo is one of four major islands of the great Indo-Malayan Archipelago (along with Sumatra, Java and Sulawesi), and is situated on the eastern edge of the Sunda Shelf, stretching between coordinates 04° S–07° N and 109–119° E. It is the second largest tropical island in the world (after New Guinea), covering a land area of approximately 743,380km^2. North-west of Borneo is the tip of Peninsular Malaysia, separated by the Natuna Sea. Further away is the southern tip of Indo-China, separated from Borneo by the South China Sea. East of Borneo is Sulawesi, separated by Selat Makasar, and to the north-east is a series of islands, most of which are of the Sulu Group. The Java Sea is located to the south of Borneo, extending south of the island of Java.

Politically, a major part of this continental island falls in the Indonesian part of Kalimantan (area: 539,460km^2), the rest within the east Malaysian states of Sarawak (124,450km^2) and Sabah (73,710km^2). Inset within Sarawak is the independent Sultanate of Negara Brunei Darussalam (5,760km^2).

Geology & Landscape

Described as a geological composite, Borneo sits on the eastern margin of the Sunda Shelf, a Lauresian continental plate. During the Pleistocene glaciation, sea levels fell to 120–200m below current levels, uniting the islands of the Sundas.

Several of the largest rivers of South-east Asia flow through Borneo, some forming important biogeographic limits. All of Sarawak's rivers drain into the South China Sea. Another major river in Sarawak, the Sungei Rejang, drained primary forests of northwestern Borneo (the area seriously logged at present). The important rivers of Sabah include the Segama and the Kinabatangan. Kalimantan's great rivers include the Kapuas, Barito, Kahayan, Kayan and Mahakam. Many of the coastal rivers are broad and meandering. The Kapuas, for instance, drops only 50m between Putussibau, about 900km inland from the sea.

Cave systems represented on Borneo include typically isolated limestone outcrops in coastal Sarawak, northwestern Sabah and southeastern Kalimantan. Arguably the most famous of these are the Niah Caves of the Subis massif, which have yielded 40,000-year-old human remains.

Climate

The climatic change coinciding with the last glacial maxima, about 23–18,000 years ago, is associated with changes in the Earth's vegetation as well fauna, the climate of South-east Asia being relatively more dry, cool as well as seasonal. The period coincided with lowered sea levels that joined the Sunda islands to the South-east Asian mainland, potentially permitting the emigration of biota from both directions.

Borneo lies within the tropics, and the equator crosses this great island approximately over the city of Pontianak in Kalimantan Barat. The island is characterized by high, equitable temperature and heavy rainfall spread virtually throughout the year. The relatively wetter periods are observed during the passage of the Northeast Monsoons (November–April), although the Southwest Monsoons

(April–August) also bring rainfall to the area. Within Borneo, the western and central parts are wetter than the eastern parts. The South China Sea to the north is markedly affected by the passage of the monsoon winds, while the Sulu Sea and Selat Makasar rarely experience large waves. Daytime temperatures in most parts of the low-lying areas are 30–32°C, and humidity is typically above 70 per cent. On the other hand, temperatures in the highlands, such as Gunung Kinabalu, can be low. Annual precipitation is in the range 4,000–5,000mm. However, localized effects of geomorphological features, including mountain massifs and bays, have significant effect on weather conditions, and therefore on the local biota within the island. A major influence of short-term climatic fluctuations on Borneo and across the Indo-Pacific region is the Pacific El Niño, or the Southern Oscillation Cycle, which is known to bring uncontrolled forest fires, leading to haze, water shortage, and loss of crops, land and life. The advent of the Northeast Monsoons in December–February brings much precipitation to the area (the season, in the vernacular, is referred to as 'Landas').

Vegetation

The plant life of Borneo is characterized by high species diversity and low endemism, the latter attributed to the land connection during the lowered sea levels of the Pleistocene epoch. For instance, the Kinabalu flora itself numbers as many as 5,000–6,000 species, comprising more than 1,000 genera and 200 families. The richest forests of South-east Asia are in northwestern Borneo, north of Sungei Kapuas in the west, encompassing the north-east portion of Sarawak, Brunei and the southwestern lowlands of Sabah, an area that served as a Pleistocene refugium. The region is geologically relatively recent, and has a rugged topography, comprising mostly infertile shallow soils, compared to other areas of Borneo. The dipterocarp-dominated rainforests of South-east Asia developed in the Miocene, replacing monsoon forests, which were widespread in the Sundas during the Oligocene and early Miocene epochs.

Vegetational zonation for Borneo is arguably best known from Gunung Kinabalu, which has a largely intact vegetation. The lowland forest is six layered, with emergent trees and sparse undergrowth. At about 1,200m is the upper boundary of lowland rainforest (where the majority of emergent trees comprises primarily the dipterocarps). The lower montane forest is five layered, lacking emergents. The upper limit of the lower montane forest is 2,000–2,350m, that of the upper montane forest between 2,800 and 3,000m. The upper montane forest has a dense herbaceous layer. The upper limit of the lower subalpine coniferous forests is 3,400m; it is sparse in undergrowth and lower in height. Fragmented upper subalpine forests occur at an altitude of 3,700m. Above this level is a zone of alpine rock desert, with scattered communities of alpine scrub, and it has been suggested that the tree line may coincide with the lowest elevation, where nocturnal ground frost is frequent.

Important forest types are briefly described here. These include mixed dipterocarp forest in the yellow-red soils in the uplands, dominated by either *Dryobalanops lanceolata* or *Shorea parviflora* (often also with *Shorea macroptera*) or *Anisoptera grossivenia*, as well as *Dryobalanops aromatica* and *Dipterocarpus globosus*. The lower montane and upper montane forests of Borneo are also remarkable in their structure and composition. Canopy height is reduced, sometimes to 18–30m, with few emergent trees, and buttressed trees being less common; there is an absence of large woody climbers and a greater abundance of vascular epiphytes. Moss vegetation, with an abundance of bryophytes, in addition to gnarled trees, characterizes the upper limits of montane forests. Upper montane forests also have trees with small, leathery leaves, and conifers (including *Dacrydium*), and representatives of the families Mytaceae and Ericaceae, produce acid to decay resistant litter; leaf leachates are considered to mobilize iron.

Forests associated with peat swamps are particularly distinctive, and are widespread over northwestern (Sarawak), western, southern and eastern (Kalimantan) Borneo, along coastal areas abutting estuarine plains as well as in small river valleys. Plant life is adapted to high-stress environments, including mineral-deficient substrate and poorly aeriated and acidic waters, in addition to shortage of surface water during dry periods.

Another distinctive vegetation type is the heath forests, or Kerangas (from the Iban term for forested land that will not support hill rice farming), confined to white-sand podsol soils, growing on either raised beach terraces, or sandstone ridges and plateaux. Vegetation types represented here belong to the families Myrtaceae, Theaceae and Podocarpaceae, as well as Ericaceae, Clusiaceae and Ebenaceae. The heath-forest flora is poorer than evergreen forests, but richer than peat-swamp forests. This latter forest type is especially extensive along the northwestern coast of Borneo. It lies within present or former areas of swamp forests, with either slow-flowing or nearly stagnant water. Other distinctive vegetation types within Borneo include limestone flora, which is rich in plant and invertebrate endemics. Mangroves forests, which are rich in species, especially at the mouth of Sungei Kinabatangan of Sabah, Brunei Bay, and the mouths of the major south-flowing rivers of Kalimantan, comprise yet another noteworthy vegetation type. Mangroves of the Sunda Shelf islands have been described as the most biologically diverse in the world, and trees may reach 50m in height.

Current levels of habitat loss continue to be high. Among other effects are concerns of its impact on climate change, local precipitation and loss of biological diversity.

The Reptile Fauna

This book describes all species of reptile currently (up to the end of 2024) known from the island of Borneo. It is organized along the groups described below.

Order Crocodylia includes crocodiles and their relatives. At least three and perhaps four species are known in Borneo. The group demonstrates a broad, posteriorly flattened skull; the presence of external ear openings; many conical and pointed teeth; shortening of the forelimbs; long to extremely elongated snout; skull with temporal fenestrae; developed neural spines; developed secondary palate, and a four-chambered heart.

Order Chelonia (Chelonii or Testudines of some researchers) includes turtles and tortoises. They may be the most easily recognized group among reptiles, and demonstrate lack of teeth (in living species); the fusion of dorsal ribs with pleural elements of the carapace to form a shell; an akinetic skull; temporal region complete or emarginated, no temporal fenestrae; dermal roof elements reduced; parietal foramen absent, and shoulder girdle internal to ribs and shell.

Order Squamata comprises most of the world's living reptiles (lizards and snakes). They have relatively slender bodies covered with scales; frequently a parietal foramen; paired hemipenes; numerous teeth; a single temporal arch (lost in gekkotans and snakes); movable quadrate; single upper temporal fenestra; median cranial elements, including premaxillae, frontals and parietals frequently fused, and loss of temporal arch. Snakes and lizards have only a few characteristics that separate them – snakes lack palpebrals, parietal eye and foramen, tympanum, pectoral girdle and forelimbs, and possess a spectacle covering the eye.

Invasive Species Several reptiles have become invasive on Borneo, presumably through human agencies. These have been brought in for food (Red-eared Slider *Trachemys scripta* and Chinese Softshell Turtle *Pelodiscus sinensis*), inadvertently with horticultural products (Garden Lizard *Calotes versicolor*), or probably accidentally, as escapees from the pet trade (Monocled Cobra *Naja kaouthia*). These species have been included in this book, as field naturalists are likely to encounter them.

Snake-bite Management

Although a majority of the snakes on Borneo are harmless, several dangerously venomous snakes do occur on the island. Among these are members of two families – the Elapidae (comprising cobras, kraits, coral snakes and sea snakes) and the Viperidae (vipers, including pit vipers). Venom from the first group affects the nerves – hence 'neurotoxic' venom that blocks the conduction of nerve impulses to muscles. Symptoms include loss of muscle control, manifested by drooping eyelids, loss of muscle tone in the facial features and paralysis of the diaphragm, resulting in an inability to breathe. Envenomation by vipers affects the circulatory system – hence 'haemotoxic' venom that damages the walls of blood vessels. Symptoms include severe local pain and swelling, non-clotting of blood and kidney failure. Bites from the so-called back-fanged snakes or keelbacks (members of the genus *Rhabdophis*) require medical attention.

For bites from species belonging to the families Elapidae and Viperidae, arrangements to transport the bitten person to a hospital should be made immediately. The patient needs to be calmed and reassured, and kept immobile, as movements can increase systemic absorption of venom. Most local hospitals in the region have access to appropriate antivenom and epinephrine to treat anaphylaxis. Basic facilities for treatment of venomous snake-bite include a system for assisted breathing (for serious cases of neurotoxic envenomation), and treatment for acute renal failure (caused by viper bites). Accurate identification of the snake responsible for the bite is critical for treatment.

In the case of cobra, coral snake or krait bites, a compression bandage needs to be tightly applied, although permitting the pulse of the bitten limb, in order to retard venom absorption, as localized entrapment of venom may lead to amplification of tissue damage. Viper bites can be worsened by the application of a torniquet due to the potential of the venom to cause severe local tissue damage. Antivenom needs to be administered by a qualified physician only when signs of local or systemic envenomation are evident, as a number of bites from venomous snakes are associated with dry bites (no venom transmitted). Ideally, the antivenom used should be from the same species and, if possible, from the same geographical area. In this book, information on snakes that are potentially venomous, and anti-venom that is used for their bites, is included in the descriptions for relevant species.

Using This Book

This book covers all currently valid species of reptile known in Borneo, described under their respective orders. Each description is provided under families (purple boxes), and within them under genera (green boxes). The cut-off date for inclusion was 31 December 2024. A majority of species have been illustrated. Dichotomous identification keys to species (blue boxes) are provided to facilitate identification of specimens, live as well as preserved ones. Typically, these keys present diagnostic (unique to the species) morphological features in a series of alternative choices. This requires a sound knowledge of characteristics and users are urged to familiarize themselves with the terms employed (see Glossary, p. 15, and other resources available online or in zoological dictionaries).

Ideal key characteristics apply equally to all individuals of a population, regardless of size, age and sex, and are relatively constant. A good key is one that is dichotomous, ending in two possible options. Keys therefore provide unambiguous identifications of specimens using the most clearly diagnostic character, and are therefore easier to use than other methods (such as long descriptive accounts). Nonetheless, an important disadvantage of using keys includes problems associated with the identification of specimens drawn from populations that show sexual dimorphism, or have distinct changes in morphology or colouration during growth.

Dichotomous identification keys to species in this book are provided under generic accounts, and in the case of monotypic or a few constituent species within a particular genus on Borneo, under families (like Bataguridae, Cheloniidae, Natricidae and Viperidae). Characteristics chosen in a majority of cases are external ones that can be quickly examined. In a few cases, scale counts are unavoidable, and in the case of separation of two snake families (Colubridae v Natricidae and within the Typhlopidae), an osteological and/or visceral comparison becomes essential.

Species Descriptions

Each species includes the following information:

- Recommended common English name.
- Current scientific name, authority and date.
- In parentheses, vernacular names, as encountered in the literature or those collected in the field. These tend to be biased towards the larger, more common or

venomous species.
- Maximum length attained – straight carapace length (SCL) in turtles and tortoises; total length (TL) in crocodilians, snakes and anguid lizards, and snout–vent length (SVL) in all other lizards.
- Morphological features important for identification, based on examination and published literature.
- Colouration in life, generally based on a careful examination of live individuals,
- Notes on habits and behaviour, including habitat, elevational range, habits, diet and reproduction, when known.
- Distributional range within Borneo (country-wide), and notes on occurrence in extralimital areas. For the former, numerous historical localities refer to rivers, rather than localities on terra firma, and hence imply areas adjacent to or in the proximity of the respective drainages. It is also important to mention here that for several such records, habitat loss and modification may have lead to the extinction of local populations.
- Conservation status, according to the 2022 IUCN Red List of Threatened Species, Red List Categories, Version 2023-1 (2023) (www.iucnredlist.org). Definitions of threat categories are provided below.
- For each species, a general distribution map within the island is provided, based on known localities (or presumed, in the case of what is currently considered widespread).

The local Bornean Malay spelling of river (Sungei) rather than that used in Peninsular Malaysia (Sungai) has been retained here, following dialectical difference between the two, and its widespread usage in the former area.

Abbreviations & Conventions

asl above sea level
IUCN International Union for Conservation of Nature and Natural Resources (now The World Conservation Union)
SCL straight carapace length (of turtles)
SVL snout-vent length (of squamates, other than snakes and anguid lizards)
TL total length (of crocodiles, snakes and anguid lizards)

Threat Categories

A number of reptile species are threatened, primarily directly through human activities. Recognizing and reversing these threats requires specialist knowledge and action.

The primary activity for protection of biodiversity is the recognition of such species and their threats. Appended below are the threat categories, as recognized by the IUCN Red List.

Extinct (EX) A taxon is Extinct when there is no reasonable doubt that the last individual has died. A taxon is presumed Extinct when exhaustive surveys in known and/or expected habitat, at appropriate times (diurnal, seasonal, annual), throughout its historic range have failed to record an individual. Surveys should be over a time frame appropriate to the taxon's life cycle and life form.

Extinct in the Wild (EW) A taxon is Extinct in the Wild when it is known only to survive in cultivation, in captivity or as a naturalized population (or populations) well outside the past range. A taxon is presumed Extinct in the Wild when exhaustive surveys in known and/or expected habitat, at appropriate times (diurnal, seasonal, annual), throughout its historic range have failed to record an individual. Surveys should be over a time frame appropriate to the taxon's life cycle and life form.
Critically Endangered (CR) A taxon is Critically Endangered when the best available evidence indicates that it meets any of the criteria A to E for Critically Endangered, and it is therefore considered to be facing an extremely high risk of extinction in the wild.
Endangered (EN) A taxon is Endangered when the best available evidence indicates that it meets any of the criteria A to E for Endangered, and it is therefore considered to be facing a very high risk of extinction in the wild.
Vulnerable (VU) A taxon is Vulnerable when the best available evidence indicates that it meets any of the criteria A to E for Vulnerable, and it is therefore considered to be facing a high risk of extinction in the wild.
Near Threatened (NT) A taxon is Near Threatened when it has been evaluated against the criteria but does not qualify for Critically Endangered, Endangered or Vulnerable now, but is close to qualifying for or is likely to qualify for a threatened category in the near future.
Least Concern (LC) A taxon is Least Concern when it has been evaluated against the criteria and does not qualify for Critically Endangered, Endangered, Vulnerable or Near Threatened. Widespread and abundant taxa are included in this category.
Data Deficient (DD) A taxon is Data Deficient when there is inadequate information to make a direct, or indirect, assessment of its risk of extinction based on its range and/or population status. A taxon in this category may be well studied, and its biology well known, but appropriate data on abundance and/or range are lacking. Data Deficient is therefore not a category of threat. Listing of taxa in this category indicates that more information is required and acknowledges the possibility that future research will show that threatened classification is appropriate.
Not Evaluated (NE) A taxon is Not Evaluated when it is has not yet been evaluated against the criteria.

Topographical Diagrams

TURTLES

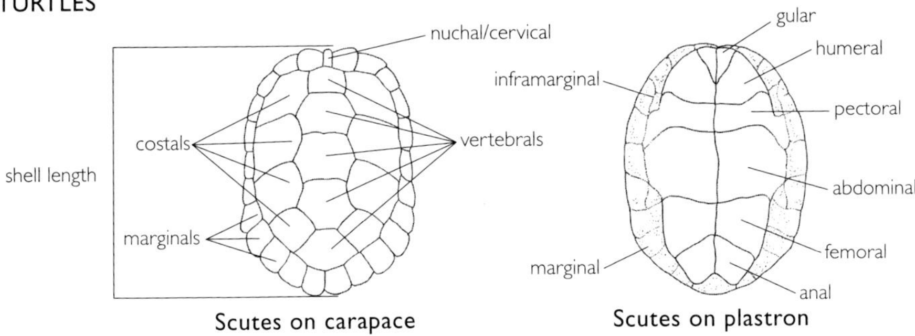

Scutes on carapace

Scutes on plastron

LIZARDS

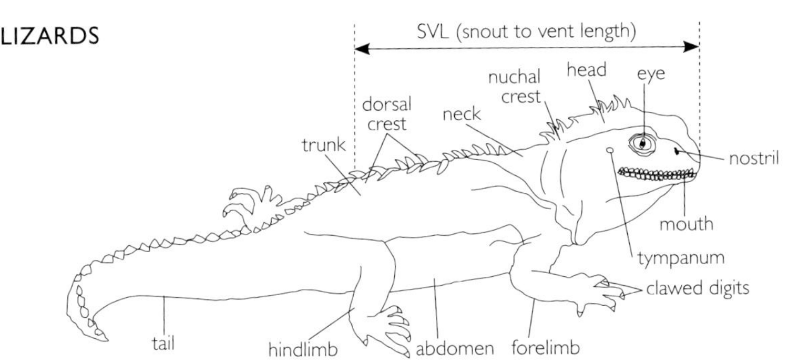

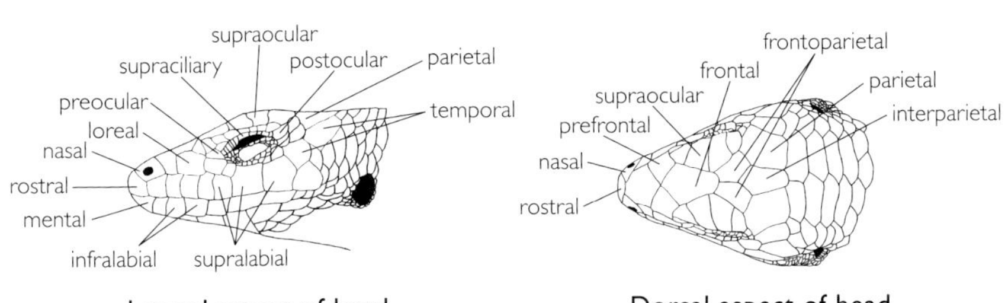

Lateral aspect of head

Dorsal aspect of head

SNAKES

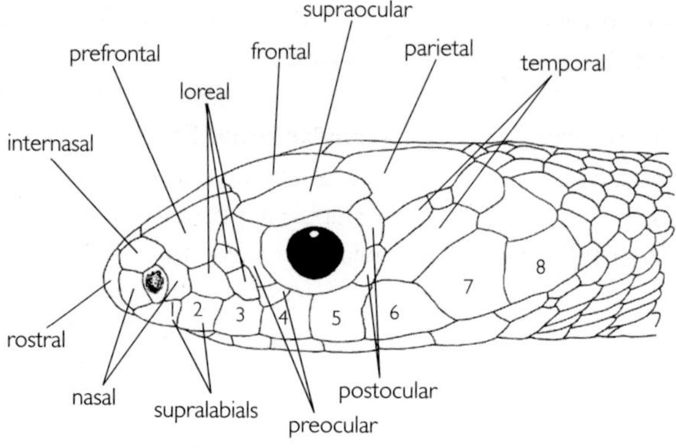

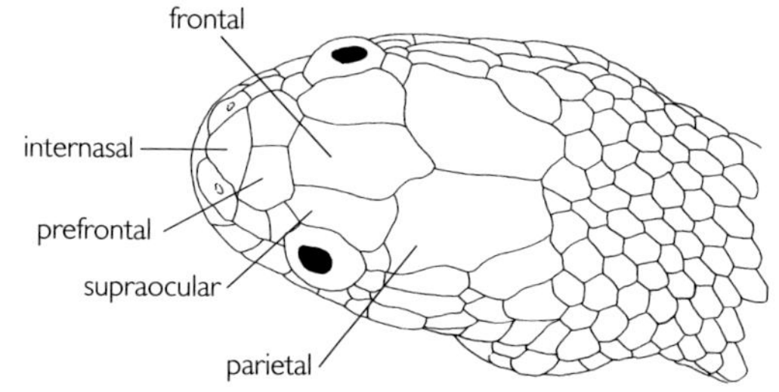

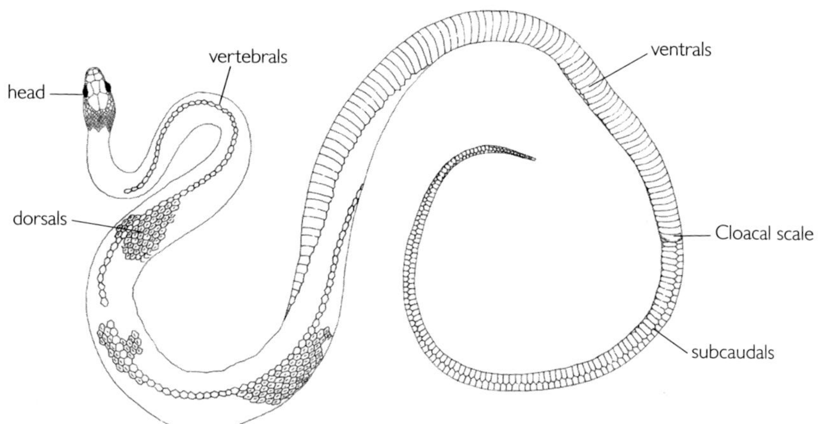

Head and body scales of a snake

Glossary

annulated Having ringed segments.
anterior Towards front of body.
arboreal Living in trees.
auricular Ear opening or position, indicated by depression, where tympanum is located.
autotomy Spontaneous or reflexive separation of a body part (typically, a tail).
axilla Armpit.
canthals Scales of canthal ridge.
canthus (plural, **canthi**) Corner of eye.
canthus rostralis Ridge from eye to tip of snout.
caudal Towards tail.
cervical Pertaining to neck.
chevron 'V'-shaped mark.
cloacal plate Terminal ventral scale or scute.
clutch Entire compliment of eggs or neonates from a single female.
concave Bent inwards, rounded.
conical scales Cone-shaped scales.
convex Bent outwards, rounded.
crests Ridges on head, composed of highly modified scales or skin.
cryptic Camouflaged or hidden.
cuneate Triangular scale lying between two labials.
cusp Toothlike-projection of a jaw.
denticulate Tooth-like.
dimorphism Difference in morphology between members of same species.
diurnal Active during the day.
dorsal Towards upper surface of body.
dorsal crest Ridge of highly modified (often conical) scales along back.
dorsum Back or dorsal surface of body.
emarginate Indented.
fangs Recurved, elongated teeth on upper jaw, through which venom passes.
femoral pore Opening on undersides of thighs.
furrow Well-defined groove.
granular scales Small, convex, non-overlapping scales, typically with a pebbly appearance.
gular Pertaining to or located on throat.
gular crest Ridge of modified scales along ventral surface of throat and chin.
gular fold Transverse fold of skin across throat.
hemipenal bulge Swelling at base of tail, on account of presence of male hemipenes.
hemipenes Paired sex organs of males (of squamates).
hood Expanded skin behind head, especially in cobras.
humeral Pertaining to upper part of forelimb.
infralabial Lower labial.
inguinal Pertaining to region around bases of hindlimbs.
iridescent Rainbow-like sheen on surface of very smooth scales.
keel Raised ridge down back, tail or scale.
keratinous Containing keratin.
knob Rounded protuberance.
labial Pertaining to lip.
lateral Pertaining to side of body.
lobe Segment of casque.
loreal Scale on side of head, between nasals and preoculars.
lumbar Pertaining to lower back.
maxillary teeth Refers to teeth on maxillary bones in upper jaws.
medial Pertaining to midline of body.
melanistic Showing abundance of dark pigment.
mental Single scale at anterior border of lower jaw.
nare Nostril.

nasal glands Refers to glands in nostrils that secrete excess salt.
nasals Scales on sides of head in which are opening of nares.
nuchal venom gland iIntegumental glands in paravertebral region of neck of several snake species.
ocellus (plural **ocelli**) Rounded, eye-like spot.
occipital Towards back of head.
occipital spines Spiny projections extending from occipital region of head.
osteoderm Bony deposit within scale or dermal layer.
oviparity Reproduction through production of eggs provided with membranes and/or shells.
ovoviviparity Reproduction through production of live young that hatch from eggs within female oviducts.
paravertebral stripe Refers to stripe on one side of midline of dorsum.
parietal Head scale behind frontal.
parietal eye Sensory structure opening through top of skull.
patagium Fold of skin on flanks and supported by ribs (in lizards of genus *Draco*).
phalange Bone of finger or toe.
postcloacal Enlarged scale behind cloaca.
posterior Towards rear of body; opposite of **anterior**.
postlabial Scale behind labial.
postmental Scale behind mental along line of chin.
postnasal Scale behind nasals and anterior to loreal.
postocular Behind eye.
precloacal Scale anterior to cloacal aperture.
precloacal groove Longitudinal groove on mid-venter before cloaca.
precloacal pore Extension of femoral pore series on to body.
prefrontal Scale anterior to frontal.
preocular Anterior to eye.
reticulation Colour pattern resembling mesh of a net.
rostral horn Annulated horn protruding from anterior tip of snout.
rostrum Literally, 'beak', most anterior part of head snout.
rugose Wrinkly or warty.
scalation Pattern of scales on body or on specific part of body.
scute Enlarged scale.
serrated Having a saw-like appearance.
sexual dimorphism Condition in which males and females have distinctly different forms.
snout–vent length Measurement between snout-tip and vent.
spatulate Shaped like a spatula, flat and rounded at tip.
spicule Tiny, pointed structure.
spinose Sharp, pointed shape like a thorn.
spur Sharp, spinous appendage.
squamation Scale arrangement.
subcaudals Scales beneath tail.
subdigital lamella Scale on ventral surface of digits in lizards.
subocular Beneath eye.
superciliary Small scale bordering orbit.
supralabial Upper labial scale.
supranasal Scale above nasal.
supraocular Scale above eye.
temporal Scale behind postocular.
total length Measurement between snout-tip and tail-tip.
transverse Cross-wise or diagonal.
tubercular scales Small, knob-like scales.
tympanum Eardrum.
ventral Towards belly or underside.
vertebrals Middorsal row of scales.
whorl Symmetrical row of enlarged scales circling a tail segment.

CROCODILES • Order CROCODYLIA

Key to Bornean Crocodylia

1a. Snout broad..Crocodylidae
1b. Snout narrow..*Tomistoma schlegelii* (p. 21)

Family Crocodylidae True Crocodiles

Crocodiles include the largest reptiles on Borneo, and are associated with larger rivers across the island as the apex predator of such ecosystems. Reports of attacks by large individuals on humans are not uncommon, and are typically attributed to the Saltwater Crocodile.

Members of the family of true crocodiles are characterized by the possession of heavy armoured skin, long snout, heavy jaws, robust body in adults, and dorsal scales with osteoderms, or small bony deposits. All are entirely aquatic, inhabiting fresh waters, with only a few occupying saltwater and brackish-water habitats such as mangroves and, less often, sea coasts and even coral reefs.

Aquatic adaptations include webbed hands and feet, nostrils capable of closing via valves, and eyes that can be covered by a transparent membrane. The osteoderms not only function as armour, but also act as heat exchangers, allowing them to rapidly raise or lower temperatures, and also neutralize acidosis, the effect of carbon dioxide accumulation in the blood from being submerged for long periods underwater. Other unusual features of crocodiles include a muscular partition separating the pectoral and abdominal cavities, alveoli in the lungs and a four-chambered heart.

Crocodilians are carnivorous, some also scavenging on land, their dietary spectrum comprising vertebrates, and in juveniles, arthropods, fish and small vertebrates associated with wetlands. A few crocodiles may pose danger to humans and livestock. The sex of crocodiles is determined by the incubation temperature of the eggs, and morphological and molecular evidence suggests a relationship of crocodiles with birds rather than with other reptiles.

Crocodylus Crocodiles

Medium-sized to gigantic (TL > 7m) crocodilians, with long, relatively broad (compared to *Tomistoma*) snout; paired ridges on snout; median nasal septum not ossified; premaxilla perforated to receive first pair of mandibular teeth; iris a shade of green.

Key to Bornean species of *Crocodylus*

1a. Longitudinal interorbital ridge present..*Crocodylus siamensis* (p. 20)
1b. Longitudinal interorbital ridge absent..**2**

2a. Transverse ventral scales 25..*Crocodylus raninus* (p. 19)
2b. Transverse ventral scales 29–35..*Crocodylus porosus* (p. 18)

SALTWATER CROCODILE *Crocodylus porosus* Schneider, 1801

(Bahasa Indonesia: Buaya Muara, Buaya Badas Kuning; Iban: Rawing; Bahasa Malaysia: Buaya Katak, Buyaya Tembaga)

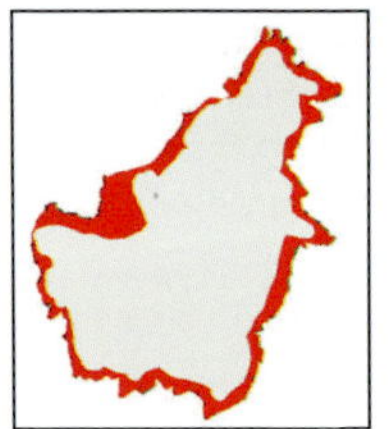

SIZE 7m (in the largest known individual); more typically 5m in males, 3m in females.

IDENTIFICATION FEATURES Large crocodilian; head large with heavy snout in adults; pair of ridges running from orbit to centre of snout; scales on back more oval than in most other crocodiles; anterior nuchals typically absent; males exceed females in mass.

COLOUR Juveniles brightly coloured, with black spots or blotches on pale yellow or grey background; adults have less contrast and greyish-olive; belly paler.

HABITAT AND BEHAVIOUR Favours mud-bottomed larger rivers and coasts, especially areas with mangrove forests, and occasionally, standing bodies of water, such as large inland lakes, marshes and oxbow lakes, more than 1,000km inland. Ambush predator. Juveniles feed on crabs, shrimps, large insects, fish, lizards and snakes; adults can take turtles, birds and mammals. Attacks on humans sometimes attributed to large adults. Females construct mound nests, where 60–80 eggs are deposited, and nest is guarded until eggs hatch.

DISTRIBUTION Sri Lanka, east coast of India, Bangladesh, Myanmar, Thailand, Cambodia, Vietnam, Peninsular Malaysia, Singapore, Sumatra, Borneo, Java, Bali, eastern Indonesia, New Guinea, the Philippines, Australia and the South Pacific. Also, Seychelles, probably as a vagrant. Widespread in appropriate habitats across Borneo. The best site for viewing it is Sungei Santubong, adjacent Santubong National Park, and many other tributaries of Sungei Sarawak, around Kuching.

IUCN THREAT STATUS Low Risk/Least Concern.

Clutch hatching

Close-up of hatchling

Close-up of head

Adult

BORNEAN SWAMP CROCODILE *Crocodylus raninus* Müller & Schlegel, 1844

(Bahasa Malaysia: Buaya Badas Hitam)

SIZE 4m

IDENTIFICATION FEATURES Large head and broad snout; lachrymal ridges weakly developed; palatine-pterygoid suture transverse; four well-developed postoccipital scales; longitudinal interorbital ridge absent; transverse ventral scale rows 25; transverse throat scale rows 38–39; dorsal scales have sharp keels, giving rough appearance.

COLOUR Adults dark green, with dark bands on tail; juveniles more yellow, also with distinct dark bands.

HABITAT AND BEHAVIOUR Apparently restricted to marshes and swamps, such as peat-swamp forests. Diet and reproductive habits unstudied.

DISTRIBUTION Central Borneo, including Sarawak and Kalimantan Barat, specific records being from the Sungei Mahakam tributary region of Kalinjau and Danau Sohuwi. Endemic to Borneo.

IUCN THREAT STATUS Not Evaluated.

NOTE The taxonomic status of the species awaits study, pending the collection of specimens with locality information.

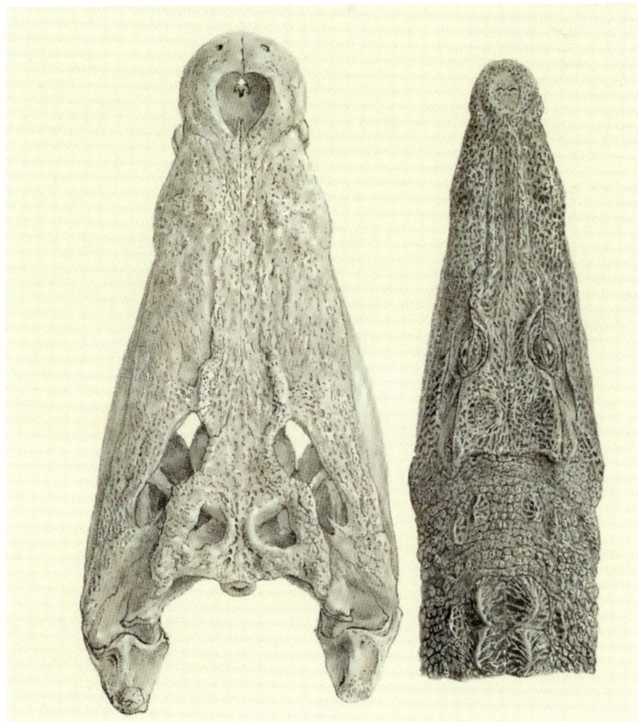

After Müller and Schlegel (1844).

Siamese Crocodile *Crocodylus siamensis* Schneider, 1801

(Bahasa Malaysia: Buaya Serunai)

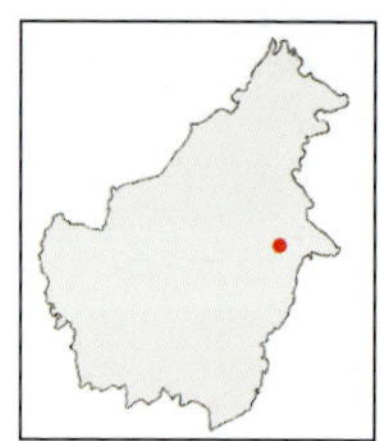

SIZE 4m in males; 3.2m in females. Most individuals under 3m.

IDENTIFICATION FEATURES Large head and broad, smooth snout, with pair of ridges running from orbit to centre of snout; longitudinal ridges anterior to eyes, raised bony crest behind eye, and webbing on feet not reaching tips of digits.

COLOUR Juveniles more brightly coloured than adults. Top of body pale olive-green, dark green, yellow or grey with black spots or blotches; adults greyish-olive, with indistinct bands.

HABITAT AND BEHAVIOUR Inhabits fresh waters, especially in association with forested habitats, and also marshes. Diet includes fish, snakes, frogs and small mammals. Mound nester; nests comprise plant debris mixed with mud, where 12–50 eggs, each 75–80 x 50mm, are laid.

DISTRIBUTION Thailand, Cambodia, Laos (probably extinct), Vietnam and possibly Java, in Indonesia. Within Borneo, only recorded from Kalimantan.

IUCN THREAT STATUS Critically Endangered.

Family Gavialidae Gharials

These crocodilian species have heavy armour and narrow, parallel-sided snouts. They inhabit fresh waters, including large rivers with sandy banks, and blackwater lakes and swamps surrounded by peat-swamp forests. Characteristics shared with true crocodiles include webbed feet, nostrils capable of closing via valves, eyes with transparent membranes, muscular partition separating pectoral and abdominal cavities, alveoli in the lungs and a four-chambered heart. The adult diet of gharials, while primarily comprising fish, includes birds, turtles and mammals up to the size of monkeys, while juveniles may ingest insects, crustaceans and frogs. Currently, this family is represented by two living species, although the fossil record shows a diverse fauna in North and South America, Africa and Asia. The taxonomy of many of these species requires further study.

Tomistoma **Malayan False Gharials**

Large (to 5.5m) crocodilians, with long, narrow snouts; no paired ridges on snout; median nasal septum not ossified; premaxillae grooved to receive first mandibular teeth; iris a shade of brown.

MALAYAN FALSE GHARIAL *Tomistoma schlegelii* (Müller, 1838)

(Bahasa Malaysia: Buaya Jenjolong; Bahasa Indonesia: Buaya Sapit; Iban: Bedai Sampit)

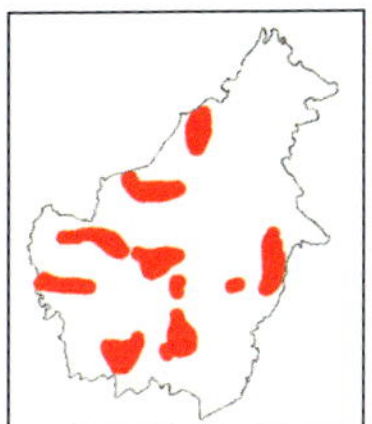

SIZE 5.5m

IDENTIFICATION FEATURES Body elongated; snout narrow, tapering gradually from skull (and ending with a rounded tip); nuchal and dorsal scutes form continuous shield of 22 transverse series.

COLOUR Top of body pale brown to dark olive, with black blotches, spots and bands, especially on flanks; these patterns sometimes fuse to form bands, especially in hatchlings; tail with broad black bands in adults; venter cream, unpatterned; juveniles bright yellow, with dark bands; iris pale brown.

HABITAT AND BEHAVIOUR Inhabits blackwater rivers, swamps and lakes overgrown with vegetation. Diet comprises fish, although much larger prey, including monkeys, has been recorded. Employs short-range, underwater acoustic signals during courtship and mating. Constructs a mound nest on peat or a mixture of peat and humus, within or close to roots of large trees, about 60cm high. Clutches comprise 20–60 eggs, measuring 70 x 100mm. Incubation period 75–90 days.

DISTRIBUTION Southern Thailand (where it is possibly extinct), Peninsular Malaysia, Sumatra, Borneo and Java. There are unverified reports from Sulawesi, Indonesia and southwestern Vietnam, and subfossil remains (c. seventeenth century) are known from China. Isolated localities in Brunei (reported by locals and in need of verification), Sabah (Sungei Klias), Sarawak (lower Sungei Baram, Loagan Bunut, Sungei Ensengai, Sungei Kelauh, Sungei Kerang, Sungei Kroh, Sungei Mayeng, Sungei Penyilam, Sungei Penjing, Sungei Samarahan and Sungei Sebangan) and Kalimantan (Gunung Palung, Danau Sentarum and Betung-Kerihun National Parks).

IUCN THREAT STATUS Endangered.

TURTLES • Order CHELONIA

Key to Bornean families of Chelonia

1a. Forelimbs flipper-like; head non-retractable **2**
1b. Forelimbs not flipper-like; head retractable **3**

2a. Shell lacks scales (except in hatchlings); both shells have distinct ridges .... Dermochelyidae
2b. Shell covered with scales; both shells lack ridges .... Cheloniidae

3a. Hindlimbs club-like .... Testudinidae
3b. Hindlimbs not club-like **4**

4a. Shell covered with scales .... Geoemydidae
4b. Shell covered with skin .... Trionychidae

Family Geoemydidae Asian Hard-shelled Turtles

Hard-shelled turtles have oval to oblong, depressed or domed carapaces bearing 11 pairs of sutured peripherals and the nuchal lacking costiform processes. The large plastron is hinged in a few species, the mesoplastron is absent, and the plastral buttresses are typically firmly articulated with the costal bones of the carapace. The neck can withdraw vertically into a shell, and the pelvic girdle articulates flexibly with the plastron. The family contains an ecological mixture of aquatic and terrestrial species. Members are typically omnivorous, and herbivorous adults tend to be carnivorous as juveniles. A few have specialized diets (e.g. snails), while many others scavenge in water or on land, some go for seasonally available food such as fruits and mushrooms. The family is widely distributed in Asia, and also in southern Europe, and central and northern South America.

Key to Bornean species of Geoemydidae

1. Hinge on plastron present **2**
1b. Hinge on plastron absent **4**

2a. Hinge on plastron can completely close shell .... *Cuora couro* (p. 24)
2b. Hinge on plastron, if present, weakly developed **5**

3a. Neck striped .... *Cyclemys dentata* (p. 25)
3b. Neck lacks stripes .... *Cyclemys enigmatica* (p. 26)

4a. Carapace with 6–7 vertebrals .... *Notochelys platynota* (p. 28)
4b. Carapace with 5 vertebrals **5**

5a. Carapace with black longitudinal lines .... *Batagur borneoensis* (opposite)
5b. Carapace without black longitudinal lines **6**

6a. Carapace with pattern .... *Heosemys spinosa* (p. 27)
6b. Carapace unpatterned black **7**

7a. Carapace smooth in adults, unicarinate in juveniles; forearms without enlarged scales .... *Orlitia borneensis* (p. 29)
7b. Carapace with faint keels in adults; tricarinate in juveniles; forearms with enlarged scales .... *Siebenrickiella crassicollis* (p. 30)

Batagur Terrapins

Hard-shelled turtles with oval shells that are smooth; posterior neurals elongated; plastron slightly kinetic; neural plates short sided in front; five vertebral scutes; broad triturating surfaces of jaws; denticulated middle ridge; digits fully webbed; forelimbs four or five clawed.

PAINTED TERRAPIN *Batagur borneoensis* (Schlegel & Müller, 1845)

(Bahasa Malaysia: Beluku; Iban Beluku)

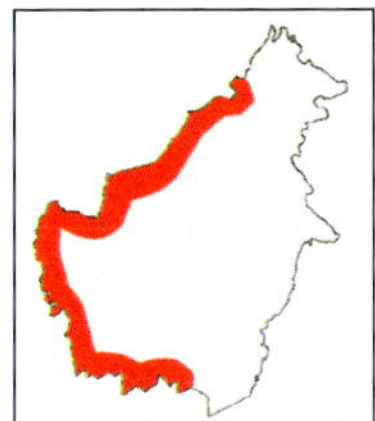

SIZE 600mm

IDENTIFICATION FEATURES Carapace oval, flattened; marginals smooth; in juveniles, shell with a vertebral keel distinct; plastral lacks hinge; forelimbs have five claws; snout slightly upturned.

COLOUR Carapace light brown or olive, with three black longitudinal stripes; heads of females olive; non-breeding males have grey heads; adult males in breeding season have white forehead and red stripe between eyes; plastron unpatterned cream.

HABITAT AND BEHAVIOUR Inhabits coastal areas and estuaries. Herbivorous, consuming aquatic macrophytes; shrimps, crabs and fishes sometimes eaten. Nesting migration occurs, adults travelling as far as 3km downriver to nest on sea beaches. Eggs elongated, each 68–76 x 36–44mm. Clutch size 10–16, and two clutches are produced annually. Incubation period 61–82 days.

DISTRIBUTION Southern Thailand, Peninsular Malaysia, Sumatra and Borneo; records from Myanmar remain unverified. Bornean records are from coastal Sarawak, Brunei and Sabah. The best site to encounter the species is Samunsam Wildlife Sanctuary in western Sarawak, where a hatchery to augment wild populations exists.

IUCN THREAT STATUS Critically Endangered.

Adult male in breeding colours

Forehead and face of breeding male

Cuora Box Turtles

Hard-shelled turtles with oval, smooth shells in adults, and weak neural and pleural keels in juveniles; plastron completely closes shell through the development of a hinge between the hyoplastron and hypoplastron; neural plates short sided behind; five vertebral scutes; bridge indistinct; digits fully webbed; forelimbs five clawed.

Malayan Box Turtle *Cuora couro* (de La Tour, in Schweigger, 1812)

(Bahasa Malaysia: Kura-kura, Kura-kura Patah)

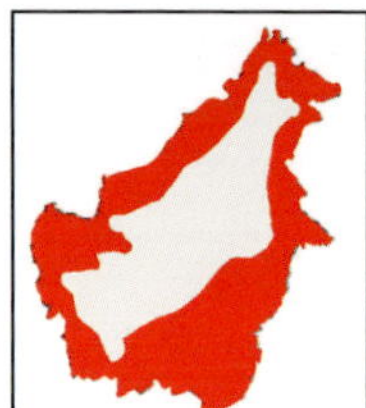

SIZE 216mm

IDENTIFICATION FEATURES Carapace high-domed and smooth; weak vertebral keel in adults; marginals not well defined; juveniles have two additional keels laterally; plastron has a transverse, movable hinge behind pectoral and abdominal scutes; posterior anal notch absent on plastron.

COLOUR Carapace olive, brown or nearly black; plastron yellow or cream, with a single black blotch; face has yellow longitudinal stripes.

HABITAT AND BEHAVIOUR Inhabits standing and slow-flowing waterbodies, including rivers, lakes, marshes and mangrove swamps, as well as agricultural areas; juveniles more aquatic than the terrestrial adults. Diet includes water plants, fungi, worms and aquatic insects, and the species is known to wander on land during the fruiting season. Eggs elongated, clutch size 1–6, two clutches laid annually, eggs each 40–55 x 25–34mm. Incubation period 1.5–3 months.

DISTRIBUTION Peninsular Malaysia, Singapore, Borneo, Sumatra, Java, Bali. Bornean records are from across the island, recent records including Kuching, Kota Samarahan, Miri, Baleh National Park and the Tinbarap Palm Oil Estate.

IUCN THREAT STATUS Endangered.

View of carapace

Facial view

View of plastron

Cyclemys Leaf Turtles

Hard-shelled turtles with oval shells; carapace tricarinate; carapace and plastron attached by ligament; hinge between hypplastron and hypoplastron in adults; neural plates short sided behind; five vertebral scutes; digits fully webbed; forearms have oval scales; forelimbs five clawed.

Common Leaf Turtle *Cyclemys dentata* (Gray, 1831)

(Bahasa Malaysia/Brunei: Bulus; Iban Beluku)

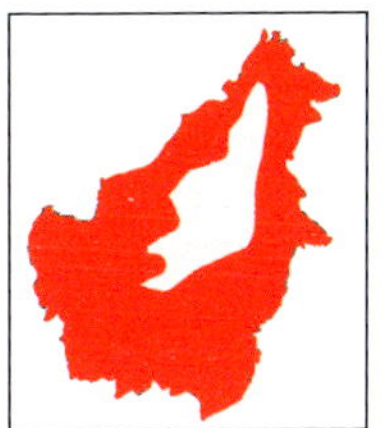

SIZE 210mm

IDENTIFICATION FEATURES Carapace oval, depressed, bearing a vertebral and two lateral keels; enlarged scales on forehead; plastron with hinge in adults around 23–25cm in shell length; femoral midseam shorter than anal midseam; anal notch narrow, acute angled.

COLOUR Carapace dark brown, unpatterned or with fine black lines; forehead dark speckled; temples and neck have dark stripes; plastron yellow, unpatterned or with dark radiating pattern; plastron of juveniles mottled with black.

HABITAT AND BEHAVIOUR Inhabits plains and low hills, and associated with small rivers, streams and sometimes ponds. Omnivorous, feeding on figs and invertebrates. Nests dug on the ground; clutches comprise 2–4 elongated hard-shelled eggs. Incubation period 75 days.

DISTRIBUTION Southern Peninsular Malaysia, Sumatra, Borneo, Java, Palawan and the Sulu Archipelago; introduced into Leyte (the Philippines). Bornean records are from across the island.

IUCN THREAT STATUS Near Threatened.

Sunda Leaf Turtle *Cyclemys enigmatica* Fritz, Guicking, Auer, Sommer, Wink & Hundsdörfer, 2008

(Bahasa Malaysia: Bulus; Iban Beluku)

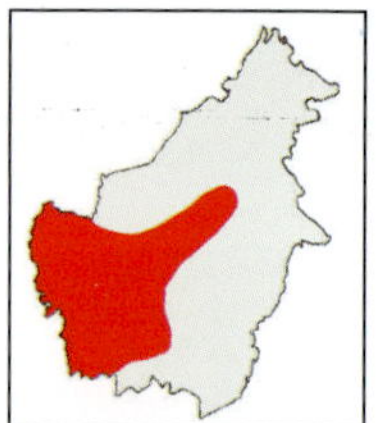

SIZE 235mm

IDENTIFICATION FEATURES Carapace oval, depressed, bearing a vertebral and two lateral keels; enlarged scales on forehead; plastron with a hinge in adults; femoral midseam greater or equal to anal midseam; anal notch wide, obtuse angled.

COLOUR Carapace dark brown, typically with reddish tinge; forehead copper to brown, lighter than temporal region; neck dark, lacking stripes; plastron dark brown to black, sometimes with dark radiating lines.

HABITAT AND BEHAVIOUR Inhabits plains and low hills, and associated with streams and sometimes ponds. Omnivorous, feeding on fallen fruits and invertebrates. Reproductive habits unknown.

DISTRIBUTION Southern Peninsular Malaysia, Sumatra, Borneo and Java. Due to its similarity to the Common Leaf Turtle (p. 25), specific records from Borneo are only from Sarawak and western Kalimantan.

IUCN THREAT STATUS Near Threatened.

Plastra of male (left) and female (right)

Heosemys Hill Turtles

Hard-shelled turtles with oval shells that are flattened on top, tricarinate, with a weak neural and pleural keels; spinous process on marginals, especially in juveniles; horizontal temporal arch absent; neural plates short sided behind; five vertebral scutes; digits partially webbed; and forelimbs five clawed.

SPINY HILL TURTLE *Heosemys spinosa* (Gray, 1831)

(Bahasa Malaysia: Kura Kura Bukit Duri)

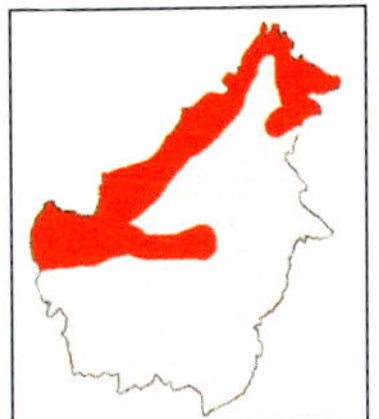

SIZE 220mm

IDENTIFICATION FEATURES Carapace oval, and arched in adults; strong vertebral keel; marginals of adults smooth; those of juveniles have greatly expanded marginals bearing distinct spines; plastral with weak hinge in females; fingers and toes partially webbed; hindlimbs club shaped. Adult males possess relatively longer and thicker tails than adult females.

COLOUR Carapace brown; small, yellowish-orange or red spot behind eye; plastron brown, each scute with radiating lines.

HABITAT AND BEHAVIOUR Inhabits tropical forests, from the lowlands to middle elevations. Diet comprises both plants and animals, and includes mushrooms; also known to scavenge on land. Three elongated, hard-shelled eggs produced. Hatchlings measure 63mm.

DISTRIBUTION Southern Myanmar, Thailand, Peninsular Malaysia, Singapore, Sumatra, Pulau Bangka, Pulau Natuna, the Batu Archipelago, Borneo, the Sulu Archipelago and Tawi-Tawi (the Philippines). Bornean records exist for Bako National Park, Kota Samarahan, Serian, Deded Krian National Park and Kubah National Park in western Sarawak; Ulu Temburong in Brunei, and Deramakot and Danum Valley Conservation Area in Sabah.

IUCN THREAT STATUS Endangered.

Notochelys Flat-shelled Turtles

Hard-shelled turtles with oval shells; carapace and plastron joined by ligament; hinge between hyoplastron and hypoplastron in adults; plastral buttress absent in adults; neural plates short sided in front; 6–7 vertebral scutes; digits fully webbed; forelimbs five clawed.

Malayan Flat-shelled Turtle *Notochelys platynota* (Gray, 1834)

(Bahasa Malaysia: Kura Kura Punggung Datar)

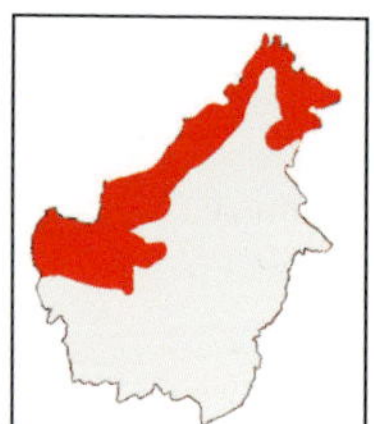

SIZE 400mm

IDENTIFICATION FEATURES Carapace flat, with low, interrupted vertebral keel; six or seven vertebrals; weak transverse hinge on plastron; narrow, crescentic scales on forearm; toes fully webbed.

COLOUR Carapace olive, yellowish-brown or brick-red, and bright green in hatchlings; head brown; juveniles have two yellow longitudinal stripes; plastron yellowish-orange, each scute with black blotch.

HABITAT AND BEHAVIOUR Inhabits shallow, macrophyte-dominated waterbodies, including swamps, marshes and forest streams. Diet comprises aquatic macrophytes. Three hard-shelled eggs laid, each 56 x 27–28mm.

DISTRIBUTION Southern Thailand, Vietnam, Peninsular Malaysia, Singapore, Sumatra, Borneo and Java. Recent records exist for Deramakot in Sabah and across central and northern Sarawak and Kalimantan.

IUCN THREAT STATUS Vulnerable.

Orlitia Giant Turtles

Hard-shelled turtles with oval, smooth shells; plastral buttress robust; neurals elongated, short sided in front; five vertebral scutes; large nasopalatine foramen; upper triturating surface of jaw surface broad posteriorly, with sharp middle ridge; digits fully webbed; forelimbs five clawed.

Malayan Giant Turtle *Orlitia borneensis* Gray, 1873

(Bahasa Indonesia: Baning Dayak)

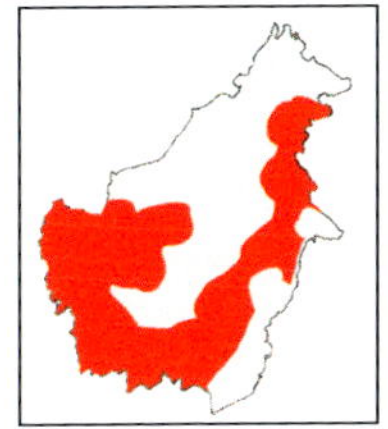

SIZE 800mm

IDENTIFICATION FEATURES Carapace humped in juveniles, turning smooth with growth; adults have relatively narrow shells; plastral lacks hinge; head large; glandular skin from eye to tympanum; band-like scales on outer faces of forelimbs; fingers and toes extensively webbed.

COLOUR Carapace unpatterned black, brown or grey; plastron yellowish-orange or brown, sometimes with dark flecks; head dark, with pale line from mouth to back of head.

HABITAT AND BEHAVIOUR Inhabits large rivers and lakes, especially blackwater habitats. Diet in the wild unknown, but captives are omnivorous. Eggs elongated, with brittle hard shells, measuring 80 x 40mm. Hatchlings 60mm, with rough shell texture and markedly serrated marginals at carapace posterior.

DISTRIBUTION Peninsular Malaysia, Sumatra and Borneo. A middle Pleistocene fossil from the Solo River confirms its occurrence on Java. Recent records for Borneo exist for Sungei Sarawak near Batu Kawa, west of Kuching, Danau Sentarum National Park and Mesangat in Kalimantan, and a few other localities in these states.

IUCN THREAT STATUS Critically Endangered.

Close-up of head

Siebenrockiella Marsh Turtles

Hard-shelled turtles with oval, depressed shells; carapace tricarinate in juveniles; plastral buttress moderately developed; triturating surfaces of jaws moderately narrow and lack distinct midridge; nasopalatine foramen small; neurals short sided in front; five vertebral scutes; digits fully webbed; forelimbs five clawed.

Black Marsh Turtle *Siebenrockiella crassicollis* (Gray, 1831)

(Bahasa Malaysia: Kura Kura Kolam)

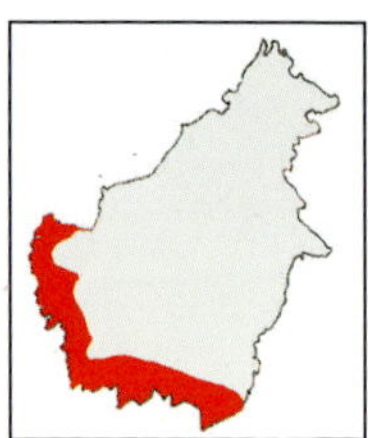

SIZE 200mm

IDENTIFICATION FEATURES Carapace oval, with serrated posterior marginals; vertebral region of adults flattened; juveniles have a vertebral and two lateral keels; adults have single keel; plastral lacks hinge.

COLOUR Carapace dark grey or nearly black; plastron paler grey; juveniles have light head spots, which are retained by adult females, these markings fading with growth in males.

HABITAT AND BEHAVIOUR Inhabits standing or sluggish waterbodies, including blackwater marshes, swamps, ponds, streams and lakes. Diet includes macrophytes, aquatic animals such as worms, snails, shrimps and frogs; also known to scavenge. Lays 3–4 clutches of 1–2 eggs, each 45 x 19mm. Hatchlings measure 52mm.

DISTRIBUTION Southern Myanmar (Tenasserim, Tanintharyi Division), Thailand, Laos, Cambodia, Vietnam, Peninsular Malaysia, Singapore, Sumatra, Borneo and Java. Recent record for Borneo is from Deramakot in Sabah.

IUCN THREAT STATUS Endangered.

Close-up of head

Family Cheloniidae Sea Turtles

Sea turtles have a large, non-retractable head; smooth carapace bearing scutes; large plastron; secondary palate; elongated, flipper-like forelimbs, and short, rounded hindlimbs. The family includes all the world's living marine turtles (except the sole living member of the Dermochelyidae). They are cosmopolitan in the world's oceans, and exclusively aquatic; only the adult females come ashore, though in some parts of the world, adult males may ascend beaches to bask. The diet of sea turtles includes marine organisms, from jellyfish to corals. Eggs are laid in holes excavated in sandy beaches. Most sea-turtle populations are threatened by harvesting for their flesh, accidental capture, sand mining and modifications of their nesting habitats, including light pollution.

Key to Bornean species of Cheloniidae

1a. One pair of prefrontals .... *Chelonia mydas* (p. 33)
1b. Two pairs of prefrontals **2**

2a. Nuchal not contacting costals .... *Eretmochelys imbricata* (p. 34)
2b. Nuchal contacts costals **3**

3a. Three enlarged inframarginals on bridge .... *Caretta caretta* (p. 32)
3b. Four enlarged inframarginals on bridge .... *Lepidochelys olivacea* (p. 35)

Caretta Loggerhead Turtles

Hard-shelled marine turtles with oval shells; carapace covered with large scales and lacking ridges; large head; interprefrontal scale present; flippers two clawed.

Loggerhead Sea Turtle *Caretta caretta* (Linnaeus, 1758)

(Bahasa Malaysia: Penyu Karah, Penyu Sisek Tempurong)

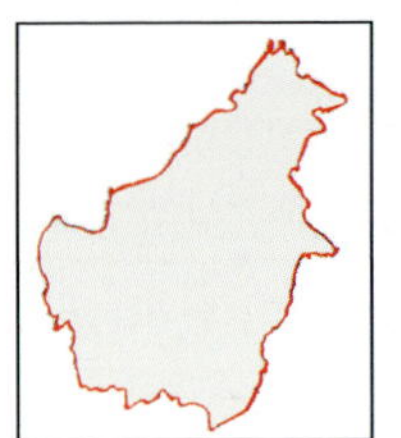

SIZE 1.2m

IDENTIFICATION FEATURES Carapace elongated, with a tapering end; five pairs of costals, costal 1 contacting nuchal; head massive; three or four infralabial scutes that lack pores; 13 marginal scutes; bridge has three inframarginals.

COLOUR Carapace reddish-brown; plastron yellowish-brown or yellowish-orange.

HABITAT AND BEHAVIOUR Associated with warm, subtropical seas, bays, lagoons and estuaries. Diet comprises molluscs and crustaceans. Eggs each 34.7–55.2mm, with 23–178 eggs produced. Incubation period 49–80 days.

DISTRIBUTION Cosmopolitan. Bornean records are uncertain, given its historical confusion with the Olive Ridley Turtle (p. 35).

IUCN THREAT STATUS Vulnerable.

Chelonia Green Turtles

Hard-shelled marine turtles with oval shells; carapace covered with large, non-overlapping scales and lacking ridges; marginals not serrated; jaws rounded; flippers single clawed.

GREEN TURTLE *Chelonia mydas* (Linnaeus, 1758)

(Bahasa Malaysia: Penyu Hijau; Bahasa Brunei: Penyu Pulau; Penyu Emegit)

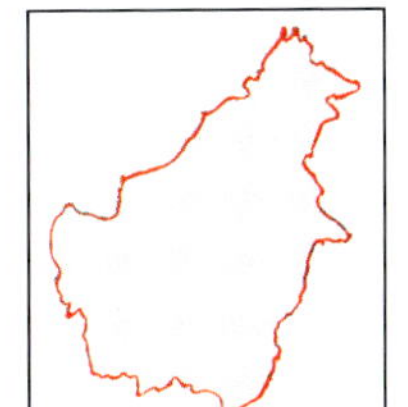

SIZE 1.4m

IDENTIFICATION FEATURES Carapace heart shaped; pair of prefrontal scales on forehead; scutes of carapace not overlapping; upper jaw without hook; forelimbs have single claw.

COLOUR Carapace olive or brown, with dark radiating pattern; plastron pale yellow. English name is for colour of their fat, once in demand for making turtle soup.

HABITAT AND BEHAVIOUR Associated with tropical regions, and common around oceanic islands and along coasts with wide sandy beaches. Juveniles carnivorous, while adults consume only sea grass and seaweed. Eggs soft shelled, spherical; a nest contains 98–172 eggs, each 41.4–42.1mm. Incubation period c. 60 days. Up to 11 nests may be laid within a nesting season.

DISTRIBUTION Cosmopolitan. Bornean records are from across the island, especially from small, rocky islands off the coast. The best place to see nesting and hatchling emergence is at the Talang-Satang National Park in western Sarawak.

IUCN THREAT STATUS Endangered.

Head and forebody

Eretmochelys Hawksbill Sea Turtles

Hard-shelled marine turtles with oval shells; carapace covered with large overlapping scales and lacking ridges; marginals serrated; jaws beak-like; flippers two clawed.

Hawksbill Sea Turtle *Eretmochelys imbricata* (Linnaeus, 1766)

(Bahasa Malaysia: Penyu Sisik; Penyu Lilin)

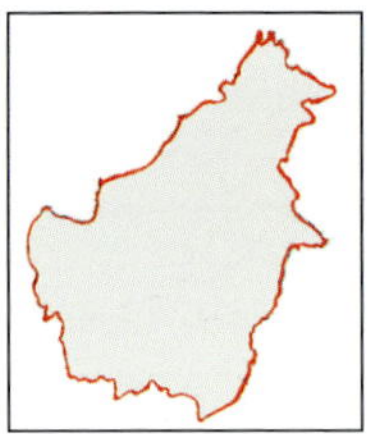

SIZE 1m

IDENTIFICATION FEATURES Carapace heart shaped; scutes of carapace have four pairs of imbricate costal scutes; two pairs of prefrontal scales; upper jaw relatively narrow, elongated; upper jaw forwards projecting, to form bird-like beak.

COLOUR Carapace olive-brown; juveniles have darker blotches than adults; plastron pale in adults, dark in juveniles.

HABITAT AND BEHAVIOUR Inhabits reefs, bays, estuaries and lagoons. Diet includes sponges, algae, corals and shellfish. Clutch size 96–177; eggs each 30–35mm. Incubation period 57–65 days.

DISTRIBUTION Cosmopolitan. Isolated and occasional records from Borneo, especially small islands off Sarawak and Sabah.

IUCN THREAT STATUS Critically Endangered.

Close-up of head

Lepidochelys Ridley Sea Turtles

Hard-shelled marine turtles with oval shells; carapace covered with large, non-overlapping scales and lacking ridges; pores near rear margin of each inframarginal; flippers with 1–2 claws.

OLIVE RIDLEY SEA TURTLE *Lepidochelys olivacea* (Eschscholtz, 1829)

(None recorded)

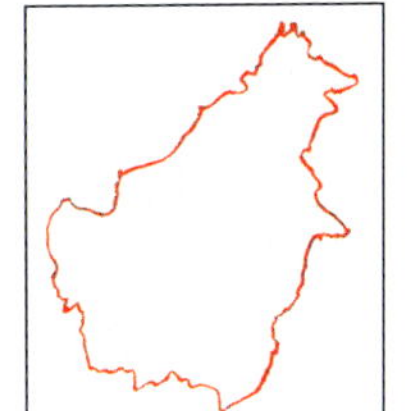

SIZE 80cm

IDENTIFICATION FEATURES Carapace broad, heart shaped, posterior marginals serrated, with juxtaposed costal scutes; 5–9 pairs of costals; bridge with four inframarginals, each with a pore; adult shell smooth; hatchling shell tricarinate; upper jaw hooked.

COLOUR Carapace olive-green or greyish-olive; plastron greenish-yellow; juveniles grey-black dorsally; venter cream.

HABITAT AND BEHAVIOUR Associated with sea beaches in vicinity of mangroves. Breeding tends to take place en masse, involving hundreds and even thousands of females, in a phenomenon referred to as 'arribadas'. Clutches comprise 50–160 eggs, each 34–43mm. Incubation period 45–60 days. Hatchlings measure 37.9–49.9mm.

DISTRIBUTION Pacific and Indian Oceans. Bornean records are relatively few, and it is known from Sarawak, Brunei and Sabah.

IUCN THREAT STATUS Vulnerable.

Family Dermochelyidae Leatherback Sea Turtles

This family consists of a single living species (with several fossil relatives), with a broad, ridged shell, lacking epidermal scutes. The dermal bones of the shell are replaced by a mosaic of small platelets; the limbs are paddle shaped, lacking claws; the forelimbs are enlarged and the hindlimbs are broadly connected to the tail via a web. The sole living species inhabits the world's oceans and feeds on jellyfish, and also crustaceans, molluscs, fish and seaweeds. It frequently wanders into cold waters of the Arctic, presumably in search of jellyfish.

Dermochelys Leatherback Sea Turtles

Turtle with an elongated shell that is leathery in adults, and covered with small, irregularly polygonal scales in hatchlings; carapace has seven distinct keels; plastron has five keels; limbs paddle-like, lacking claws; adults dark brown, blackish-blue or black.

LEATHERBACK SEA TURTLE *Dermochelys coriacea* (Vandelli, 1761)

(Bahasa Malaysia: Penyu Timbo; Penyu Belimbing; Kamban; Kambau; Agal)

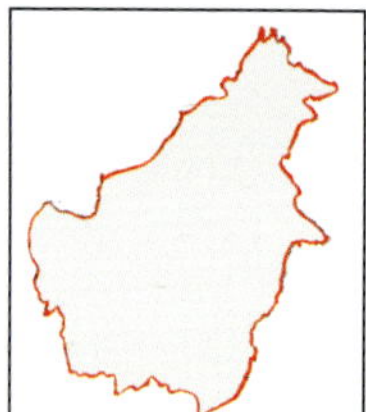

SIZE 2.5m

IDENTIFICATION FEATURES Carapace elongated, tapered towards end; seven ridges on carapace and five on plastron; shell covered with skin in adults; distinct scales present on hatchling shells; limbs paddle-like and clawless.

COLOUR Top of body black, dark brown or blackish-blue, with paler flecks of pale yellow or white; venter paler.

HABITAT AND BEHAVIOUR Inhabitant of open oceans, nesting on oceanic islands as well as on wide beaches on the mainland. Diet comprises primarily jellyfish. In northern temperate waters, capable of diving up to 1,200m below the surface in search of food. Nests more deeply compared to other sea turtles. Clutches comprise 90–130 eggs, each 50–54mm.

DISTRIBUTION Cosmopolitan. Occasional visitor to shores of Borneo.

IUCN THREAT STATUS Vulnerable.

Family Emydidae New World Hard-shelled Turtles

The sole naturalized species of this New World family in the region has an oval, domed carapace with 11 pairs of sutured peripherals around the margin; nuchal lacking costiform processes; large plastron lacking mesoplastron; plastral buttresses articulating with costals of carapace; the neck withdrawing vertically, and the pelvic girdle articulating flexibly with the plastron. It inhabits freshwater habitats, including ponds and marshes in tropical Asia. The diet of this invasive species is omnivorous, and its natural distribution includes the southern portion of North America.

Trachemys Sliders

Hard-shelled turtles with oval shell; mandible flattened ventrally; anterior edge of middle ridge of triturating surface of jaws not cusp-like; fifth toe has three phalanges; carapace has longitudinal ridges.

Red-eared Slider *Trachemys scripta* (Thunberg, 1792)

(No vernacular names recorded)

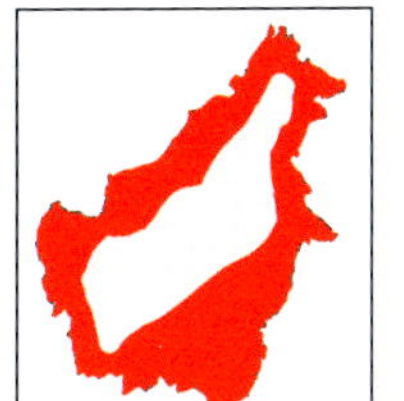

SIZE 280mm

IDENTIFICATION FEATURES Carapace rounded, with a nearly smooth outline; plastron lacks hinge; females much larger than males, which possess elongated claws on fingers.

COLOUR Carapace green with yellow lines, turning darker with growth; plastron bright yellow with large black mark on each scute; orange or red patch on temples; yellow stripes on neck.

HABITAT AND BEHAVIOUR Omnivorous, feeding on leaves, fruits, fish, frogs and carrion. Clutches comprise 2–25 eggs, each 30–42 x 19–29mm, and incubation period is 65–75 days. Hatchlings measure 30–33mm.

DISTRIBUTION Native of the Mississippi River drainage, USA. Naturalized populations known from Borneo.

IUCN THREAT STATUS Least Concern.

Hatchling

Adult

Family Testudinidae Land Tortoises

These are mostly terrestrial turtles with a high shell; head and limbs with heavy scales; columnar forelimbs; hindlimbs club shaped, and head and limbs completely retractable inside the shell. While the family reaches its greatest diversity in the world's arid regions, the South-east Asian species (with one exception) are linked to relatively moist habitats. A bulk of their diet comprises vegetation, although carrion is sometimes ingested, probably as a source of calcium and protein. Their eggs are buried in the ground or laid in leaf-litter mounds. Tortoises are cosmopolitan in distribution, except in Australia.

Manouria Asian Giant Tortoises

Hard-shelled turtles with oval shells; vertebral region slightly concave in adults; forelimbs columnar, cylindrical; hindlimbs club shaped; forelimbs five clawed; digits free of web.

Asian Giant Tortoise *Manouria emys* (Schlegel & Müller, in Temminck, 1840)

(Bahasa Indonesia: Beluku; Bahasa Malaysia: Baning; Kura-kura Anam Kaki)

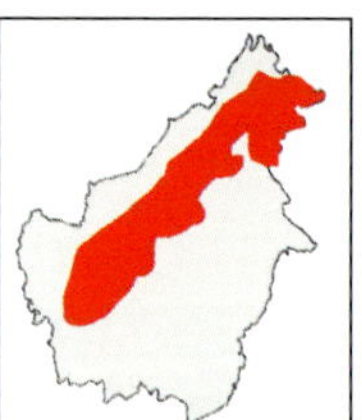

SIZE 500mm

IDENTIFICATION FEATURES Carapace low; vertebral region depressed; distinct growth rings on carapace scutes; posterior marginals weakly serrated; upper jaw hooked; outer surface of forelimbs bears large scales; pair of tuberculate scales on thighs; pectoral scutes separated or fused with each other.

COLOUR Shell blackish-brown; plastron lighter; head and limbs brown.

HABITAT AND BEHAVIOUR Associated with primary forests in the mid-hills. Diet comprises vegetation, although insects and frogs are also eaten. Constructs mound nest by sweeping leaf litter in which 23–51 hard-shelled, spherical eggs of diameter 51–54mm are deposited. Unusually for a turtle, it guards its nest, attacking presumed egg predators. Incubation period 60–75 days; hatchlings 60–66mm.

DISTRIBUTION Eastern India, Bangladesh, Myanmar, Thailand, Vietnam, Cambodia, Laos, Peninsular Malaysia, Sumatra and Borneo. There appear to be few recent records for Borneo, all from Sepilok Forest Reserve, the Danum Valley Conservation Area and Tabin Wildlife Reserve in Sabah. Historical records from Kalimantan and Sarawak.

IUCN THREAT STATUS Critically Endangered.

Family Trionychidae Softshell Turtles

These are highly aquatic turtles with skin-clad, flattened shells that have reduced bony elements; three claws on each limb, and nostrils located on a fleshy proboscis. The family includes one of the world's largest freshwater turtle species, and comprises primarily inhabitants of fresh waters, one invading brackish waters or sea coasts. The diet of these turtles is mostly small invertebrates and fish, and they are also known to scavenge.

Key to Bornean species of Trionychidae

1a. Snout shorter than eye diameter........*Pelochelys cantori* (p. 43)
1b. Snout as long as eye diameter........**2**

2a. Neck and carapace with a linear pattern........*Chitra chitra* (p. 41)
2b. Neck and carapace lack a linear pattern........**3**

3a. Carapace straight sided, with a black vertebral stripe........*Dogania subplana* (p. 42)
3b. Carapace rounded or oval, without a black vertebral stripe........*Amyda cartilaginea* (p. 40)

Amyda Asian Softshell Turtles

Softshell turtles with oval shells; one or two costal bones not separated in vertebral region by neural bones; head not large; snout not downturned; postorbital arch not edged; eyes nearer temporal region than nostrils; sides of carapace ovoid; five plastral callosities; epiplastra has elongated anterior projections; proboscis as long as eye opening.

Malayan Softshell Turtle (p. 40)

Asian Softshell Turtle *Amyda cartilaginea* (Boddaert, 1770)

(Bahasa Indonesia: Bulus; Bahasa Malaysia: Labi Labi; Kuja)

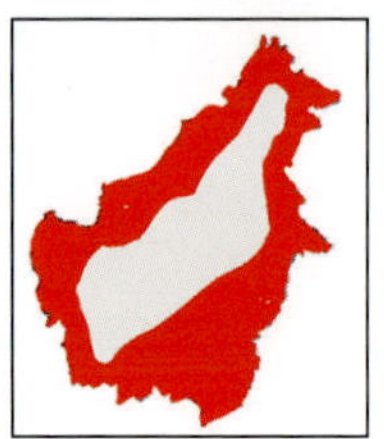

SIZE 750mm

IDENTIFICATION FEATURES Carapace rounded or oval; rounded sides to carapace; narrow head; distinct tubercles at anterior carapace margin.

COLOUR Top of body greenish-grey or olive, sometimes with yellow-bordered black spots or radiating streaks, which tend to disappear with growth; plastron cream in males, grey in females; head nearly unpatterned or with yellow spots.

HABITAT AND BEHAVIOUR Inhabits side streams of large, muddy rivers, peat swamps and marshes. Nocturnal and carnivorous, feeding on fish, frogs, shrimps and water insects. Nests excavated in river banks, and 4–8 eggs, each 21–33mm, laid. Incubation period 130–140 days. Hatchlings 37.2–49.5mm.

DISTRIBUTION Eastern India, Myanmar, Thailand, Laos, Cambodia, Vietnam, Peninsular Malaysia, Singapore, Sumatra, Borneo, Java, Bali, Lombok and Sulawesi (Indonesia), the latter two localities probably based on introductions. Bornean records are from across the island, the species being known in Loagan Bunut National Park, Long Saan in upper Baram, Baleh National Park, Tinbarap Palm Oil Estate and the Serian blackwater region of Sarawak.

IUCN THREAT STATUS Vulnerable.

Chitra Narrow-headed Softshell Turtles

Softshell turtles with oval shells; one or two costal bones not separated in vertebral region by neural bones; head not large; snout not downturned; postorbital arch not edged; eyes nearer nostrils than temporal region; sides of carapace ovoid; proboscis as long as eye opening.

South-East Asian Narrow-headed Softshell Turtle

Chitra chitra Nutaphand, 1986

(Bahasa Indonesia: Bulus Besar and Labi Labi Bintang)

SIZE 1.2m

IDENTIFICATION FEATURES Carapace rim smoothly joins cartilaginous part to skin of neck; head small, with tiny proboscis; eyes situated close to snout-tip; neural bones narrow; carapace does not end abruptly; lower arm has 2–4 scales.

COLOUR Top of body brown or yellowish-brown, with bright yellow or tan, dark-edged stripes, including inverted chevron-like marking on anterior of carapace and neck; neck has five lines; plastron cream or pale pink. Pale colouration; midline and lateral vertebral carapace stripes; bell-shaped mark on carapace anterior indistinct; no 'X'-shaped figure between eyes; ocelli in eye region; indistinct dark speckling and ocelli on chin; broad costal markings.

HABITAT AND BEHAVIOUR Inhabits medium to large, fast-flowing rivers with sandy bottoms. Piscivore. Clutches comprise 60–117 eggs, each 34–40mm. Incubation period 62–66 days. Hatchings 40–42mm.

DISTRIBUTION Thailand, including Khwae, Mae Klong and Mae Ping River systems, and isolated localities in Peninsular Malaysia, and central and eastern Java, including Pasuruan and Solo Rivers. The only Bornean record is from Danau Sentarum, in western-central Borneo.

IUCN THREAT STATUS Critically Endangered.

Frontal view

Dorsal view

Dogania Malayan Softshell Turtles

Softshell turtles with oval shells; costal bones separated in vertebral region by neural bones; four plastral callosities; head large; snout long, slightly downturned; postorbital arch reduced to an edge; eyes nearer temporal region than nostrils; sides of carapace straight; proboscis as long as eye opening.

MALAYAN SOFTSHELL TURTLE *Dogania subplana* (Geoffroy Saint-Hilaire, 1809)

(Bahasa Indonesia: Kuja; Bahasa Malaysia and Bahasa Brunei: Labi Labi; Layi Layi)

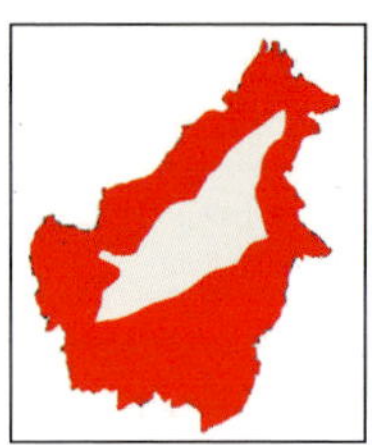

SIZE 350mm

IDENTIFICATION FEATURES Carapace flat, oval, with distinctly straight sides; head large, bearing a downturned snout; adults have carapace hinge.

COLOUR Carapace dark olive or brown, with dark median stripe and 2–3 pairs of black-centred, eye-like spots, a pattern most distinct in juveniles and fading with growth; plastron cream or grey; juveniles have reddish-brown blotch behind eyes.

HABITAT AND BEHAVIOUR Associated with clear rocky, shallow streams with sandy bottoms, in mid-hills as well as lowlands. The large head is considered adaptive for cracking shells of aquatic snails; fish, prawns and crabs also eaten.

DISTRIBUTION Southern Myanmar (Tennaserim and Mergui Archipelago), Thailand, Peninsular Malaysia, Singapore, Java, Sumatra, Pulau Singkep, Pulau Natuna, Borneo, Java and Palawan (the Philippines). Bornean records are from Tanjung Datu National Park, Bako National Park, Santubong National Park, Kubah National Park, Lambir Hills National Park, foothills of Gunung Penrissen, Baleh National Park and Kampung Sebako at foothills of Gunung Pueh in Sarawak, and Deramakot in Sabah.

IUCN THREAT STATUS Least Concern.

Frontal view

Dorsal view

Close-up of head

Pelochelys Giant Softshell Turtles

Softshell turtles with oval shells; one or two costal bones not separated in vertebral region by neural bones; seven plastral callosities; head not large; snout not downturned; postorbital arch not edged; eyes nearer temporal region than nostrils; sides of carapace ovoid; proboscis under half eye opening.

Asian Giant Softshell Turtle *Pelochelys cantorii* Gray, 1864

(Bahasa Malaysia: Labi Labi)

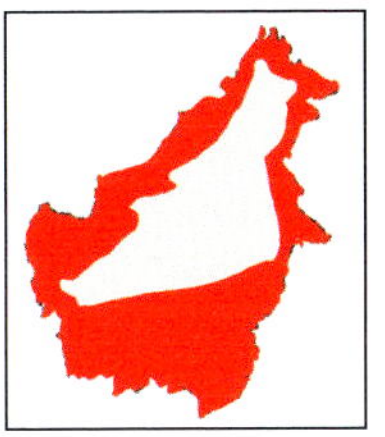

SIZE 1.5m

IDENTIFICATION FEATURES Carapace flattened, relatively more elongated in juveniles and oval in adults; head short; eyes close to tip of snout; juveniles have numerous tubercles on carapace and low vertebral keel; proboscis very short and rounded.

COLOUR Carapace olive or brown, spotted or streaked with lighter or darker shades, with lighter outer edge; plastron unpatterned white.

HABITAT AND BEHAVIOUR Mainly associated with coasts and large rivers. Diet comprises fish, shrimps, crabs and molluscs, sometimes aquatic plants. Clutches comprise 24–70 eggs each of diameter 30mm. Hatchlings measure 42mm.

DISTRIBUTION India, Bangladesh, Myanmar, Thailand, Laos, Cambodia, Vietnam, Peninsular Malaysia, Singapore, Sumatra, southern China, the Philippines and possibly Sulawesi. Records from isolated Bornean localities in Sabah (Marak Parak, Gunung Kinabalu), Sarawak (Sungei Rajang), and Kalimantan (Kutai, Sungei Berau and Maruwai).

IUCN THREAT STATUS Critically Endangered.

Dorsal view

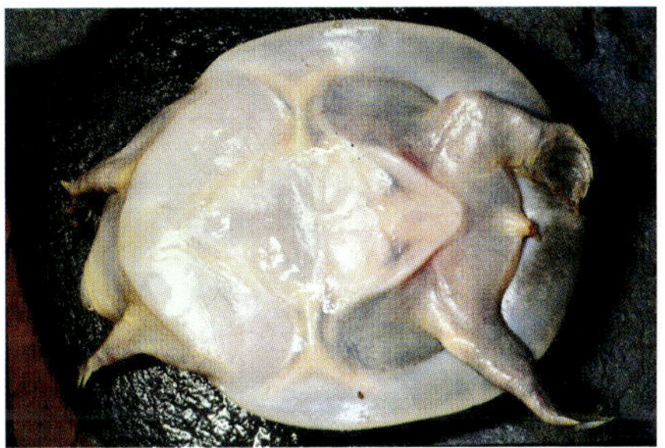

Plastron

Frontal view

Forehead

Pelodiscus Chinese Softshell Turtles

Softshell turtles with oval shells; one or two costal bones not separated in vertebral region by neural bones; seven plastral callosities; head not large; snout not downturned; postorbital arch not edged; eyes nearer temporal region than nostrils; sides of carapace ovoid; proboscis as long as eye opening.

CHINESE SOFTSHELL TURTLE *Pelodiscus sinensis* Wiegmann, 1835

(Bahasa Malaysia: Labi Labi)

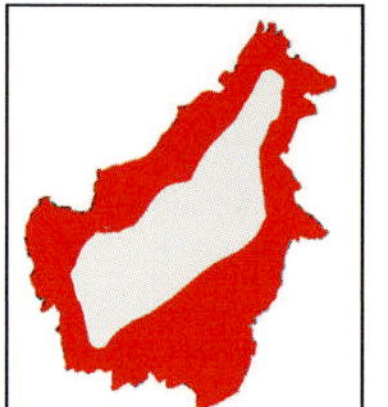

SIZE 250mm

IDENTIFICATION FEATURES Carapace oval, slightly longer than wide, smooth in adults and with longitudinal rows of low tubercles in juveniles; preneural bones absent.

COLOUR Top of body unpatterned olive to greyish-green; juveniles have light-bordered spots; fine black radiating lines around eyes; plastron unpatterned cream or yellow in adults; juveniles have large black blotches.

HABITAT AND BEHAVIOUR Inhabits ponds, marshes and rivers, in lowlands and mid-hills, up to at least 600m asl. Basking is rare. Diet comprises insects, crustaceans and fish; aquatic macrophytes also consumed. Clutches of 9–28 eggs produced, diameter 20–24mm. Incubation period 40–80 days. Hatchlings measure 27mm.

DISTRIBUTION Native to southern China, Vietnam, Korea, Japan and Russia, and introduced into Thailand, Peninsular Malaysia, Singapore, Borneo, Sumatra, Japan, Timor, Hawaii and Bonin Islands. There is an established population around Kuching in western Sarawak.

IUCN THREAT STATUS Vulnerable.

Hatchling

Adult

LIZARDS • Order SQUAMATA SAURIA

Key to Bornean families of lizards

1a. Eyes concealed; snout covered with single large scale....................Dibamidae
1b. Eyes exposed; snout covered with several small scales..................**2**

2a. Tongue long, deeply bifid..**4**
2b. Tongue short...5

3a. Limbs absent, body with lateral folds..Anguidae
3b. Limbs present, body without lateral fold **5**

4a. Ear opening absent; lower eyelid with transparent window; gular fold absent...Lanthanotidae
4b. Ear opening present; lower eyelid without transparent window; gular fold present...Varanidae

5a. Forehead with granules or small scales...**6**
5b. Forehead with large symmetrical shields **8**

6a. Teeth acrodont; body compressed laterally ....Agamidae
6b. Teeth pleurodont; body depressed **7**

7a. Eyelids present; egg-shell parchment type..Eublepharidae
7b. Eyelids absent; eggs brittle shelled ....Gekkonidae

8a. Ventrals differentiated from dorsals; femorals or inguinal pores present...Lacertidae
8b. Ventrals and dorsals undifferentiated; femoral and inguinal pores absent...Scincidae

Family Agamidae Agamid or Dragon Lizards

These are moderate to large (SVL 45–350cm), arboreal (on Borneo) lizards.Their charactristics include: large, overlapping body scales; osteoderms absent on body; pectoral girdle with a 'T'-shaped or cruciform interclavicle; clavicles curved, rod shaped; well-developed tail and limbs; acrodont dentition; pterygoid lacking teeth; vomer flat or convex; well-developed crests on head and body; many with keeled scales, especially in adult males; throat flaps present; fracture plane in caudal vertebrae absent; pupils rounded; oviparous; having leathery egg-shells.

Key to Bornean genera of Agamidae

1a. Wing-like dermal expansion on sides of body .... *Draco*
1b. No wing-like dermal expansion on sides of body **2**

2a. Tympanum concealed; inner lining of mouth blue **3**
2b. Tympanum exposed; inner lining of mouth not blue **4**

3a. Dorsal crest present; large unicarinate scales laterally .... *Aphaniotis*
3b. Dorsal crest reduced or absent; large tubercular scales laterally .... *Pelturagonia*

4a. Males with long compressed rostral appendage; females with small knob on snout-tip .... *Harpesaurus*
4b. Males without rostral appendage; females without knob on snout-tip **5**

5a. Gular fold present; dorsal scale unequal .... *Gonocephalus*
5b. Gular fold absent; dorsal scale equal **6**

6a. Nuchal crest absent **7**
6b. Nuchal crest present **8**

7a. Oblique fold in front of shoulder .... *Pseudocalotes*
7b. No oblique fold in front of shoulder .... *Bronchocela*

8a. Head and throat with large scales below tympanum; SVL > 130mm .... *Hypsicalotes*
8b. Head and throat lack large scales below tympanum; SVL < 100mm .... *Calotes*

Aphaniotis Shrub Lizards

Small (SVL up to 72mm), arboreal agamids, with compressed, slender bodies; tympanum hidden; dorsals heterogenous; ventrals larger than dorsals, keeled; nuchal and dorsal crests weakly developed or absent; gular fold absent; precloacal and femoral pores absent; limbs long and slender; fifth toe longer than first; inner lining of mouth blue.

Key to Bornean species of *Aphaniotis*

1a. Snout-tip with a fleshy appendage .... *A. ornata* (p. 49)
1b. Snout-tip without a fleshy appendage **2**

2a. Snout pointed; convex scale above rostral projecting anteriorly; flanks with transverse bands; irises of males brow .... *A. acutirostris* (opposite)
2b. Snout rounded; convex scale above rostral not projecting anteriorly; flanks without transverse bands; irises of males blue or green .... *A. fusca* (p. 48)

Long-snouted Shrub Lizard *Aphaniotis acutirostris* Modigliani, 1889

(No vernacular names recorded)

SIZE 72mm

IDENTIFICATION FEATURES Body slender, compressed; snout acute, much longer than eye diameter; projecting convex scale above rostral; limbs long; fifth toe longer than first; dorsal scales small, with scattered larger scales; row of tuberculate scales on paravertebral region, parallel to vertebrae; dewlap weak; males have weak nuchal crest; precloacal and femoral pores absent; ventral scales keeled.

COLOUR Dorsal surface brown with darker variegation or with green bands; dark radiating lines from orbit of eye; throat sometimes has dark spots; rest of ventral surface pale brown; males have distinct yellow dewlap; inner lining of mouth blue.

HABITAT AND BEHAVIOUR Inhabits lowland rainforests, and found on shrub-like vegetation. Diet presumably comprises small insects and other invertebrates. Typically, two eggs laid every 30–100 days, each 7 x 12mm. Incubation period 48–63 days.

DISTRIBUTION Sumatra, the Mentawai Islands, Pulau Simuleur, Pulau Weh and Pulau Nako. A single Bornean record from Kalimantan Barat.

IUCN THREAT STATUS Least Concern.

Brown Shrub Lizard *Aphaniotis fusca* (Peters, 1864)

(No vernacular names recorded)

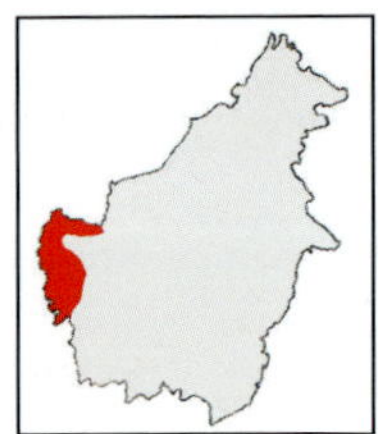

SIZE 67mm

IDENTIFICATION FEATURES Body slender, compressed; snout rounded, not longer than diameter of eye; no convex scale projecting above rostral; limbs long; fifth toe longer than first; dorsal scales small, with scattered larger scales; nuchal crest reduced, composed of short, triangular scales; dewlap weak, especially in females; precloacal and femoral pores absent; limbs long and slender; tail long, slender, ending in blunt tip.

COLOUR Dorsal surface dark brown or brownish-olive; flanks more olive; venter pale olive-brown; two dark interorbital bars; lips cream; inner lining of mouth dark blue, tongue pale blue; pupil black, rounded; ocular ring bright yellow; iris blue in males, brown in females and juveniles; gular sac in adult males black with oval yellow spots, with cream-yellow blotches.

HABITAT AND BEHAVIOUR Inhabits primary and lightly disturbed forests in the lowlands and mid-hills. Diurnal and arboreal; sometimes found in social groups. Diet comprises caterpillars, ants, beetles, millipedes, cockroaches, termites and small lizards. Clutch size 1–2 eggs, measuring 7 x 18mm; each hatchling SVL 23–43mm.

DISTRIBUTION Southern Thailand, the Malay Peninsula, Borneo, Sumatra, Simalur, Nias (Mentawai Islands), Singkep and the Natuna Islands. Western Borneo, including western Sarawak (Bau, Gunung Gading, Santubong, Bako National Park, Gunung Matang) and Kalimantan (Taman Nasional Betung Kerihun).

IUCN THREAT STATUS Least Concern.

Threat display

Ornate Shrub Lizard *Aphaniotis ornata* (van Lidth de Jeude, 1893)

(No vernacular names recorded)

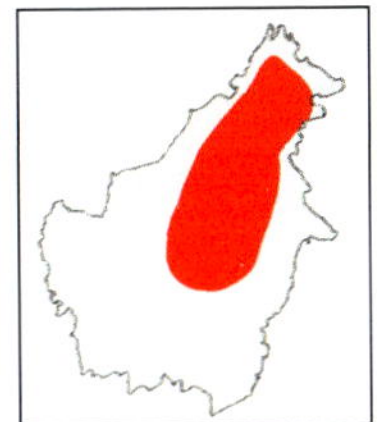

SIZE 57mm

IDENTIFICATION FEATURES Body slender and compressed; snout as long as eye diameter; snout-tip with fleshy conical appendage that lies backwards at tip of snout, narrowest at base and covered with imbricate keeled scales; tympanum not exposed; canthus rostralis sharp in males, rounded in females; supralabials 8–9; infralabials 8–9, both keeled; males have low nuchal crest comprising seven erect scales, with two rows of smaller erect scales at base; limbs long; fifth toe longer than first; dorsal scales small, with scattered larger scales; forehead scales keeled, those on occiput and supraocular largest; dewlap and fold weak; precloacal and femoral pores absent.

COLOUR Dorsal surface medium brown to brownish-red; nape, thighs and waist pale green when not stressed; pupil black with pale yellow ring; small yellow spot on each side of eyelids; supercilliaries pale yellow; labials yellowish-brown; rostral appendage brown; dark radiating lines from eyes to upper lips; oral cavity black; lining of mouth pale blue; gular sac yellow-cream, unpatterned; limbs have indistinct dark bands; venter yellowish-cream; tail banded with yellow, more distinct apically.

HABITAT AND BEHAVIOUR Inhabits lowland rainforests and mid-hills of northern, northeastern and northwestern Borneo, up to 900m, and associated with saplings and tree trunks. Two eggs produced, each 7 x 15mm.

DISTRIBUTION Endemic to northern and central Borneo, including Sarawak (Upper Baloi); Brunei Darussalam (Bukit Pangar, Tutong); Sabah (Kinabalu Park, Danum Valley Field Centre, Deramakot, Tawau Hills Park, Sungei Kallang, Serinsim, Sandakan Bay) and Kalimantan (Long Blu, on upper reaches of Sungei Mahakam).

IUCN THREAT STATUS Least Concern.

Ex-situ individual

Close-up of head

Lateral view

Dorso-lateral view

Bronchocela Crested Lizards

Medium-sized to large (SVL 80–130mm) arboreal agamids; body slender with long limbs and tail; dorsals and laterals homogenous; dorsal crest present; dewlap absent or weakly developed; eggs spindle shaped.

Key to Bornean species of *Bronchocela*

1a. Six to 10 upper dorsal scale rows pointed backwards and upwards .... *B. cristatella* (below)

1b. Two to four upper dorsal scale rows pointed backwards and upwards .... *B. jubata* (p. 52)

Crested Green Lizard *Bronchocela cristatella* (Kuhl, 1820)

(Bahasa Indonesia: Londok; Bahasa Malaysia: Balingkasa)

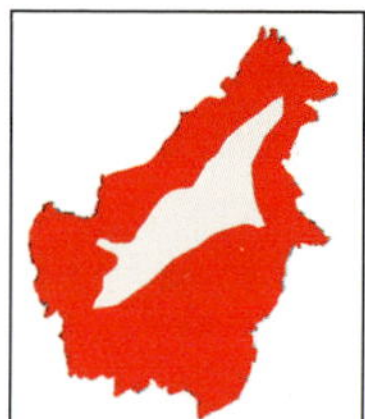

SIZE 130mm

IDENTIFICATION FEATURES Body slender; head long; snout slightly longer than orbit, tapering and rounded; body distinctly compressed; nuchal crest weak, comprising lanceolate scales; dorsal crest somewhat distinct, not extending to tail; scales in middorsal region slightly large and strongly keeled; no dewlap or oblique skin fold on shoulder and no swollen tail-base in adult males; scales unequal, smooth; toe IV equal to toe V; tympanum diameter half that of orbit; ventral scales nearly five times as large as dorsal scales; middorsal scale rows 60–100.

Close-up of head

COLOUR Dorsal surface bright green when stressed, sometimes with white or light blue spots that may form bars, or with white bars, changeable to brown, with indistinct dark bands on body and tail; pale grey stripe across eyes; tympanum greyish-brown; venter yellowish-green; pupil rounded, black, with narrow yellow ring; eyelids greenish-yellow; anterior half of tail green with indistinct grey bands; posterior half brownish-grey.

HABITAT AND BEHAVIOUR Widespread in the lowlands and mid-hills up to 1,700m. Diurnal and arboreal. Forest-edge species; also invades parks and gardens, and associated with shrubs and tree trunks. Diet includes bugs, beetles, flies and ants, in addition to small lizards. Eggs 30–35.8 x 8.1–11mm; 1–4 eggs in a clutch.

DISTRIBUTION Southern Myanmar, Thailand, the Malay Peninsula, Sumatra, Borneo, Java, to the Lesser Sundas, Makulu and the Philippines; old records from New Guinea presumably based on introductions. Widespread on Borneo.

IUCN THREAT STATUS Least Concern.

Sleeping adult

Maned Forest Lizard *Bronchocela jubata* Duméril & Bibron, 1837

(Bahasa Indonesia: Londok)

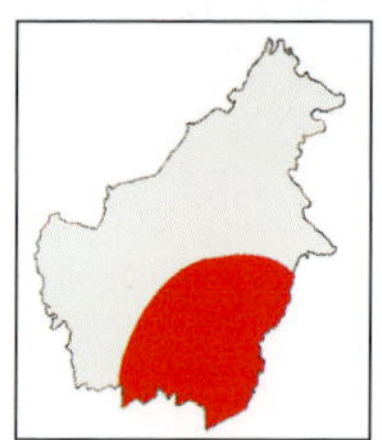

SIZE 150mm

IDENTIFICATION FEATURES Body relatively robust; limbs relatively long; nuchal and dorsal crest comprising lanceolate scales that point posteriorly; dewlap reaches pectoral region; axillary fold absent; tympanum half or more than half diameter of orbit of eye; forehead scales keeled; supralabials 9–10; infralabials 8–9; midbody scale rows 43–53; row of large scales along chin, parallel to labials; ventral scales large, strongly keeled; fingers III and IV subequal.

COLOUR Dorsal surface green, capable of changing to brown or black, with yellow or red spots or vertical bars; venter cream.

HABITAT AND BEHAVIOUR Inhabits lowland rainforests, and also human-altered habitats. Arboreal and diurnal; active on tree trunks. Insectivore. Clutches of two eggs, each 43–53 x 9.5–10mm, produced. Hatchlings 162mm.

DISTRIBUTION Java, Bali, Borneo, Singkep, Sulawesi, and the Karakelang and Salibabu Archipelagos. Within Borneo, Bankajang (Kalimantan Barat), Balikpapan (Kalimantan Timur) and Banjarmasin (Kalimantan Selatan).

IUCN THREAT STATUS Least Concern.

Calotes Garden Lizards

Body stout in adults, laterally compressed; dorsal scales large, rhomboidal, homogeneous; enlarged spines on back of head; no large scales below exposed tympanum; dorsal crest developed in adults; precloacal and femoral pores absent; tail long, and swollen at base in adult males.

GARDEN LIZARD *Calotes versicolor* (Daudin, 1802)

(No vernacular names recorded)

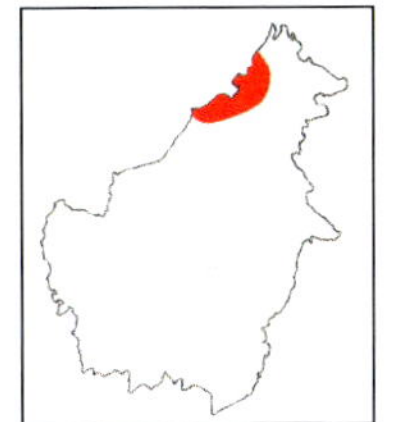

SIZE 500mm

IDENTIFICATION FEATURES Body relatively robust, laterally compressed; head large; nuchal and dorsal crest comprising continuous scales; body scales point backwards and upwards; fold in front of shoulder absent; two sets of spines above tympanum; gular pouch in breeding males; scales around body 35–52.

COLOUR Dorsal surface variable, from greyish-olive and pale brown to bright orange or red, particularly on anterior; dark patch on throat of displaying males; dark radiating lines from orbit region; juveniles and females have pale yellow dorsolateral stripes.

HABITAT AND BEHAVIOUR Inhabits open habitats such as parks and gardens. Distributed in coastal areas on Borneo, and on mainland Asia, up to 1,000m asl. Diet comprises invertebrates and small vertebrates. Oviparous, digging a hole under trees where about 12 eggs are laid.

DISTRIBUTION Eastern Iran, Afghanistan, east through Bangladesh, Bhutan, India, Nepal, Pakistan, Sri Lanka, to Hainan and Hong Kong in the east, and Indo-China, the Malay Peninsula and Sumatra in the south-east). Introduced into the Andaman Islands, Singapore, Réunion and Mauritius, in the Mascarene Archipelago and Florida in the USA. Within Borneo, around Bandar Seri Begawan in Brunei Darussalam; Miri in Sarawak), and Kota Kinabalu in Sabah.

IUCN THREAT STATUS Least Concern.

Draco Flying Lizards

Medium-sized to large (SVL to 139mm) arboreal agamids; body slender; two large, spreadable patagia, supported by the last 5–7 elongated and movable ribs; neck lappets present; dewlaps typically present in both sexes; tail long; femoral and precloacal pores absent.

Key to Bornean species of *Draco*

1a. Patagial ribs five 2
1b. Patagial ribs six 4

2a. Nostril directed dorsally 7
2b. Nostril directed laterally 5

3a. Patagium shades of yellow, green or brown, with 5–6 pairs of dark bands; dewlap grey in males, yellow in females; dewlap with large scales .... *D. obscurus* (p. 61)
3b. Patagium black or dark; dewlap without large scales 6

4a. Patagium black with orange or yellow spots; dewlap black in males, grey, black or red in females, basally with a white spot .... *D. melanopogon* (p. 60)
4b. Patagium brown with dark marbling that may form bands; dewlap bright yellow in males, red or orange in females, both basally with a dark spot .... *D. haematopogon* (p. 58)

5a. Supraciliary edge with thorn-like scale; SVL 97–132mm (males), 95–119mm (females 6
5b. Supraciliary edge without thorn-like scale; SVL 70–90mm (males), 79–88mm (females) .... *D. cristatellus* (p. 56)

6a. Dewlap white with a yellow tip .... *D. punctatus* (p. 62)
6b. Dewlap orange .... *D. fimbriatus* (p. 57)

7a. Nostril directed dorsally; supraciliary edge without thorn-like scale 8
7b. Nostril directed laterally; supraciliary edge with thorn-like scale 9

8a. Patagium orange with black bands; toe IV lamellae 17–23; SVL to 106mm .... *D. quinquefasciatus* (p. 63)
8b. Patagium green with black bands; toe IV lamellae 28–34; SVL to 139mm .... *D. maximus* (p. 59)

9a. Patagium red with black spots or green with dark stripes; dark spot on interorbital and nuchal regions .... *D. cornutus* (opposite)
9b. Patagium black with yellow spots; no dark spot on interorbital or nuchal regions .... *D. sumatranus* (p. 64)

HORNED FLYING LIZARD *Draco cornutus* Günther, 1864

(Bahasa Indonesia: Cicak Terbang; Bahasa Indonesia: Cicak Terbang)

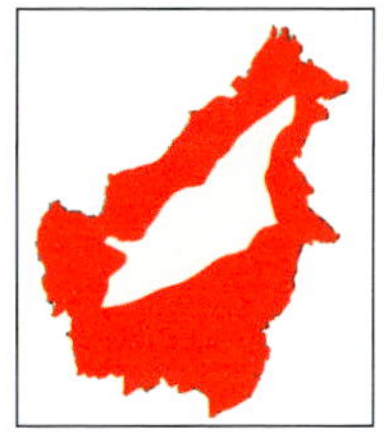

SIZE 89mm

IDENTIFICATION FEATURES Body slender; thorn-like scale on supraciliary edge longer than 0.6mm; tympanum typically without scales; dewlap triangular, covered with small scales; nostrils oriented laterally; neck lappets have large scales; males and females have reduced nuchal crest consisting of triangular scales, a dorsal crest being absent in the species; supralabials 8–10; lamellae under toe IV 20–27; dorsals smooth, sometimes intermixed with some larger scales; posterior edges of thighs have fringe-like scales; ventrals subequal to dorsals, keeled.

COLOUR Dorsal surface bright green to olive-brown, sometimes with indistinct dark bands in males, changeable to dull brown; tan or light brown with extensive darker areas in females; patagium reddish-orange, becoming more green towards body in males and brown in females, with dark spots that may fuse to form bands; wide, dark interrupted margin; black interorbital spot and spots on sides of neck; dewlap and inner neck lappet in males yellowish-orange or greenish-white, the former tipped with orange and with blue base, lacking black dots basally; in females, these are white with yellowish-brown speckling; venter bluish-grey in males, yellowish-cream or white in females, with blackish-grey blotches; under surfaces of limbs and tail pale blue.

HABITAT AND BEHAVIOUR Inhabits coastal areas and lowlands up to the mid-hills, to 700m. Diurnal and arboreal. Diet comprises small black ants. Lays 2–3 three eggs. Hatchlings 18mm.

DISTRIBUTION Range includes Sumatra, Java, Banguran Islands, Borneo and the Sulu Archipelago. Widespread in coastal and inland localities on Borneo, including Labuan, Sabah, Sarawak, Brunei Darussalam and Kalimantan. The best places for seeing it include Bako National Park in Sarawak, and Ulu Temburong, Brunei, and Kiau, Kinabalu Park in Sabah.

IUCN THREAT STATUS Least Concern.

Crested Flying Lizard *Draco cristatellus* Günther, 1872

(Bahasa Indonesia: Cicak Terbang; Bahasa Indonesia: Cicak Terbang)

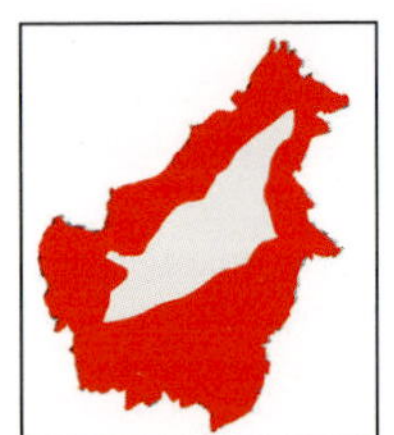

SIZE 90mm

IDENTIFICATION FEATURES Body slender; patagial ribs 5–6; nostrils oriented laterally or slightly obliquely; tympanum large; dewlap triangular; tapering gradually, covered with small scales; males have low but continuous caudal crest, formed by triangular, pointed scales; supralabials 9–10; subcaudals and adjacent scales subequal.

COLOUR Dorsal surface brownish-grey to greenish-olive with dark transverse bars or spots; patagium dark brown with lighter lines; venter blue with dark spots; gular sac white, dark streaked, with bright yellow tip and base.

HABITAT AND BEHAVIOUR Associated with canopy of lowland rainforests and mid-hills. Diet comprises small insects. Reproductive biology unstudied.

DISTRIBUTION Range includes the Malay Peninsula. Known from isolated localities on Borneo, including Gunung Santubong, Sungei Baram, Sungei Seran, Kapit Division, Sungei Mengiong in Sarawak; Poring, Sandakan, Dewhurst Bay, Sungei Kuamut in Sabah, and Rantau and Upper Mahakam in Kalimantan.

IUCN THREAT STATUS Data Deficient.

Fringed Flying Lizard *Draco fimbriatus* Kuhl, 1820

(Bahasa Indonesia: Cicak Terbang; Bahasa Indonesia: Cicak Terbang)

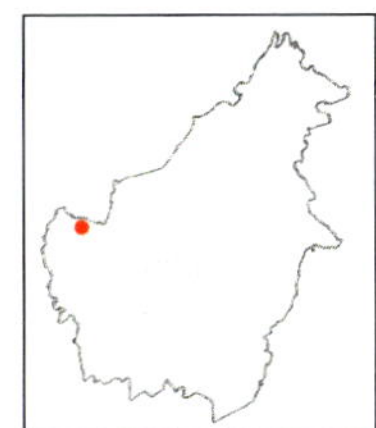

SIZE 132mm

IDENTIFICATION FEATURES Body relatively robust; spinous projection on supraciliary region; six patagial ribs; nostrils oriented laterally; dewlap triangular, with small scales; tympanum large, typically without scales; supralabials 11; infralabials 9–12; males have low nuchal sail and caudal crest comprising small, triangular scales; females have low nuchal crest; posteriors of thighs and tail-base have fringe-like scales; top of body generally smooth, with some intermixed keeled scales.

COLOUR Top of body dark grey-brown, lacking distinct bands; forehead and back of head grey to brownish-grey, with indistinct dark marks between eyes; sides of head and jaws greyish-cream, with darker mottling; narrow grey interorbital bar; supralabials and infralabials have darker areas, forming indistinct bars; digits and forelimbs and hindlimbs have diffuse brownish-grey bands on a greyish-cream background; throat yellowish-grey, with black speckling anteriorly; outer aspect of throat lappet yellow-grey with black spots; inner aspect pinkish-orange or salmon-pink; gular sac salmon-pink; venter yellow; patagium pale grey, lacking a radiating pattern, dorsally heavily suffused with dark areas; ventrally, patagium pale grey, lacking dark speckles on anterior edge; tail with narrow, indistinct dark bands; intervening areas of tail grey; tail-base unpatterned yellow.

HABITAT AND BEHAVIOUR Inhabits lowland forests, and associated with large trees. Dorsal pattern similar to lichen-covered tree bark, presumably offering camouflage when at rest. Two to four eggs constitute a clutch, with each egg measuring 16–17 x 10–11mm, and hatching taking place 40 days later.

DISTRIBUTION Range extends through the Malay Peninsula, including Singapore, Sumatra, the Mentawai Archipelago and Borneo. On Borneo, only confirmed record is from Kubah National Park in Sarawak. This species is often confused with the Spotted Flying Lizard (p. 62).

IUCN THREAT STATUS Least Concern.

Red-bearded Flying Lizard *Draco haematopogon* Boie, in Gray, 1831

(Bahasa Indonesia: Cicak Terbang; Bahasa Indonesia: Cicak Terbang)

SIZE 94mm

IDENTIFICATION FEATURES Body slender; five patagial ribs; caudal crest absent; nostrils oriented dorsally; tympanum large, covered with skin; dewlap elongated, triangular, tapering gradually, covered with small scales; row of keeled scales on snout; no thorn-like scale on supraciliary edge; supralabials 11–12; subcaudals 1.2–1.6 times length of adjacent scales; dorsals 148–184.

COLOUR Top of body olive-grey to brownish-grey, with indistinct lighter and darker spots; patagium black with yellow spots; males have black spot at base of red dewlap; in females, grey-brown spot at base of dewlap; limbs and tail medium brown with indistinct brownish-grey bands; pupils rounded, black; area around eyes yellowish-brown.

HABITAT AND BEHAVIOUR Inhabits mid-hills and submontane forests; also offshore islands once connected to hill ranges. Diurnal and active on tree trunks; diet assumed to comprise ants and other small arthropods. Clutch size 2–3 eggs.

DISTRIBUTION Range includes the Malay Peninsula, Sumatra, Java and Borneo. Rather rare hill-forest species of flying lizard, and reported from isolated localities throughout Borneo. The best places to see it include Gunung Penrissen and Gunung Mulu National Park in Sarawak, and Sepilok Forest Reserve and Pulau Gaya in Sabah.

IUCN THREAT STATUS Least Concern.

LARGE FLYING LIZARD *Draco maximus* Boulenger, 1893

(Bahasa Indonesia: Cicak Terbang; Bahasa Indonesia: Cicak Terbang)

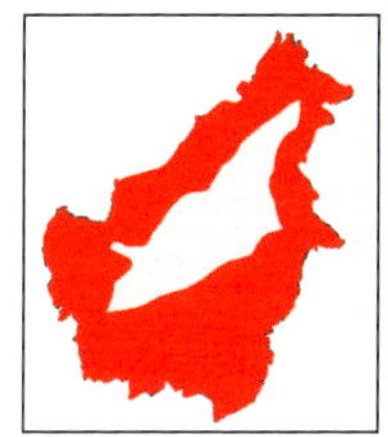

SIZE 139mm

IDENTIFICATION FEATURES Relatively robust; six patagial ribs; spinous projection on supraciliary absent; dewlap elongated, triangular, covered with small scales, much larger in males than in females; neck lappets have slightly larger scales on edges; nostrils oriented upwards; tympanum partially or completely covered with scales; supralabials 14–15; dorsals mostly homogenous, with small series of large scales at base of patagium; fringe-like scales on posterior margins of thighs and tail-base; ventrals larger than dorsals, keeled; males have a nuchal sail and relatively larger dewlaps than females.

COLOUR Top of body green or olive-green, with brownish-olive, wavy pattern of bands; patagium black with discontinuous olive-brown lines; lower surface of patagium almost unpigmented cream with yellow flush; inner surface of neck lappets pale grey in males, dark grey in females; dewlap black basally, with white dots, the distal portion reddish-brown in males, olive-brown in females; pectoral and abdominal areas cream with a pink flush; under surfaces of limbs and tail cream with some pale yellow areas.

HABITAT AND BEHAVIOUR Found on large trees in riparian forests, from the lowlands up about 1,000m asl. Clutches comprise 1–5 eggs, each 11 x 18mm.

DISTRIBUTION Range includes the Malay Peninsula, Sumatra, Borneo and the Natuna Islands. Widespread on Borneo, with records from all parts. The best sites for viewing it include Gunung Gading, Kuching Division and Gunung Matang in Sarawak, and Tawau Hills Park, Poring in Kinabalu Park, Danum Valley Field Centre, Maliau Basin, Sabah and Maruwai and Nasional Bentung-Kerihun in Kalimantan.

IUCN THREAT STATUS Least Concern.

Black-bearded Flying Lizard *Draco melanopogon* Boulenger, 1887

(Bahasa Indonesia: Cicak Terbang; Bahasa Indonesia: Cicak Terbang)

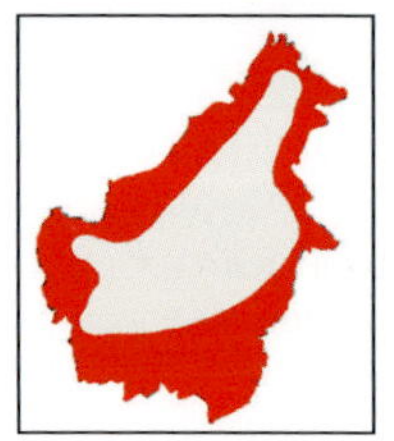

SIZE 93mm

IDENTIFICATION FEATURES Body slender; no spinous projections on supraciliary; tympanum large, partially or completely covered with small scales; five patagial ribs; nostrils oriented upwards; dewlap long and triangular; neck lappets have large scales; dewlap elongated, tapering only basally, scales covering dewlap slightly enlarged; supralabials 12–13; dorsals smooth or weakly keeled, heterogenous; lateral row of large scales; fringe-like scales on posterior of thighs and tail-base; ventrals keeled, larger than dorsals.

COLOUR Top of body olive or green with dark markings, comprising brownish-grey bands or diamond-shaped spots, with lighter intervening areas; patagium dorsally black with scattered yellow-orange spots, the pattern visible ventrally; dewlap in males black, with white edge at posterior aspect of base; in females, dewlap black to brownish-red, white basally; distal portion of inner aspect of neck lappet black, basal portion white; pectoral and abdominal region cream with pink flush; under surfaces of limbs and tail greyish-yellow.

HABITAT AND BEHAVIOUR Associated with non-riparian portions of primary and secondary lowland dipterocarp and non-dipterocarp forests, up to about 600m asl. Found on tree trunks. Diet comprises ants, small beetles, millipedes, isopods and termites. Clutches comprise two eggs, each 6.5–7.6 x 14–15.7mm.

DISTRIBUTION Range extends from southern Thailand to the Malay Peninsula, including Singapore, Sumatra, Borneo and the Natuna Islands. Widespread on Borneo. The best places to see it include Gunung Santubong, Gunung Matang, Lambir Hills National Park and Gunung Mulu National Park in Sarawak; Ulu Temburong District and Bukit Patoi in Brunei Darussalam, and Poring, Tawau Hills Park in Sabah. Also known from Kalimantan.

IUCN THREAT STATUS Least Concern.

Obscure Flying Lizard *Draco obscurus* Boulenger, 1887

(Bahasa Indonesia: Cicak Terbang; Bahasa Indonesia: Cicak Terbang)

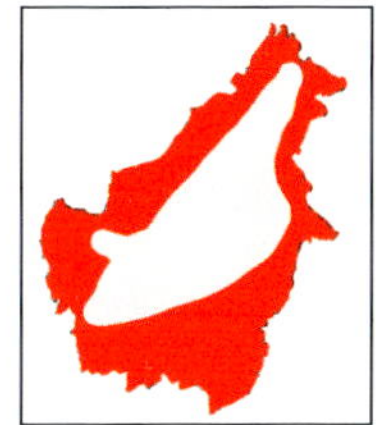

SIZE 113mm

IDENTIFICATION FEATURES Body slender; spinous projections on supraciliary absent; five patagial ribs; nostrils oriented upwards; dewlap oval-elongated, being tapered distally, but expanded at tip, slightly constricted at base, its distal portion with large scales; tympanum entirely or partially covered with scales; supralabials 9–10; dorsals heterogenous, mostly smooth; lateral row of large scales; posterior aspect of thighs and tail-base with fringe-like scales; ventrals larger than dorsals, keeled.

COLOUR Top of body grey-brown with scattered dark brown flecks; five cyan butterfly-shaped marks across vertebral region; patagium yellowish-olive with 5–6 broad dark bands that enclose large yellowish-olive areas; back of forehead has small, black butterfly-shaped mark; narrow dark bar across back of head that joins orbits; nape has two oval black spots; dewlap grey in males, olive-yellow in females; inner side of neck lappet maroon in males, pale pink in females; patagium olive-grey margined; forehead olive-brown with dark brown flecks; upper surfaces of forelimbs and hindlimbs grey-brown, with dark brown bands; tail grey-brown with dark brown bands; most distinct basally; venter unpigmented olive-yellow, except throat, which has dark grey flecks; venter of patagium olive.

HABITAT AND BEHAVIOUR Associated with lowland dipterocarp forests and mid-hills, up to about 1,000m asl. Diurnal and arboreal, and active on low tree trunks; diet presumably ants. Between one and five eggs laid at a time.

DISTRIBUTION Range extends across southern Thailand, the Malay Peninsula, Sumatra, Borneo and the Natuna Islands. Known from isolated localities across Borneo. Best seen in locations including Gunung Matang and Niah National Park in Sarawak; Poring and Kiau, within Gunung Kinabalu Park, and Tawau Hills Park in Sabah. Also recorded in Kalimantan.

IUCN THREAT STATUS Least Concern.

SPOTTED FLYING LIZARD *Draco punctatus* Boulenger, 1900

(Bahasa Indonesia: Cicak Terbang; Bahasa Indonesia: Cicak Terbang)

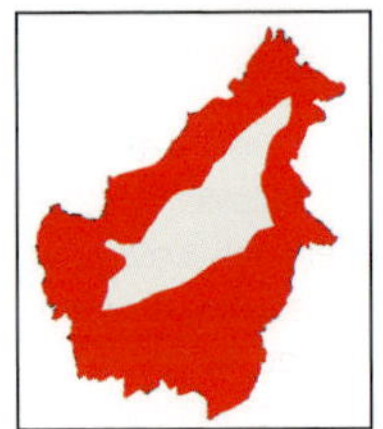

SIZE 78mm

IDENTIFICATION FEATURES Body relatively robust; spinous projection on supraciliary region; five patagial ribs; nostrils oriented laterally; dewlap triangular, with small scales; tympanum large, typically without scales; supralabials 11; infralabials 9–12; males have low nuchal sail and caudal crest comprising small, triangular scales; females have low nuchal crest; posteriors of thighs and tail-base have fringe-like scales; top of body generally smooth, with some intermixed keeled scales.

COLOUR Top of body has four dark grey-brown bands with poorly defined borders; intervening areas greyish-cream; forehead and back of head brownish-grey, with dark oval mark between eyes; sides of head and jaws greyish-cream, with darker mottling; interorbital bar greyish-cream; supralabials and infralabials have darker areas, forming bars; digits and forelimbs and hindlimbs have diffuse dark brown bands on a greyish-cream background; throat yellowish-grey, with black speckling anteriorly; outer aspect of throat lappet yellow-grey with black spots; inner aspect pinkish-orange or salmon-pink; gular sac salmon-pink, unpigmented in males and with lighter anterior and lower margins in females; rest of venter light grey in males; yellow-cream with dark brown speckles in females; patagium dark grey, with greyish-yellow lines, forming radiating pattern, dorsally heavily suffused with dark areas, apart from areas adjacent to body; ventrally, patagium yellowish-grey, with dark speckles on anterior edge; tail has 16 dark brown bands; intervening areas of tail greyish-cream; tail-base yellow-cream with black speckles.

HABITAT AND BEHAVIOUR Inhabits lowland and mid-elevation forests, and associated with large trees, often at heights of 20m above substrate. Dorsal pattern remarkably similar to that of tree bark, presumably offering camouflage when at rest. Reproductive habits unknown, given the history of misidentification (with *Draco fimbriatus*).

DISTRIBUTION Range extends through the Malay Peninsula, including Singapore, Sumatra, the Mentawai Archipelago, Borneo, and the archipelagos of Batu and Banya. Widespread on Borneo, the best localities to observe the species is Niah National Park in Sarawak and Poring Hot Springs in Sabah. This species is often confused with *D. fimbriatus*.

IUCN THREAT STATUS Least Concern.

At rest

Lateral view

FIVE-BANDED FLYING LIZARD *Draco quinquefasciatus* Hardwicke & Gray, 1827

(Bahasa Indonesia: Cicak Terbang; Bahasa Indonesia: Cicak Terbang)

SIZE 110mm

IDENTIFICATION FEATURES Body slender; spinous projections on supraciliary absent; six patagial ribs; nostril oriented upwards; dewlap elongated, slender, crescentic, tapering to narrow tip and covered with small scales; males have low nuchal sail; tympanum small, entirely covered with small scales; neck lappets have large scales; supralabials 12–14; dorsals heterogenous, mostly smooth; lateral row of enlarged scales; posterior margin of thighs and tail-base have fringe-like scales; ventrals subequal to dorsals, keeled.

COLOUR Top of body bright green in males, brownish-olive in females, with dark speckling; patagium yellow or orangish-red above, with five dark brown, dark olive or black cross-bars, and transverse row of light spots in the middle; dewlap and inner side of neck lappets bright yellow in males, and of the same colour, but duller, in females; limbs and tail brownish-olive, with pale bands.

HABITAT AND BEHAVIOUR Inhabits forests, from the lowlands to the mid-hills, in both riparian and non-riparian situations, and also known from peat swamps. Diurnal and arboreal. Diet comprises ants. Clutches include 2–4 eggs, each 15.7–17.2 x 9.6–10.2mm.

DISTRIBUTION Range extends over southern Thailand and the Malay Peninsula, including Singapore, Sumatra, Pulau Sinkep, Pulau Belitung and Borneo. Known from all of Borneo. The best places to see it include Gunung Santubong, Ranchan Pool Forest, Serian, Lambir Hills National Park, Loagan Bunut National Park in Sarawak; Sungei Ingei, Ulu Temburong District and Bukit Patoi in Brunei Darussalam; Sepilok, Gunung Kinabalu and Danum Valley Field Centre in Sabah, and Gunung Kenepai and Batu Jorong in Kalimantan.

IUCN THREAT STATUS Least Concern.

COMMON FLYING LIZARD *Draco sumatranus* Schlegel, 1844

(Bahasa Indonesia: Cicak Terbang; Bahasa Indonesia: Cicak Terbang)

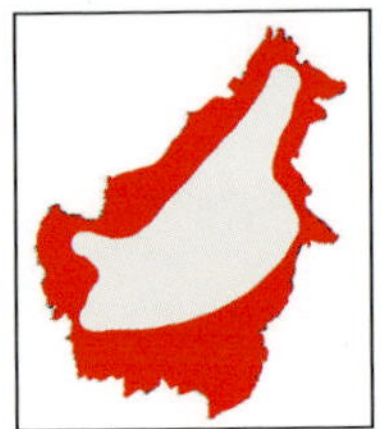

SIZE 85mm

IDENTIFICATION FEATURES Body slender; patagium with six (rarely five or seven) ribs; caudal crest absent; nostrils oriented laterally; dewlap triangular, covered with small scales; nuchal crest present, comprising 6–20 compressed, triangular or keeled scales in males, and up to 14 in females; supralabials 6–12; dorsals 103–166; tympanum typically covered with smooth skin; 2–4 large, keeled scales between tympanum and eye.

COLOUR Adult males have powder-blue forehead when displaying; ground colour of top of body light brown with dark brown blotches; patagium dorsally brownish-grey, with obscure yellowish-orange blotches, isolated or confluent with each other, those on the edges, especially the anterior, brighter yellow; under surface of patagium yellow with black mottling, especially on anterior edge; large, pale cream patch below eye at angle of jaws; snout-tip dark grey-brown; pale brown interorbital bar; black spots on interorbital and sagittal regions; rest of head and neck brown, variegated with darker and lighter spots; ocular region dark brown, with yellow radiating lines on upper and lower eyelids; chin cream with dark grey mottling; gular sac bright yellow, with black dots at base in males, pale blue distally, grey, mottled with grey-brown, basally; venter bluish-grey.

HABITAT AND BEHAVIOUR Inhabits open forests and plantations; commonly seen in parks and gardens. Diurnal and arboreal. Diet comprises ants, termites and beetles. Clutches of 1–5 eggs are produced.

DISTRIBUTION Range extends from Thailand and the Malay Peninsula, including Singapore, Sumatra, the Mentawai and Riau Archipelagos, to Borneo and Palawan. Widespread on Borneo. The best places to view it include Kuching city, the Matang Range, Kota Samarahan in Sarawak; Sandakan in Sabah, and Banjarmasin, Gunung Kenepai and Sinkawang in Kalimantan.

IUCN THREAT STATUS Least Concern.

Gonocephalus Angle-headed Lizards

Medium-sized to large (SVL to 160mm) arboreal agamids; body strongly compressed; distinct angular bony ridge along snout from eye to nostril; dewlap, nuchal crest, shoulder fold and tympanum present; forehead covered with small, granular scales; dorsals and laterals subequal, typically intermixed with large scales; femoral and precloacal pores absent.

Key to Bornean species of *Gonocephalus*

1a. Supraciliary border strongly raised, angular posteriorly .... *G. doriae* (p. 67)
1b. Supraciliary border not strongly raised, rounded posteriorly **2**

2a. Dorsal scales equal, without large scales; females have nuchal sails, not crests .... *G. grandis* (p. 68)
2b. Dorsal scales unequal, with large scales; females have nuchal crests **3**

3a. Dorsal crest reduced, ridge-like; enlarged dorsolateral scales forming oblique rows .... *G. mjobergi* (p. 70)
3b. Dorsal crest distinct; dorsolateral scales not enlarged **4**

4a. No enlarged scales above and below tympanum; flanks have light yellow oval spots within network of dark reticulations; iris of males brown or light blue, surrounding skin not orange ...... *G. bornensis* (p. 66)
4b. Enlarged scales above and below tympanum; flanks with or without light rounded spots; no dark reticulations; iris of males bright blue, surrounding skin orange .... *G. liogaster* (p. 69)

Bornean Angle-headed Lizard (p. 66)

Bornean Angle-headed Lizard *Gonocephalus bornensis* (Schlegel, 1848)

(No vernacular names recorded)

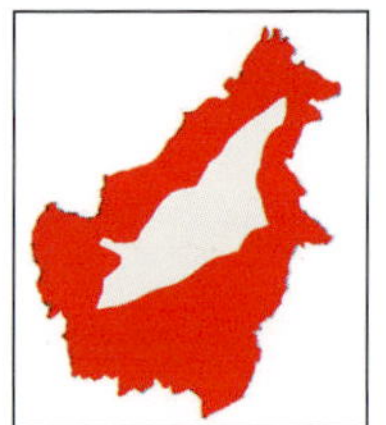

SIZE 136mm

IDENTIFICATION FEATURES Body robust; tail thick, nearly rounded; nuchal and dorsal crests continuous, highly developed in males, comprising elongated, lanceolate scales that run up to level of tail and decreasing in height posteriorly; females have high nuchal crest but lower dorsal crest; dorsal crest fused in juvenile males; in females, nuchal crest distinctly longer than dorsal crest; supraciliary border rounded; dorsals intermixed with larger scales; no large scales below orbit of eye; ventrals smooth, more convex on pectoral region.

COLOUR Top of body bright green with darker markings, comprising five dark transverse bands; dark nuchal-gular collar encircling throat; sides of head and flanks have dark green spots, each with pale green centre; lower jaw yellow with dark brown longitudinal stripes, those near throat converging; tympanum unpatterned green; body flanks have light oval spots; nuchal and body crest brown and yellow; dewlap pale with dark, broken stripes; labials have brown bars; brown radiating lines from orbit of eye; anterior edge of sclera bright blue; limbs green with dark bands; venter yellow with grey-brown transverse bands entering from flanks; tail yellow-cream with 12 brown-black bands.

HABITAT AND BEHAVIOUR Inhabits primary rainforests in the lowlands and mid-hills, up to about 1,100m asl. Diurnal and arboreal. Diet includes ants and spiders. Clutches comprise four eggs, each 4.2–4.3 x 9.0–9.9mm; hatchlings 35–37mm.

DISTRIBUTION Endemic to Borneo. Widespread on Borneo. The best sites for the species include Kubah National Park, Gunung Gading National Park, Lambir Hills National Park and Gunung Mulu National Park in Sarawak; Melilas in Belait District and Ulu Temburong in Brunei Darussalam; the lower reaches of Gunung Kinabalu, Sepagaya Forest Reserve and Deramakot, Kinabatangan in Sabah, and Maruwai and Taman Nasional Gunung Palung in Kalimantan.

IUCN THREAT STATUS Least Concern.

Marquis Doria's Angle-headed Lizard *Gonocephalus doriae*

(Peters, 1871)
(No vernacular names recorded)

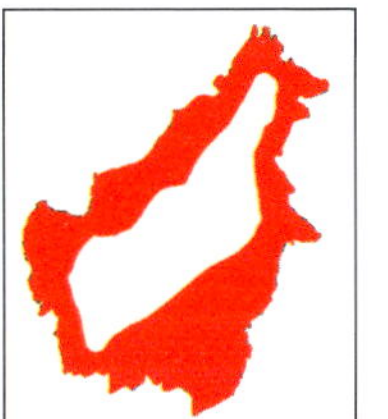

SIZE 163mm

IDENTIFICATION FEATURES Body robust; tail thick, nearly rounded; nuchal crest composed of low, overlapping and slightly crescentic scales; superciliary border strongly raised and angular posteriorly; no spine-like scales on crest in adult males.

COLOUR Top of body of males bright green, with an indistinct grey, wavy pattern, or with large areas of orange, changeable to reddish-brown with dark and light flecks; gular sac yellow, orange or grey with six bluish-grey stripes; lower flanks have eight transverse bars; top of female and juvenile body green with dark blotches; venter bright yellow or cream; limbs and tail bright green, with pale green or grey and orange bands; under surfaces of tail and limbs pale yellow.

HABITAT AND BEHAVIOUR Inhabits lowland rainforests, and associated with low tree trunks, branches and shrubs. Diurnal. Insectivorous and oviparous.

DISTRIBUTION Endemic to Borneo. Isolated records from across Borneo. Sites where the species has been seen recently include Gunung Gading, Gunung Buri, Batang Ai National Park in Sarawak; Deramakot, Kinabatangan District, Danum Valley Field Centre, Tawau Hills Park in Sabah, and Banjarmasin and Barabai in Kalimantan.

IUCN THREAT STATUS Least Concern.

GIANT ANGLE-HEADED LIZARD *Gonocephalus grandis* (Gray, 1845)

(No vernacular names recorded)

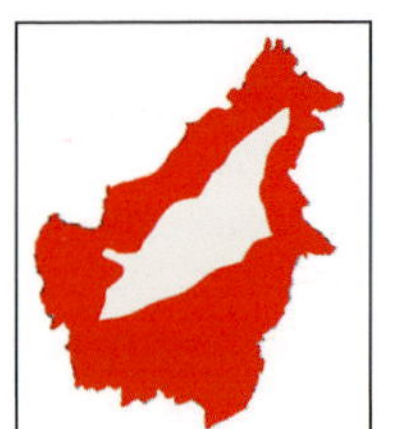

SIZE 160mm (males), 137mm (females)

IDENTIFICATION FEATURES Body robust, laterally compressed; tail thick, nearly rounded; males have high nuchal and dorsal crests, comprising lanceolate scales, with dorsal crest separated from nuchal crest in males; females lack dorsal crest but have a nuchal sail; supraciliary border rounded; dorsals subequal, not with larger scales; ventrals smooth.

COLOUR Top of body greenish-brown, through olive to nearly black; flanks blue, with yellow spots in males; dewlap yellow to red, striped with blue; females and juveniles brownish-green to black, with light, 'V'-shaped bands; flanks of females similar to those in adult males, while those of juveniles have dark reticulations on a pale background; limbs grey-brown with indistinct dark bands; tail grey-brown with dark bands. Males have yellowish-orange throat, with three indistinct stripes, two of which meet at base; pupils black with yellow ring; rest of eyes grey-brown; females have broad postocular stripe, extending to axilla, followed by four dark blotches. Top of body of juveniles brownish-grey, with pale brown, thin interorbital bar; upper lip yellowish-brown, unbarred; lower lip cream; pale brown radiating lines in supraciliary region; throat cream with thin, dark brown transverse bars; broad, dark brown postocular stripe extends to axilla, meeting the one on throat, narrowing ventrally and broadening dorsally; four paired bands on top of body of same colour; lower flanks medium brown with yellow-cream spots; limbs and tail dark banded dorsally; venter of body, limbs and tail dirty grey, with darker and lighter variegation.

Close-up of juvenile head

HABITAT AND BEHAVIOUR Inhabits primary and disturbed forests in the plains and mid-hills, up to around 1,400m asl, near streams and small rivers. Diurnal and arboreal. Diet comprises arthropods, including caterpillars, beetles, grasshoppers, ants, flies, cockroaches and spiders. Clutches comprise 1–6 eggs, each 10–11 x 21–26mm; hatchlings 32–35mm.

Close-up of adult head

DISTRIBUTION Range extends from Peninsular Thailand, the Malay Peninsula, offshore islands like Pinang, Pulau Tioman, Sumatra and associated islands, including Pulau Nias, Pulau Nako and Pulau Sipura of the Mentawai Archipelago, to Borneo. Widespread on Borneo. On Borneo, can be encountered in Gunung Matang, Gunung Santubong, Gunung Gading, Lambir Hills National Park in Sarawak; Tasek Merimbun, Tutong District in Brunei Darussalam; Gunung Kinabalu Park, Sandakan, Tawau Hills Park and Danum Valley Field Centre in Sabah, and upper reaches of Sungei Bulangan, Maruwai, Taman Nasional Bentuang Karimun in Kalimantan.

IUCN THREAT STATUS Least Concern.

Lateral view of adult head

Blue-eyed Angle-headed Lizard *Gonocephalus liogaster* (Günther, 1872)

(No vernacular names recorded)

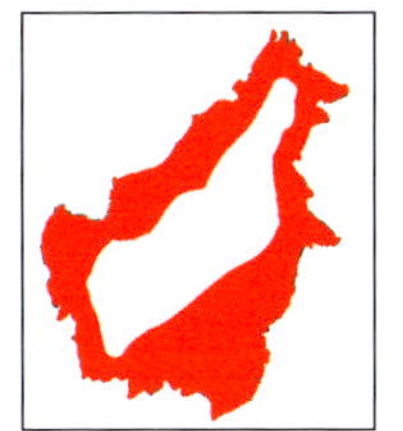

SIZE 140mm

IDENTIFICATION FEATURES Body robust; tail thick, nearly rounded; nuchal and dorsal crests continuous, comprising lanceolate scales that are longest anteriorly; in females, nuchal crest long, dorsal crest low, ridge-like; supraciliary border rounded; dorsals intermixed with larger scales; row of large scales below orbit of eye.

COLOUR Top of body of males brown, olive or green, with dark reticulated pattern on flanks; females have yellow cross-bars; pupils black, rounded, and irises of males bright blue; skin surrounding orbit reddish-orange or bright yellow; in females, eyes are brown; two or three rows of bright yellow scales between upper labials and orbit; throat grey with blackish-grey blotches; limbs and tail brownish-olive with blackish-grey bands.

HABITAT AND BEHAVIOUR Inhabits lowland forests and peat-swamp forests. Arboreal and diurnal; active on trunks of trees, near streams and small rivers. Diurnal and insectivorous. Clutches comprise 1–4 eggs, each 11 x 23mm.

DISTRIBUTION Range includes Peninsular Malaysia, Sumatra and Borneo. Widespread on Borneo. Most likely to be seen in Samunsam Wildlife Sanctuary, Gunung Gading, Gunung Santubong, Gunung Matang, Semenggoh Wildlife Sanctuary, Niah National Park in Sarawak; upper reaches of Sungei Ingei, Belait District and Sungei Rampayoh, Belait District in Brunei Darussalam; Danum Valley Field Centre, Maliau Basin, Tawau Hills Park and Poring, Gunung Kinabalu Park in Sabah, and Pamukan Bay, Sambas and Taman Nasional Kayan Mentarang in Kalimantan.

IUCN THREAT STATUS Least Concern.

Threat display

Threat display

Juvenile

Adult

Mjöberg's Angle-headed Lizard *Gonocephalus mjobergi* Smith, 1925

(No vernacular names recorded)

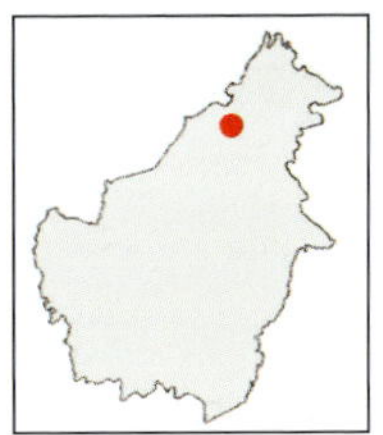

SIZE 88mm

IDENTIFICATION FEATURES Body large, robust; tail thick, nearly rounded; snout longer than eye diameter; supraciliary border not raised; tympanum equal to eye diameter; nostril within a single nasal; forehead scales feebly keeled; supralabials 8–9; infralabials eight; single flat, large scale below tympanum; dewlap small, its edge feebly serrated and covered with small scales; nuchal crest comprising 11 lanceolate scales; dorsal crest a small ridge, not continuous with nuchal crest; enlarged dorsolateral scales forming oblique rows; dorsal scales reduced, feebly keeled; scales on tail feebly keeled.

COLOUR Top of body of adults grass-green in life (greyish-blue in preservative); venter pale green.

HABITAT AND BEHAVIOUR Species known only from type locality, at 2,134–2,250m, the highest altitudinal record for a member of the genus. Nothing else known of natural history of this montane species, although like its other congeners, it is suspected to be an arboreal insectivore.

DISTRIBUTION Endemic to Borneo. Restricted to Gunung Murud in Sarawak.

IUCN THREAT STATUS Data Deficient.

Holotype of *Gonocephalus mjobergi* Smith, 1925 BMNH RR 1946.8.13.87 (ex-BMNH 1924.8.28.8), from Mt Murud, Borneo....7000 feet altitude.

Harpesaurus Horned Lizard

Medium-sized (SVL 59mm) arboreal agamids; body slender, compressed; scales subequal; tympanum distinct; males have single, cylindrical elongated rostral appendage; dewlap present; transverse gular fold absent.

Bornean Horned Lizard *Harpesaurus borneensis* (Mertens, 1924)

(No vernacular names recorded)

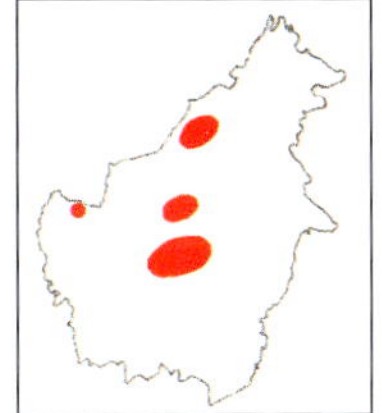

SIZE 59mm

IDENTIFICATION FEATURES Body slender, compressed; single cylindrical rostral appendage (measuring 7.0mm in a 51.5mm SVL individual) in males that is surrounded by four large, petal-like scales; nuchal crest well developed, dorsal and caudal crests present in males, dewlap present; gular scales acute, with keeled tips; supralabials seven; infralabials seven; tympanum oval, 2/3 orbit diameter; dorsal scales and scales on limbs smooth; ventral scales keeled and precloacal and femoral pores absent; tail long, wider at base and becoming more slender apically, prehensile, capable of being coiled upwards.

COLOUR Top of body reddish-brown, with black spots arranged laterally in seven oblique series; yellow-brown oblique bands on top of body; sides of head have dark brown, radiating lines from orbit of eye; supralabials cream with dark blotches; rostral appendage greyish-cream; tympanum black; upper surfaces of forelimbs and hindlimbs brown with yellow-brown bands; tail ringed with black and reddish-brown bands; base of crest, and throat and venter white.

HABITAT AND BEHAVIOUR Inhabits lowland rainforests; associated with tree trunks. Diurnal; diet includes ants. Ovoviviparous, producing live young.

DISTRIBUTION Endemic to Borneo. Has been sighted at Kubah National Park, Niah National Park, Batang Baleh and Bukit Sarang in Sarawak. Also known from central area of the drainage of Sungei Kapuas in Kalimantan.

IUCN THREAT STATUS Data Deficient.

Hypsicalotes Bornean Crested Dragon Lizard

Large (SVL to 145mm) arboreal agamid; body robust; limbs and tail long; sides of face with large scale approximating the size of orbit below tympanum; nuchal and dorsal crests present and separate from each other, continuing on to tail; dewlap distinct in adult males, its scales small; distinct shoulder fold; dorsal scales heterogenous; tail swollen basally, its posterior strongly compressed.

Kinabalu Crested Dragon *Hypsicalotes kinabaluensis* (de Grijs, 1937)

(No vernacular names recorded)

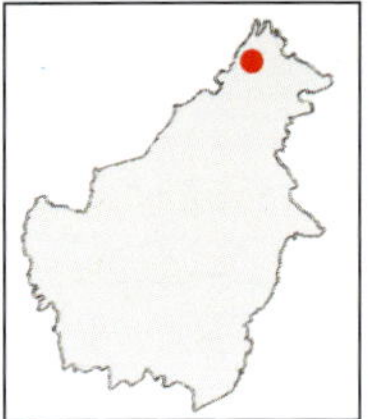

SIZE 145mm

IDENTIFICATION FEATURES Large species. Body relatively robust, and slightly compressed; two large, spinose scales on top of and to back of tympanum; scales imbricate, keeled; midbody scale rows 51–54; nuchal and dorsal crests present, separated from each other; large lanceolate scales forming median row on gular pouch in males, which has reduced, heterogenous scales; lower portion of jaw corner has smooth scale, about twice as large as tympanum, which is slightly smaller than orbit of eye; short oblique fold in front of shoulder; finger IV slightly longer than finger III.

COLOUR Top of body green; trunk has chocolate-brown and black spots that form bands; gular pouch pale red with black and white stripes in anterior margin; venter brown with green spots;

Male

Female

tail brown with small green areas, bearing narrow pale bands; forelimbs and hindlimbs have pale narrow bands; pupil black with a yellow ring; rest of eye brown. In stressed individuals, the green body is changeable to brown, and the labials and chin shields turn yellow.

HABITAT AND BEHAVIOUR Inhabits highland dipterocarp to submontane forests, at 900–1,600m asl. Arboreal, with some terrestrial activity. Diet and reproductive habits unknown.

DISTRIBUTION Endemic to Borneo, and known only from Kinabalu Park in Sabah.

IUCN THREAT STATUS Data Deficient.

Threat display

Female

Pelturagonia Bornean Shrub Lizards

Small (SVL to 84mm) arboreal agamids; body short, squat; head large; tympanum not visible externally; dewlaps absent or reduced; nuchal crest low; rostral divided into two or more scales; spiny tail-base of males swollen and depressed dorsally; inner lining of mouth blue; oviparous.

Key to Bornean species of *Pelturagonia*

1a. Large supraocular spines present..*P. spiniceps* (p. 78)
1b. Supraocular spines absent..**2**

2a. Nasal separated from supralabials; continuous vertebral crest.......*P. nigrilabris* (p. 77)
2b. Nasal in contact with supralabials; vertebral crest indicated by widely spaced vertebral scute..**3**

3a. Two rows of enlarged subcaudals..*P. cephalum* (p. 76)
3b. Four rows of enlarged subcaudals..**4**

4a. Nuchal crest comprises 3–8 scales; area between caudal crest flat..*P. borneensis* (opposite)
4b. Nuchal crest comprises 10–12 scales; area between caudal crest not flat..*P. anolophium* (below)

CRESTED SHRUB LIZARD *Pelturagonia anolophium* Harvey, Larson, Jacobs, Shaney, Streicher, Hamidy, Kurniawan & Smith, 2019

(No vernacular names recorded)

SIZE 193mm

IDENTIFICATION FEATURES Body stout; supercilium lacks long spinose scale; enlarged middorsal scales widely spaced, not forming continuous dorsal crest; nuchal crest consisting of 10–12 triangular scales; paravertebrals on neck and scales of pectoral gap pointing backwards; area between caudal dorsolateral crests not flat, with enlarged projecting scales; four rows of enlarged subcaudals near tail-base; gulars sharply keeled and mucronate, 31–36; scales around midbody 92–109.

COLOUR Top of body green with contrasting yellow tubercles and reddish-brown to brick-red bands; gular region lacks dark pigmentation or with black reticulation without black lines extending from labials.

HABITAT AND BEHAVIOUR Inhabits hills, presumably the upper reaches of hill dipterocarp and submontane forests at 832–1,090m asl, and associated with shrubs. Diet unstudied. Clutches of about 2–4 produced.

DISTRIBUTION Endemic to Borneo. Only known from Gunung Lumut, Paser in East Kalimantan.

IUCN THREAT STATUS Not Evaluated.

Bornean Shrub Lizard *Pelturagonia borneensis* (Inger, 1960)

(No vernacular names recorded)

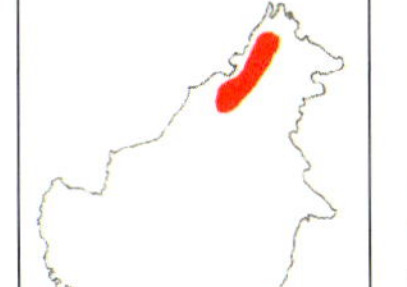

SIZE 155mm

IDENTIFICATION FEATURES Body short, squat; supraciliary spine absent; nasal in contact with supralabial; continuous row of infraorbitals; gular scales sharply keeled; nuchal crest has four conical scales that are thick in males, compressed in females; oblique axillary fold; incomplete transverse gular fold; scales on sides of tail sharply keeled; four rows of keeled subcaudals near tail-base;

COLOUR Top of body olive-grey, greenish-olive, medium brown or greyish-brown, with three yellowish-tan, narrow, transverse and oblique bands between neck and waist; faint narrower bars in intervening regions; two interorbital bars of the same colour; forehead pale or olive-green; upper lip orange-yellow; sides of head orangish-yellow with olive-grey areas, including radiating lines from orbit of eye, postocular stripes meet at back of head; enlarged scales at back of head orange; seven small scales comprise low nuchal crest; these are bicoloured orange and grey; upper surfaces of limbs and tail olive-grey with paler bars; gular region has orange-yellow and olive-green oblique stripes; rest of venter greyish-yellow, with pale grey median line and pale grey blotches; under surface of limbs and tail greyish-yellow, with pale grey blotches; inner lining of mouth deep greyish-black.

HABITAT AND BEHAVIOUR Inhabits montane forests at 1,300–1,800m asl. Diurnal, and associated with low vegetation. Diet unstudied. Clutches comprise two large, elliptic eggs.

DISTRIBUTION Endemic to Borneo. Known from sites across the island, including Bukit Batu Song and Camp 2, Gunung Mulu National Park in Sarawak; Gunung Kinabalu Park, Long Pasia, Crocker Range Park in Sabah.

IUCN THREAT STATUS Least Concern.

LARGE-HEADED SHRUB LIZARD *Pelturagonia cephalum* Mocquard, 1890

(No vernacular names recorded)

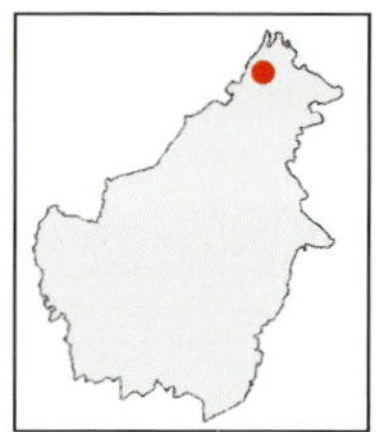

SIZE 84mm

IDENTIFICATION FEATURES Body short, squat; supraciliary spines absent; nuchal crest comprises 7–8 thick, conical scales; vertebral scale row comprises a few weakly elevated, large, widely separated scales posteriorly; nasal typically in contact with supralabials; two continuous rows of infraorbitals; gular scales obtusely keeled or smooth; five large scales on each side of expanded tail-base; enlarged spinous scales on outer edge of femur and forelimbs; ventrals keeled; two rows of large, keeled subcaudals.

COLOUR Top of body pale green, with dark green or greyish-green wavy bands, changeable to dark brown; large scales on nape yellow; tympanum region dark grey-green when stressed; two enlarged cream spines on sides of lower jaws; venter cream with dark green interscale markings joining to form stripes on greenish-cream throat; upper surfaces of limbs, dark streak on each axilla; digits and tail dark green with faint grey bands; outer edges of thighs beige with dark grey-black stripe up to cloaca; inner lining of mouth deep blue.

HABITAT AND BEHAVIOUR Inhabits submontane and montane forests at 1,300–2,100m asl. When threatened, opens its mouth, revealing the blue inner lining of the mouth; in addition, exhibits death feigning. On the ground, moves via a series of short hops. Diet unknown. Clutch size four, eggs each 15.2–16.2 x 8.8–9.5mm.

DISTRIBUTION Endemic to Borneo. Records are from the Crocker Range Park and Kinabalu Park, with one unconfirmed record from Sarawak.

IUCN THREAT STATUS Least Concern.

Black-lipped Shrub Lizard *Pelturagonia nigrilabris* (Peters, 1864)

(No vernacular names recorded)

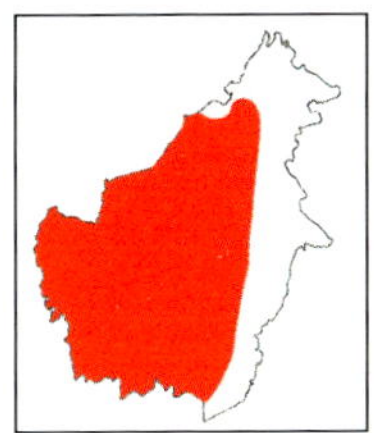

SIZE 58mm

IDENTIFICATION FEATURES Body short, squat; supraciliary spine absent; nuchal crest comprises 6–12 compressed scales; vertebral scale rows have continuous series of large, raised scales posteriorly; nasal separated from supralabials; two continuous rows of infraorbitals; gular scales distinctly keeled; four rows of large scales near tail-base.

COLOUR Top of body of adult males pale brown with four dark brown cross-bars, the first forming an incomplete collar around nape; lips pale green up to temporal region; large scales on chin and jaws pale yellow; gular sac yellow; flanks pale green or pale blue, band-like in intervening areas between the cross-bars; females and juveniles brown to olive; limbs pale brown with broad, dark grey bands; tail pale brown dorsally, with dark brown bands; chin dark grey; throat unpatterned yellow; venter yellowish-cream, with dark brown vermiculation; inner lining of mouth pale blue.

HABITAT AND BEHAVIOUR The only lowland representative of the genus. Diurnal, and active on low tree trunks and shrubs. Diet includes small arthropods. Reproductive habits unknown.

DISTRIBUTION Near endemic to Borneo; also recorded in Pulau Sirhassen, Natuna Archipelago. Most likely to be seen at Ranchan Pool in Serian, Bako National Park, Gunung Santubong, Gunung Matang and Gunung Gading in Sarawak, and Pulaumajang, Sambas and Taman Nasional Bentuang Karimun in Kalimantan.

IUCN THREAT STATUS Least Concern.

Close-up of head and forebody

Night colouration

SPINY-HEADED SHRUB LIZARD *Pelturagonia spiniceps* Smith, 1925

(No vernacular names recorded)

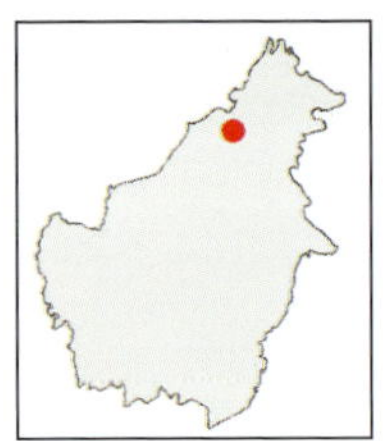

SIZE 60.3mm

IDENTIFICATION FEATURES Body short, squat; snout shorter than eye diameter; supraciliary spine three-quarters eye diameter; nasal in contact with second supralabial; infraorbitals in single continuous row; gular scales keeled; nuchal crest comprises three scales separated by 2–6 small, keeled scales; oblique fold in front of shoulder; vertebral scale row has widely separated large scales posteriorly; two rows of enlarged, keeled subcaudals; tail compressed laterally.

COLOUR Dorsal surface pale green or greenish-grey, with large brown patches, changeable under stress to brown, with thin, pale transverse lines that meet at vertebral region; spinose enlarged scales on top of body yellowish-brown; head variegated with bright green, yellow and cream scales; throat has dark transverse lines; venter cream spotted with grey; dorsal surfaces of limbs bright green with black bands; top of tail olive, turning grey posteriorly; dewlap bright yellow; inner lining of mouth black.

HABITAT AND BEHAVIOUR Inhabits high altitudes of northern Borneo at 1,200–1,800m asl, in moss and oak-laurel forests, at a height of c. 1m from substrate, on tree trunks. When threatened, opens its mouth, revealing deep blue mouth lining. On substrate, moves via a series of short hops. Diet unknown. Lays two eggs, each 12 x 7mm.

DISTRIBUTION Endemic to Borneo. Found in Gunung Murud, Limbang Division in Sarawak; Long Pasia in Sabah, and Gunung Buduk Rakik in northern Kalimantan.

IUCN THREAT STATUS Least Concern.

Pseudocalotes Forest Lizards

Medium-sized (SVL to 92mm) arboreal agamids; body compressed; tail twice as long or longer than head-body length; no dermal fold from angle of jaws to shoulder; no spines from eye to above tympanum; nuchal spines smooth or absent; nuchal crest composed of compressed scales; dorsal crest low; lateral scales smooth or weakly keeled, typically as large as or smaller than ventrals; eggs oval.

Key to Bornean species of *Pseudocalotes*

1a. Temporal region lacking spine; nuchal crest absent .... *P. nigrigularis* (below)
1b. Temporal region with two spines; nuchal crest comprises seven scales .... *P. saravacensis* (p. 80)

KINABALU FOREST LIZARD *Pseudocalotes nigrigularis* (Ota & Hikida, 1991)

(No vernacular names recorded)

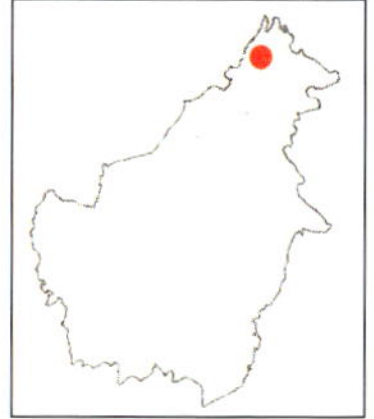

SIZE 76mm

IDENTIFICATION FEATURES Body relatively robust, short; snout tapering, rounded; tail relatively long; distinct gular pouch in males; slight oblique fold in front of shoulder; midbody scale rows 60; scales on upper 3–4 rows directed backwards, the rest directed obliquely downwards; middorsal crest slightly large, strongly keeled, numbering 37 from occipital region to above level of cloaca; anterior nine scales of crest elongated and compressed; upper half slightly keeled; lower half unkeeled; lamellae under toe IV 29.

COLOUR Dorsal surface of body dark brownish-grey, with two broad transverse white bands on trunk between forelimbs and hindlimbs; white spots beneath eyes; neck and shoulder mottled with white; dewlap black with pair of white spots near distal end; limbs have several indistinct white bands; tail has 11 broad transverse pale grey bands; venter white, with distinct dark-mottled pattern.

HABITAT AND BEHAVIOUR Inhabits submontane forests. Arboreal and diurnal. Diet and reproductive habits unstudied.

DISTRIBUTION Endemic to Borneo. Known only from Kinabalu Park in Sabah.

IUCN THREAT STATUS Data Deficient.

Bornean Long-headed Lizard *Pseudocalotes saravacensis*

Inger & Stuebing, 1994

(No vernacular names recorded)

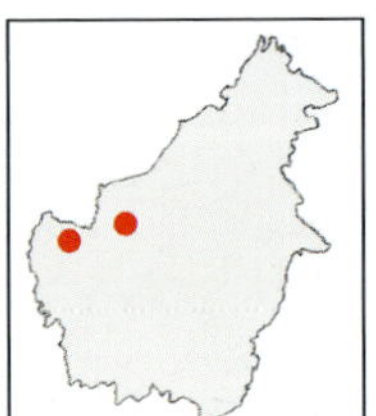

SIZE 82mm

IDENTIFICATION FEATURES Body relatively robust; trunk cylindrical; head large, with obtuse snout; tail fairly long, and distinctly swollen a short distance behind base; distal two-thirds of tail laterally compressed; two projecting, compressed, spine-like scales on temporal region, larger subequal in diameter to tympanic scale; nuchal crest comprises seven tall, flexible scales; prefrontal scales slightly large, forming inverted 'V'.

COLOUR Dorsal surface dark brown with ill-defined green-coloured areas, head slightly lighter; lacks dark, radiating lines from eye; nuchal crest and throat have orange-red wash; venter cream; tail with faint, dark blotches or bands, its base green.

HABITAT AND BEHAVIOUR Associated with lowland and hill dipterocarp forests, at elevations of 240–1,000m. Arboreal and diurnal. Diet and reproductive habits unknown.

DISTRIBUTION Endemic to Borneo. Known only from Lanjak-Entimau Wildlife Sanctuary and Gunung Penrissen in Sarawak.

IUCN THREAT STATUS Data Deficient.

Family Anguidae Glass Snakes

Anguids are small to large lizards (SVL 5.5–130cm). Their characteristics include: skull with paired nasals, postorbitals and a single fused parietal; dentition pleurodont; limbs absent in the single Bornean species; body heavily armoured with largely non-overlapping scales underlain by rectangular osteoderms and with a longitudinal ventrolateral fold that separates the dorsal and ventral armour on each side; short tail; tail autotomy; leathery egg-shell.

Dopasia Asian Glass Snakes

Large (SVL to 125mm) terrestrial anguids; snake-like, with Bornean species lacking limbs; head slightly distinct from neck; body has large rhombic scales in longitudinal and transverse rows that are attached to ossified plates; extensible lateral folds, with reduced, unossified scales; eyes with movable eyelids; ear opening present; tip of tongue thin, bifid; pterygoid teeth present; tail shows autotomy.

BORNEAN GLASS SNAKE *Dopasia buettikoferi* (van Lidth de Jeude, 1905)

(No vernacular names recorded)

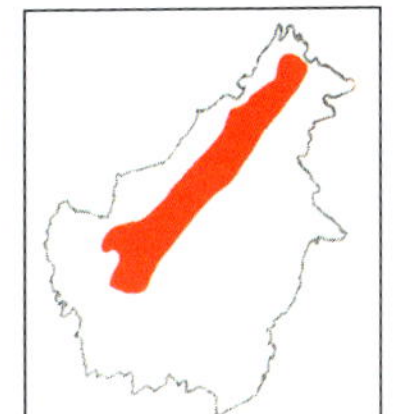

SIZE 125mm

IDENTIFICATION FEATURES Body slender; limbs externally absent; frontal and interparietal subequal and form a wide suture; supraoculars five; tympanum distinct; dorsals in 16–18 longitudinal rows and 98–105 transverse rows; central 12 rows of dorsals keeled to form unbroken straight lines; ventrals smooth, in 10 longitudinal rows; caudals keeled; ear opening distinct; teeth conical.

COLOUR Dorsally brown, with dark lateral band, edged with light above, continuing to tail; small blue interparietal spot; front portion of back has irregular transverse series of blue spots, edged anteriorly with black; five oblique dark lines across labials; labials and venter pale yellow or pale green.

HABITAT AND BEHAVIOUR Associated mostly with montane and submontane forests and rarely, with lowland regions, at 305–1,268m elevation. Terrestrial and also probably semi-fossorial, diurnal and insectivorous. One individual examined had the remains of a cockroach and the fragment of leg of a rat, the latter presumably obtained through scavenging. Reproductive habits unknown. Some members of the family are known to guard eggs, coiling around them and threatening small animals that approach the nest.

DISTRIBUTION Endemic to Borneo. Records exist from isolated localities such as Batang Lemanak, a tributary of the Sungei Lupar, Lantjak-Entimau Wildlife Sanctuary, Sela'an Linau in the Upper Baram and Gunung Mulu National Park in Sarawak; Gunung Kinabalu Park, Gunung Trus Madi and Long Pasia in Sabah, and Bukit Liang Kubung in Sungei Kapuas basin and Taman Nasional Kayan Mentarang in Kalimantan.

IUCN THREAT STATUS Least Concern.

Family Dibamidae Worm Lizards

Small to medium-sized (SVL 8–20cm) lizards. Their characteristics include: single, fused parietal bone; dentition pleurodont; pterygoid lacking teeth; limbs absent in females; males have small, flap-like hindlimbs; external ear openings absent; vestigial eyes covered by a scale; large, plate-like scales on snout and on mandible; body covered with smooth, cycloid scales; osteoderms absent; more than 26 presacral vertebrae; both ends of body blunt; tail short; tail autotomy; egg-shell highly calcareous, and a near-term embryo had two large premaxillary teeth that are thought to represent egg-teeth.

Dibamus Worm Lizards

Small (SVL to 200mm) fossorial squamates; cephalic scales large; no nasal-prefrontal contact; vomer elongated, extending posteriorly one half or more the length of maxillary tooth row; middorsal scale rows unmodified; females limbless; males have reduced limbs forming flaps.

Key to Bornean species of *Dibamus*

1a. Postocular one .... *D. leucurus* (p. 84)
1b. Postoculars two **2**

2a. Nuchal collar present; frontonasal divided; labial suture absent ..... *D. vorisi* (p. 85)
2b. Nuchal collar absent; frontonasal fused; labial suture present .... *D. ingeri* (opposite)

INGER'S WORM LIZARD *Dibamus ingeri* Das & Lim, 2003

(No vernacular names recorded)

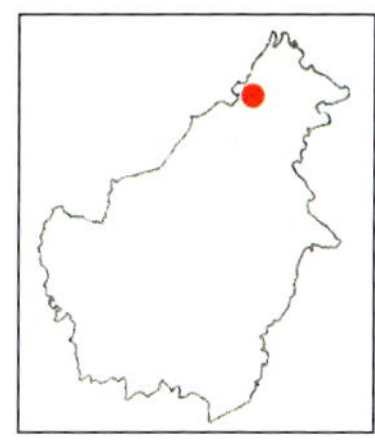

SIZE 96mm

IDENTIFICATION FEATURES Body stout; snout bluntly rounded, distinctly conical; rostral pad has a large number of evenly distributed sensory papillae; nasal and labial sutures complete, extending from ocular to nostril; frontonasal divided; interparietal single, not large, narrower than frontonasal and frontal, posteriorly bordered by three slightly smaller nuchal scales; postoculars two; supralabial single, elongated, bordering ocular ventrally; infralabial lanceolate; transverse scale rows posterior to head 24, at midbody 20, and anterior to vent 20; ventrals 163; subcaudals 36; hindlimbs reduced; flattened left hindlimb 4.7mm, covered with three scales basally, three pairs of scales along its length, and terminating in a single scale; large median scale on precloacal region, overlapped by those on sides; precloacal pores absent; postcloacal scales not reduced.

COLOUR Unknown in life; in preservative, dorsal surface tan brown, unpatterned; posterior half of body has pale linear blotches; venter slightly paler; snout-tip, sides of head, including supralabials, throat, hindlimbs and precloacal region yellowish-cream; wide nuchal band cream; back of head has a few brown linear blotches.

HABITAT AND BEHAVIOUR Inhabits submontane forests at 1,180m. Terrestrial and found inside rotting logs. Nothing else known of its natural history.

DISTRIBUTION Endemic to Borneo. Known only from Mendolong Camp, Sipitang District in Sabah.

IUCN THREAT STATUS Data Deficient.

Holotype of *Dibamus ingeri* Das & Lim, 2003 (FMNH 239756) from 'km 13.9, Mendolong Camp, Sipitang District, Sabah, East Malaysia (Borneo)'.

White-tailed Worm Lizard *Dibamus leucurus* (Bleeker, 1860)

(No vernacular names recorded)

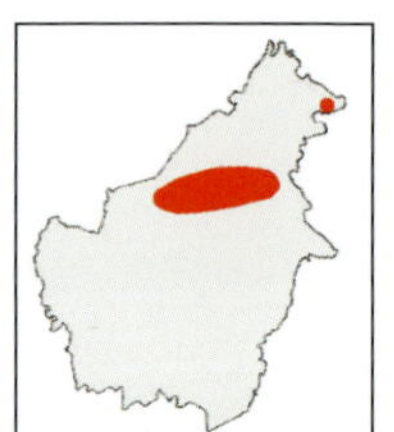

SIZE 136mm

IDENTIFICATION FEATURES Body slender; postocular single; nasal suture complete; rostral sutures incomplete; midbody scale rows 20–23; subcaudals 41–52; frontal greater than frontonasal; three small scales bound infralabials; single supralabial; large mental scale; presacral vertebrae 106–128; postsacral vertebrae 21–28; tail 16–20 per cent SVL; tail-tip rounded.

COLOUR Dorsal surface mid-brown, lacking nuchal or body bands; body scales dark edged; tip of snout pale brown.

HABITAT AND BEHAVIOUR Inhabits lowland hill forests. Fossorial, 4–10cm under soil surface. Diet and reproductive habits unstudied.

DISTRIBUTION Sumatra, Pulau Weh, Borneo and the Philippines. Within Borneo, reported from Upper Baram in Sarawak, Datu District in Sabah, and Madjalan in Kalimantan.

IUCN THREAT STATUS Least Concern.

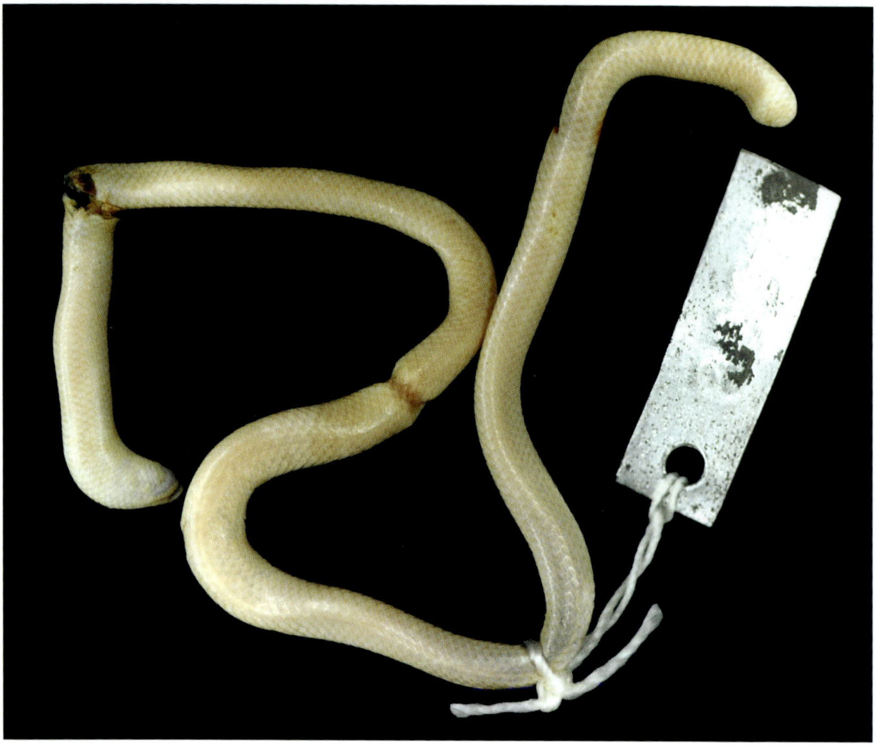

Dibamus leucurus (Bleeker, 1860) (USNM 82252) from Madjalan, Batang Sungei Kayan, Kalimantan Timur Propinsi.

VORIS'S WORM LIZARD *Dibamus vorisi* Das & Lim, 2003

(No vernacular names recorded)

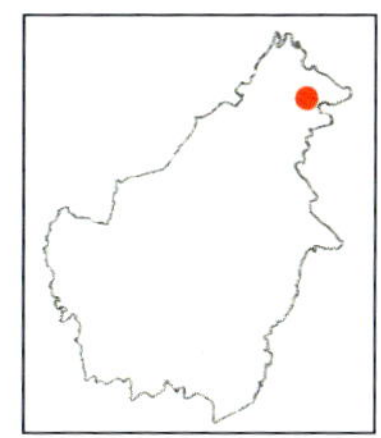

SIZE 89mm

IDENTIFICATION FEATURES Body slender; snout bluntly rounded, projecting beyond jaws; rostral pad has large number of evenly distributed sensory papillae, rostral suture incomplete, nasal suture incomplete, extending from ocular to nostril; labial suture absent; posterior border of rostral nearly straight; frontal single; frontonasal entire; interparietal single, narrower than frontonasal and frontal, posteriorly bordered by three slightly smaller nuchal scales; postoculars two; supralabial single, elongated, bordering ocular ventrally; infralabial lanceolate, separated by smaller, trapezoid mental; scales bordering posterior edge of infralabial, three bilaterally; head slightly distinct from neck and from body; tail short; transverse scale rows posterior to head 22, at midbody 20, and anterior to vent 18; ventrals 147; subcaudals 11–33; flattened hindlimb covered with three scales basally, three pairs of scales along its length, and terminating in single scale; large median scale on precloacal region, overlapped by those on sides; precloacal pores absent; postcloacal scales not reduced

COLOUR Colour in life unknown; in preservative, dorsal surface brown, unpatterned; venter slightly paler; snout-tip, sides of head, including supralabials, throat, hindlimbs and precloacal region yellowish-cream; no nuchal band; pale brown body band on anterior half of body both dorsally and ventrally; tail-tip cream.

HABITAT AND BEHAVIOUR Inhabits lowland forests. Fossorial, 5cm below soil surface around tree buttresses, and leaf litter. Diet includes beetle larvae. Reproductive habits unstudied.

DISTRIBUTION Endemic to Borneo. Only known from Danum Valley Field Centre, Lahad Datu District of Sabah.

IUCN THREAT STATUS Data Deficient.

Holotype of *Dibamus vorisi* Das & Lim, 2003 (FMNH 230187) from 'Danum Valley Field Centre (05° 01' N; 118° 03' E), Lahad Datu District, Sabah, East Malaysia (Borneo)'.

Family Eublepharidae Eyelid Geckos

Medium-sized lizards (SVL 4.5–15.5cm). Their characteristics include: skull with paired premaxillaries; supratemporal bone present (as in *Aeluroscalabotes*) or absent in some other members; vertebrae procoelous; single parietal; eyelids present, spectacle absent; skin soft with numerous small, juxtaposed scales; forelimbs and hindlimbs well developed; digits narrow; subdigital setae absent; tail autotomy; eggs elongated, with shells leathery or parchment-like, rather than brittle.

Aeluroscalabotes Cat Geckos

Medium-sized (SVL to 122mm) arboreal lizards; body covered with small, juxtaposed flat scales; postmentals large; axillary pockets absent; digits short, cylindrical basally and compressed distally; claw between dorsal and two lateral shields; transversely widened lamellae restricted to bases of digits, pupils vertical; males have precloacal pores.

Cat Gecko *Aeluroscalabotes felinus* (Günther, 1864)

(No vernacular names recorded)

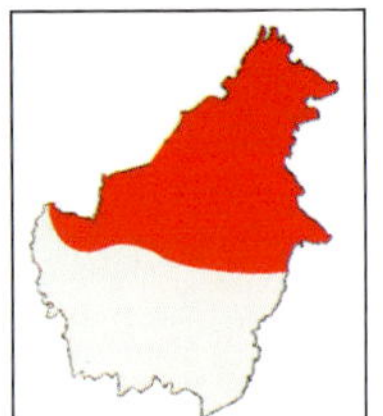

SIZE 122mm

IDENTIFICATION FEATURES Body slender; eyelids fleshy; tail rounded and capable of being curled laterally; postmentals large; axillary pockets absent; large retractable claw on each digit between a dorsal and two lateral scutes; transverse enlarged lamellae restricted to bases of digits. Several species are suspected to hide under the complex.

COLOUR Dorsal surface reddish-brown, more brightly coloured in juveniles, with oval or elongated white spots, which may be fused to form a complete or incomplete vertebral stripe on body and tail; small cream speckles on flanks and sides of head; sometimes, tail-tip is white; forehead pale brown; canthal region dark brown;

supralabials and infralabials pale unpatterned yellow, extending to limit of tympanum; limbs medium brown, with cream speckles; pupil elliptic; eyelids pale brown.

HABITAT AND BEHAVIOUR Inhabits lowland rainforests, submontane forests and peat-swamp forests. Associated with low vegetation of saplings and dead logs at night. Diet comprises large insects, including crickets and cockroaches. Clutches comprise 1–2 eggs with parchment shells, each 9.5–13 x 17.0–21.2mm. Incubation period 35–45 days. Hatchlings 78–81mm.

DISTRIBUTION Peninsular Thailand, the Malay Peninsula, Sumatra, Borneo, and reportedly also Pulau-Pulau Sula, east of Sulawesi, although the Pulau Sula record is doubtful. On Borneo, sites associated with it include Gunung Matang, Gunung Gading, Gunung Penrissen and Kota Samarahan in Sarawak; Tawau Hills Park, Crocker Range Park, Danum Valley Field Centre and Maliau Basin in Sabah, and Long Petah, Maruwai, and Taman Nasional Bentuang Karimun in Kalimantan.

IUCN THREAT STATUS Least Concern.

Family Gekkonidae True Geckos

Small to medium-sized lizards (SVL 14–370mm). Their characteristics include: forelimbs and hindlimbs well developed; single parietal; vertebrae amphicoelous; skin soft with numerous small, usually juxtaposed scales; eyes large; scales on skin soft; each eye covered with spectacle, the skin covering the eye forming a clear spectacle and lacking temporal arches; subdigital setae typically present; tail autotomy in all Bornean species; egg-shell highly calcareous.

Key to Bornean genera of Gekkonidae

1a. Digits slender or slightly dilated..**2**
1b. Digits moderately to broadly dilated **6**

2a. Pupil vertical..*Cyrtodactylus*
2b. Pupil rounded..*Cnemaspis*

3a. Terminal phalanges rising from dilated portion..**3**
3b. Terminal phalanges united with dilated portion to distal end and extending beyond..**4**

4a. Digits broadly dilated distally; scansor narrow; on toe IV covering distal half; terminal scansors, distinctly wedge shaped......**11**
4b. Digits uniformly dilated; scansors wide; on toe IV covering over distal half; terminal scansors oval or squarish............*Hemidactylus*

5a. First inner digit developed, broadly dilated..*Gehyra*
5b. First inner digit vestigial, not dilated ....*Hemiphyllodactylus*

6a. Exposed margin of scansor notched or some distal ones divided ....*Lepidodactylus*
6b. Scansor entire; distal ones with straight, exposed margin or with shallow notch..*Gekko*

Cnemaspis Day Geckos

Small to medium-sized (SVL to 80 mm) diurnal and arboreal gekkonids; body depressed; scales on top of body granular or tubercular; tail cylindrical; digits slender, clawed, not dilated; two distal phalanges compressed, forming an angle with basal portion of digits; raised eyelid-like scales distinct around eyeball; pupils rounded; adult males with or without precloacal and femoral pores.

Key to Bornean species of *Cnemaspis*

1a. Postmentals 3 .... *C. kendallii* (p. 90)
1b. Postmentals > 3 2

2a. Ventrolateral caudal tubercles absent 3
2b. Ventrolateral caudal tubercles present 4

3a. Enlarged median subcaudal tubercles .... *C. sirehensis* (p. 96)
3b. Non-enlarged median subcaudal tubercles .... *C. lagang* (p. 91)

4a. Forearm scales smooth .... *C. dringi* (below)
4b. Forearm scales keeled 5

5a. Precloacal pores ≥ 14 .... *C. nigridia* (p. 94)
5b. Precloacal pores ≤ 14 6

6a. Median row of subcaudals keeled .... *C. matahari* (p. 93)
6b. Median row of subcaudals smooth 7

7a. Ventrolateral caudal tubercle anteriorly .... *C. leucura* (p. 92)
7b. Ventrolateral caudal tubercle not anteriorly .... *C. paripari* (p. 95)

Dring's Day Gecko *Cnemaspis dringi* Das & Bauer, 1998

(No vernacular names recorded)

SIZE 46mm

IDENTIFICATION FEATURES Body slender; five postnasals; postmentals reduced; pectoral and abdominal scales distinctly elongated, imbricate and smooth; supralabials 11; infralabials 11; midventrals 39–40; median subcaudals tricarinate; males have three pairs of precloacal pores, lacking femoral pores; no precloacal groove; no postcloacal spur.

COLOUR Dorsal surface pale brown, with dark brown, parallel series along paravertebral region from nape to beyond caudal constriction; irregular series of markings along vertebral midline; pectoral and abdominal regions dark pigmented; flanks of body have distinct white patches.

HABITAT AND BEHAVIOUR Inhabits lowland rainforests. Ecology unstudied.

DISTRIBUTION Endemic to Borneo. Recorded from Labang Camp and Sungei Segaham, Kapit Division of central Sarawak.

IUCN THREAT STATUS Data Deficient.

Holotype of *Cnemaspis dringi* Das & Bauer, 1998 (FMNH 148588 (FMNH 19914), from 'Labang Camp (03° 20'N; 113° 29'E), Bintulu District, Fourth Division, Sarawak, East Malaysia, Borneo'

CAPTAIN KENDALL'S DAY GECKO *Cnemaspis kendallii* (Gray, 1845)

(No vernacular names recorded)

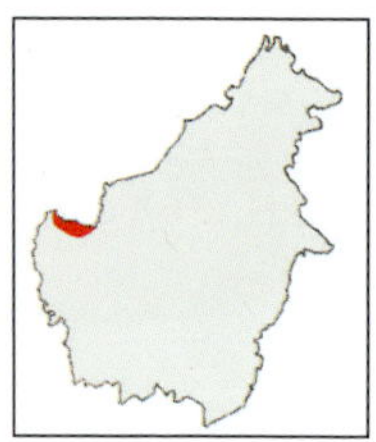

SIZE 80mm

IDENTIFICATION FEATURES Body slender; canthal ridge well developed, extending backwards over orbit; postnasals six; postmentals reduced; pectoral and abdominal scales distinctly elongated, imbricate and unicarinate; tail has median row of pointed, semi-erect scales below; supralabials 10–12; infralabials 8–11; interorbital scale rows nine; midventrals 40; top of body has larger and small scattered scales; median subcaudals tricarinate; males lack precloacal or femoral pores or precloacal groove; two postcloacal spurs.

COLOUR Dorsal surface of body pale brown or greyish-brown, with black or dark brown, oblong spots forming seven interrupted bands, each composed of three spots, across dorsal surface of torso between nuchal and caudal regions; olive-brown areas of dorsal surface reticulated with lighter shades of yellow-brown; flanks and dorsal surfaces of limbs have pale yellow suffusion; two black postocular stripes, directed to nuchal region, which has black drop-like spot medially; cream tips of two rows of tubercles that extend beyond and below tympanum; tail dorsal surface has 9–10 dark brown bands, with narrower olive-yellow intervening areas; tops of limbs pale brown or greyish-brown, with dark brown bands; phalanges have greyish-brown and olive-brown bands; venter of body and limbs unpatterned greyish-cream; manus and pes have dark grey smudges.

HABITAT AND BEHAVIOUR Inhabits lowland dipterocarp forests, peat-swamp forests and mid-hills. Associated with tree trunks, mostly in the shade. When alarmed, curls tail over back. Diet includes earthworms, beetles, millipedes and ants. Two eggs produced at a time.

DISTRIBUTION Endemic to northwestern Borneo. Widespread in western Sarawak. The best places to encounter it include Bako National Park, Gunung Matang, Gunung Penrissen, Gunung Santubong, Gunung Gading, the Ranchan Pool Forest, Serian, Semenggoh Wildlife Sanctuary; besides Semitau in Kalimantan.

IUCN THREAT STATUS Least Concern.

GUNUNG MULU ROCK GECKO *Cnemaspis lagang* Nashriq, Davis, Bauer & Das, 2022

(No vernacular names recorded)

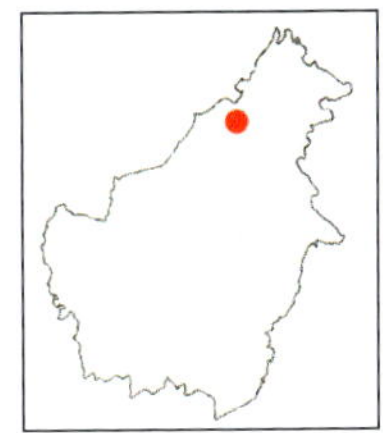

SIZE 46mm

IDENTIFICATION FEATURES Body slender; postmentals 9–10; males have 7–8 discontinuous, pore-bearing, precloacal scales with round pores, arranged in a chevron, separated at midline by 2–4 poreless scales; paravertebral and lateral row of caudal tubercles present; ventrolateral caudal tubercles absent; caudal tubercles not encircling tail; subcaudals keeled, not bearing row of enlarged median subcaudal scales; 1–2 postcloacal tubercles on each side of tail-base; 13–14 supralabials; 10–11 infralabials; pale brown and bright yellow caudal bands anteriorly; tail immaculate posteriorly.

COLOUR Forehead and dorsal surface of body pale brown; pale blue area between eyes; ground colour of nape and shoulder region mid-brown, with dark brown lines from orbit; pair of mid-brown spots on shoulder; yellow flecks on flank, forelimbs and hindlimbs; small, scattered brown spots between limb insertion; rows of brown lines along vertebral column; proximal half of tail yellow, bearing faint brown bands; ventral surfaces cream or grey; tail yellow, its posterior immaculate.

HABITAT AND BEHAVIOUR An obligate of limestone karst surfaces such as caves within lowland forests below 100m. Nocturnal. Diet and reproductive habits unstudied.

DISTRIBUTION Endemic to Borneo. Restricted to Gunung Mulu National Park in Sarawak, with records from the base of Gunung Api (90m) and Lagang Cave (100m).

IUCN THREAT STATUS Not Evaluated.

Penrissen Rock Gecko *Cnemaspis leucura* Kurita, Nishikawa, Matsui & Hikida, 2017

(No vernacular names recorded)

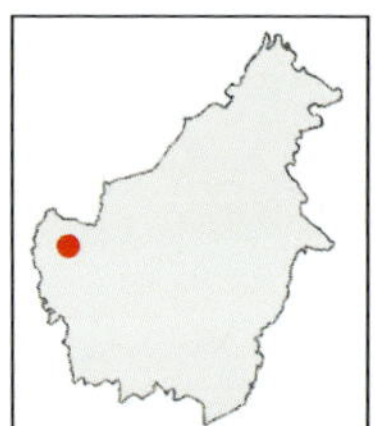

SIZE 62.3mm

IDENTIFICATION FEATURES Body slender; postnasals 5–6; postmentals 4–7; pectoral scales small, raised, juxtaposed and weakly keeled; abdominal scales elongated, arrowhead shaped, strongly raised, imbricate and keeled; supralabials 10–13; infralabials 8–12; paravertebral tubercles 24–31; no or several subcaudals enlarged; median subcaudals smooth; males have 4–9 precloacal pores, lacking femoral pores; no precloacal groove; postcloacal tubercles 3–8.

COLOUR Dorsal surface greyish-yellow, changeable to darker shade of grey under stress; dark grey postocular stripe; paired black axillary spot; midbody has yellow or white spots; dorsal surface of limbs greyish-yellow; anterior half of tail dark grey, with faint grey bands; posterior lighter.

HABITAT AND BEHAVIOUR Associated with hill dipterocarp forests and mostly restricted to sandstone formations, including edges of hill streams. Diet and reproductive habits unstudied.

DISTRIBUTION Endemic to Borneo. Confined to Gunung Penrissen in Sarawak.

IUCN THREAT STATUS Data Deficient.

Serian Rock Gecko *Cnemaspis matahari* Nashriq, Davis, Bauer & Das, 2022

(No vernacular names recorded)

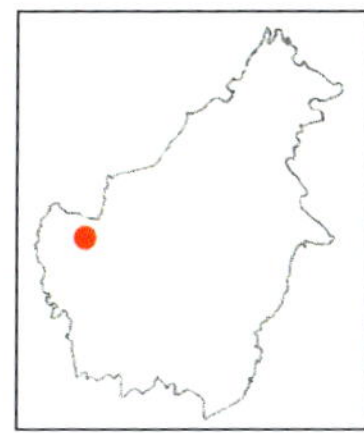

SIZE 56mm

IDENTIFICATION FEATURES Body slender; postmentals 5–9; ventral scales keeled; males have discontinuous row of 6–12 pore-bearing, precloacal scales with round pores arranged in a chevron, separated at midline by 2–4 poreless scales; paravertebral and lateral row of caudal tubercles present; ventrolateral caudal tubercles absent; caudal tubercles not encircling tail; subcaudals keeled, bearing enlarged median row of keeled subcaudal scales; 2–7 postcloacal tubercles on each side of tail-base; no enlarged femoral or subtibial scales; submetatarsal scales of first toe enlarged; supralabials 11–14.

COLOUR Dorsal surface yellow; nape and shoulder region grey with paired, rounded brown spots; white flecks on nape to shoulder region; faint yellow bands on body; front half of tail bluish, with faint brown bands; posterior half unpatterned; ventral surfaces grey; posterior half of tail white.

HABITAT AND BEHAVIOUR Obligate of limestone hills and caves, within lowland forests, at 50m. Nocturnal. Diet and reproductive habits unstudied.

DISTRIBUTION Endemic to Borneo. Confined to the Serian-Tebedu, Siburan and Jambusan-Semadang hills in Sarawak.

IUCN THREAT STATUS Not Evaluated.

Gunung Gading Rock Gecko *Cnemaspis nigridia* (Smith, 1925)

(No vernacular names recorded)

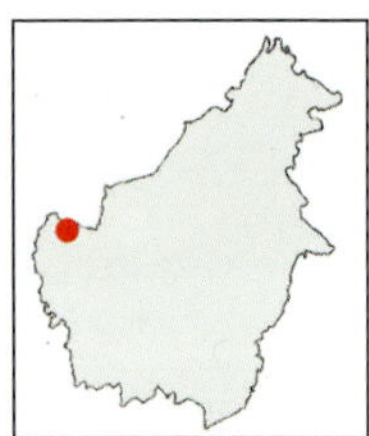

SIZE 70mm

IDENTIFICATION FEATURES Body robust; five postnasals; postmentals large; pectoral and abdominal scales imbricate and tricarinate; skin relatively hard, not prone to tearing; tail without median row of pointed, semi-erect scales below; supralabials 11; infralabials 12; midventrals 68; median subcaudals smooth; top of body has small granular scales and scattered large keeled tubercles; males typically lack precloacal or femoral pores, or precloacal groove; one or two postcloacal spurs.

COLOUR Dorsal surface brownish-olive or greenish-olive, with black blotches, especially on scapular region; two pairs of dark brown or black elongated spots on nape and axilla; single elongated spot on vertebral region; some scales on middorsum green; venter unpatterned grey; top of limbs and tail olive with dark brown bands.

HABITAT AND BEHAVIOUR Associated with granite hills and granite scree regions at 500–1,100m, within lowland and hill dipterocarp forests. Diurnal, and associated with shaded areas of tree trunks and granitic boulders. Diet includes spiders and presumably other small arthropods. Tail curling over back is known for this species. Clutches of two eggs laid communally: up to 15 eggs found at a single site; eggs each 9.8–11.2 x 8.4–10.2mm.

DISTRIBUTION Endemic to Borneo and Natuna Island. Sarawak records include Gunung Gading, Gunung Pueh and Gunung Beremput. Also recorded in Anjungan in Kalimantan.

IUCN THREAT STATUS Least Concern.

FAIRY CAVES ROCK GECKO *Cnemaspis paripari* Grismer & Chan, 2009

(No vernacular names recorded)

SIZE 50.7mm

IDENTIFICATION FEATURES Body slender; postnasals small, granular; postmentals 6–7; pectoral and abdominal scales slightly raised and keeled; supralabials 12; infralabials 10–11; paravertebral tubercles 25–31; median subcaudals enlarged, bead-like; males have one precloacal pore (sometimes absent), and lack femoral pores; no precloacal groove; two postcloacal spurs.

COLOUR Dorsal surface pale brownish-yellow, turning bright yellow in breeding males; head bright yellow with small brown flecks; pale brown postorbital bar; axillary region grey; 3–4 dark blotches on trunk; narrow yellow transverse markings on flanks; limbs nearly unpatterned yellow; venter grey; under surface of posterior half of tail cream or white.

HABITAT AND BEHAVIOUR Restricted to karst landscape of the Bau region of western Sarawak, and associated with limestone caves, including walls within the twilight zone and limestone scree of the area. Diet unstudied. Two eggs produced at a time; these are laid on depressions on walls of caves. Incubation period was more than 90 days in one instance.

DISTRIBUTION Endemic to Borneo. Confined to the Wind and Fairy Caves in western Sarawak.

IUCN THREAT STATUS Critically Endangered.

GUA SIREH ROCK GECKO *Cnemaspis sirehensis*

Nashriq, Davis, Bauer & Das, 2022

(No vernacular names recorded)

SIZE 49mm

IDENTIFICATION FEATURES Body slender; postmentals 6–7; subtibial scales keeled; ventrolateral caudal tubercles absent; paravertebral and lateral row of caudal tubercles present; caudal tubercles do not encircle tail; subcaudals keeled, bearing enlarged median row of weakly keeled scales; no enlarged femoral or subtibial scales; submetatarsal scales of first toe enlarged; supralabials 12–13.

COLOUR Forehead, top of body and limbs mid-brown; head yellow, bearing small occipital flecks; nape and shoulder region has brown, irregularly shaped blotches; grey patches on paravertebral; small, scattered yellow flecks between limb insertions; irregular yellow bands on forelimbs and hindlimbs; tail has brownish-olive and yellow banding from tail-base to tip; tail yellow.

HABITAT AND BEHAVIOUR Obligate of limestone hills and caves, within lowland forests as well as edges of plantations, at 50m. Nocturnal. Diet and reproductive habits unstudied.

DISTRIBUTION Endemic to Borneo. Known only from Gua Sireh, Serian District in Sarawak.

IUCN THREAT STATUS Not Evaluated.

Cyrtodactylus Bent-toed Geckos

Small to medium-sized (SVL to 125mm) arboreal geckos; body compressed dorsally; scales on dorsum granular, with interspersed tubercles; fingers and toes recurved, slender and non-dilated, their arched terminal phalanges compressed laterally; basally, digits with widened lamellae; males have precloacal and/or femoral pores; a few species also have precloacal groove and vertical pupils.

Key to Bornean species of *Cyrtodactylus*

1a. Dorsum with narrow pale bands **2**
1a. Dorsum without narrow pale bands **6**

2a. Forehead without pale reticulated pattern .... *C. limajalur* (p. 106)
2b. Forehead with pale reticulated pattern **3**

3a. Femoral scales enlarged; dorsum lacks isolated dark spots on vertebral region **4**
3b. Femoral scales not enlarged; dorsum has series of isolated dark spots on vertebral region .... *C. malayanus* (p. 107)

4a. Maximum SVL 116mm .... *C. consobrinus* (p. 100)
4b. Maximum SVL ≤ 116mm **5**

5a. Mean paravertebral tubercles 47.1; mean longitudinal rows of ventral scales 57.6 .... *C. kapitensis* (p. 105)
5b. Mean paravertebral tubercles 43; mean longitudinal rows of ventral scales 67 .... *C. hutan* (p. 103)

6a. Supralabials 8–9 .... *C. cavernicolus* (p. 99)
6b. Supralabials ≥10 .... **7**

7a. Males with femoral pores and large femoral scales .... *C. baluensis* (p. 98)
7b. Males without femoral pores and large femoral scales **8**

8a. Males with preanal groove **9**
8b. Males without preanal groove **10**

9a. Precloacal pores present .... *C. pubisulcus* (p. 111)
9b. Precloacal pores absent **11**

10a. Paravertebral tubercles 47–57 .... *C. miriensis* (p. 109)
10b. Paravertebral tubercles 37–48 .... *C. hantu* (p. 102)

11a. Dorsum with black blotches .... *C. ingeri* (p. 104)
11b. Dorsum with dark cross-bars **12**

12a. Preanal pores forming narrow angular series; expanded subdigital lamellae .... *C. yoshii* (p. 112)
12b. Preanal pores forming wide angular series; non-expanded subdigital lamellae **13**

13a. Tubercles present on upper arm; ventrals 48–51 .... *C. matsuii* (p. 108)
13b. Tubercles absent on upper arm; ventrals 39–46 .... *C. hamidyi* (p. 101)

Kinabalu Bent-toed Gecko *Cyrtodactylus baluensis* (Mocquard, 1890)

(No vernacular names recorded)

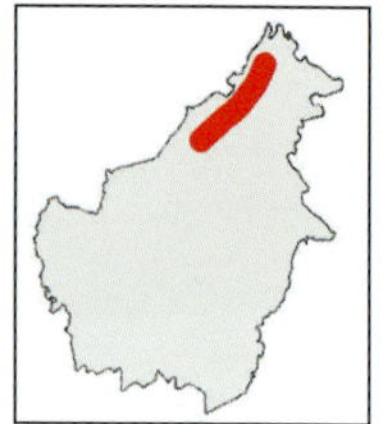

SIZE 86mm

IDENTIFICATION FEATURES Body slender; dorsals small, with larger tubercles in 21–24 rows; midventrals 40–45; precloacal groove absent; precloacal pores 9–10, forming narrow angular series within a distinct depression and separated by femoral pores; femoral pores 6–9; sharp boundary of scale size between large ventral scales and posterior scales that are granular; lamellae under toe IV 21–23, the largest of these are basal and four times as large as adjacent scales; tail ends in narrow point.

COLOUR Top of body pale olive, brown or yellowish-brown with irregular dark spots that typically form six dark cross-bars between nuchal and caudal regions, which extend to midlevel of flanks; head typically has 'V'-shaped, dark brown postocular stripe, extending as a lateral band from snout-tip to nape; labials pale, unbarred; dark preocular region; golden smudges in subocular region; forehead has dark brown irregular blotches, especially in postocular region; scattered bright yellow tubercles on forehead and dorsum of body that reach level of midventral fold, those on dorsal surfaces of limbs show less contrast; upper surfaces of limbs and phalanges olive-grey with irregular dark brown bands; manus and pes grey; tail has eight narrow, dark bands, which extend ventrally, but show reduced contrast; venter unpatterned greyish-cream.

HABITAT AND BEHAVIOUR Inhabits non-riparian sections of dipterocarp lowland rainforests to montane oak forests, at c. 150–2,500m. Nocturnal, on leaves of saplings, and more rarely, on tree trunks and buttresses. Diet presumably comprises small arthropods. Clutches of two rounded eggs, each 12 x 15mm, produced; hatchlings 31–32mm.

DISTRIBUTION Endemic to Borneo. The best places to see it include Gunung Murud and Gunung Mulu National Park in Sarawak, and Maliau Basin, Gunung Kinabalu, Gunung Trus Madi, Tawau Hills Park, Crocker Range Park and Danum Valley Field Centre in Sarawak. Also known from Kalimantan.

IUCN THREAT STATUS Least Concern.

NIAH CAVE BENT-TOED GECKO *Cyrtodactylus cavernicolus*

Inger & King, 1961

(No vernacular names recorded)

SIZE 81mm

IDENTIFICATION FEATURES Body slender; precloacal grooves containing two pairs of pores; dorsal surface covered with small, granular scales interspersed with 20–22 rows of trihedral or conical tubercles; femoral pores and large femoral scales absent; lamellae under toe IV 22–26; largest lamellae of basal phalange greater than four times as wide as adjacent scales.

COLOUR Top of body fawn-brown, with dark-edged dark brown cross-bars, changeable to brown-black under stress; dark stripe from back of eyes, meeting at nape; limbs have dark cross-bands; venter unpatterned cream; tail dark banded.

HABITAT AND BEHAVIOUR Cave and forest-dwelling species of gecko from limestone regions. Diet comprises flattened cave cockroaches and moths that dwell in guano. Reproductive habits unknown.

DISTRIBUTION Endemic to northern Borneo. Restricted to Niah National Park and Subis Forest Reserve in northern Sarawak.

IUCN THREAT STATUS Least Concern.

Peter's Bent-toed Gecko *Cyrtodactylus consobrinus* (Peters, 1871)

(No vernacular names recorded)

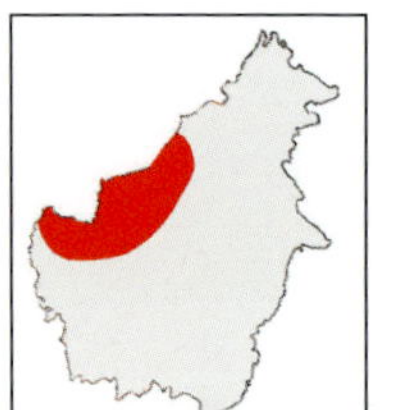

SIZE 114mm

IDENTIFICATION FEATURES Body robust; dorsal surface with paravertebral tubercles 32–44; longitudinal tubercle rows 13–16; supralabials 9–13; infralabials 8–10; midventrals 52–68; ventrals not keeled; femoral pores absent or seven; precloacal pores 5–10; preanal groove absent but a depression observed; lamellae under toe IV 23–26; single row of enlarged caudal scales.

COLOUR Forehead medium-brown, with pale yellow narrow network of reticulations; supralabials have bright yellow spots; dorsal surface of adults dark chocolate-brown with 4–8 white or yellow transverse bands with dark-edged intervening areas, between nuchal loop and caudal constriction; upper surfaces of limbs and tail have narrow pale bands edged with dark brown; gular region dark grey; pectoral, abdominal and under surfaces of limbs and tail dirty-cream or smoke-grey. Ground colour of juvenile dorsal surface lemon-yellow, with broad blackish-grey bands, which are wider than the pale intervening areas; forehead has a dark reticulated pattern; isolated yellow spots on sides of head on level below tympanum.

HABITAT AND BEHAVIOUR Inhabits lowland dipterocarp forests up to the mid-hills, to 1,100m. Diet presumably comprises arthropods. Two eggs, each 14.0–17.6mm.

DISTRIBUTION The Malay Peninsula, including the Seribuat Archipelago, Sumatra, Sinkep and Borneo. The easiest places to see it on Borneo include the caves of Bau, Gunung Santubong, Gunung Matang, Gunung Gading in Sarawak. Populations allocated to this species from other parts of Borneo, including Brunei Darussalam, Sabah and Kalimantan, need further study.

IUCN THREAT STATUS Least Concern.

Adult

Hamidy's Bent-toed Gecko *Cyrtodactylus hamidyi* Riyanto, Fauzi, Sidik, Mumpuni, Irham, Kurniawan, Ota, Okamoto, Hikida & Grismer, 2021

(No vernacular names recorded)

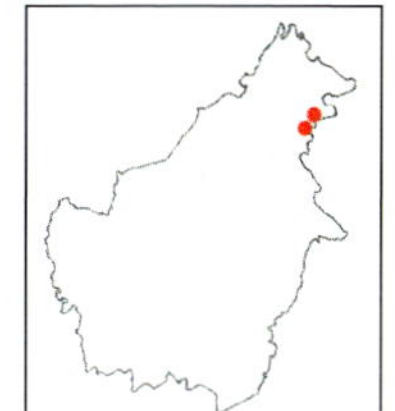

SIZE 66mm

IDENTIFICATION FEATURES Body slender; dorsal surface of upper arm lacks tubercles; tubercles present on ventrolateral body folds; paravertebral tubercles 28–30; longitudinal dorsal tubercle rows 17–20; ventral scale rows 39–46; subdigital lamellae on toe IV 17–19; precloacal pit in males; precloacal pores in males 5–7; median subcaudal not transversely arranged; enlarged median row of transverse scales.

COLOUR Top of body brownish-grey; paired crescentic mid-brown marks at back of forehead; large, paired mid-brown cross-bands between axilla and groin, edged with dark brown and bordered by a pale network forming a vertebral stripe; anterior band 'U'-shaped, and extends to eyes via postocular band; throat, belly and under surface of tail cream; six dark bands on tail.

HABITAT AND BEHAVIOUR Appears to be restricted to riparian forests. Nocturnal and aboreal; active on vegetation such as leaves and branches. Diet and reproductive habits unstudied.

DISTRIBUTION Endemic to Borneo. Known only from Tawau Hills Park in Sabah, and Samaenre Semaja village in Nunukan Regency in Kalimantan.

IUCN THREAT STATUS Not Evaluated.

Pelagus Bent-toed Gecko *Cyrtodactylus hantu* Davis, Bauer, Jackman, Nashriq & Das, 2021

(No vernacular names recorded)

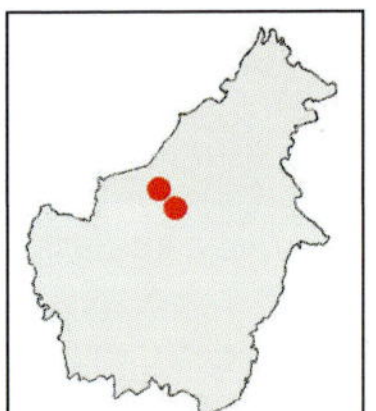

SIZE 73mm

IDENTIFICATION FEATURES Body slender; tubercles present on occiput; paravertebral tubercles 37–48; longitudinal tubercle rows 13–19; ventrolateral dermal fold present; supralabials 10–12; infralabials 9–12; midventrals 34–46; precloacal groove present; precloacal pores absent or six; dorsal tubercles not in regular longitudinal rows; subcaudals not transversely widened; lamellae under toe IV 19–22.

COLOUR Forehead, body, limbs and tail dark brown; wide, dark brown nuchal loop extends to snout-tip and edged by white line; seven dark brown bands on body, edged by narrow dark brown lines; limbs have light brown banded or blotched pattern; venter cream; tail has 10 dark bands.

HABITAT AND BEHAVIOUR Inhabits primary lowland forests, rarely entering abandoned human-made structures, and active on low-lying shrubs and tree branches. Diet and reproductive habits unstudied.

DISTRIBUTION Endemic to Borneo. Found in Ulu Baleh National Park, Pelagus National Park and around Kapit Division in Sarawak.

IUCN THREAT STATUS Not Evaluated.

Forest Bent-toed Gecko *Cyrtodactylus hutan* Davis, Nashriq, Woytek, Wikramanayake, Bauer, Karin, Brennan, Iskandar & Das, 2024

(No vernacular names recorded)

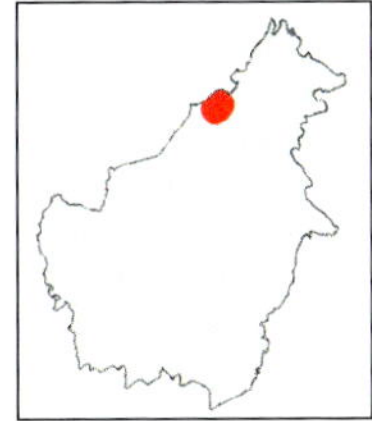

SIZE 116.3mm

IDENTIFICATION FEATURES Body fairly robust; tubercles present on occiput; paravertebral tubercles 40–49; longitudinal tubercle rows 13–20; ventrolateral dermal fold present; supralabials 9–12; infralabials 8–10; midventrals 45–80; precloacal groove absent but pit-like precloacal depression present; precloacal pores 0–11; femoral pores 0–2 in males; dorsal tubercles in longitudinal rows; subcaudals transversely widened; lamellae under toe IV 21–27.

COLOUR Top of body, forehead, limbs and tail dark brown; 6–7 light-edged, dark brown body bands, that are two times wider than pale yellow or tan interspaces; vertebral stripe absent; limbs have dark brown bands; venter unpatterned greyish-brown.

HABITAT AND BEHAVIOUR Recorded in lowland dipterocarp forests and mid-hills. Nocturnal, and found on old-growth trees. Diet and reproductive habits unknown.

DISTRIBUTION Endemic to Borneo. Known localities include Gunung Mulu National Park, Lambir Hills National Park and Niah National Park in Sarawak, and Ulu Temburong in Brunei Darussalam.

IUCN THREAT STATUS Not Evaluated.

Inger's Bent-toed Gecko *Cyrtodactylus ingeri* Hikida, 1990

(No vernacular names recorded)

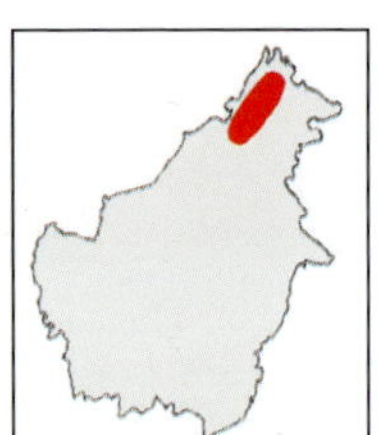

SIZE 81mm

IDENTIFICATION FEATURES Body relatively slender; dorsum has large, tuberculate scales in 17 irregular rows; rounded imbricate ventrals; supralabials 10–12; infralabials 8–10; midventrals 40–43; lamellae under toe IV 23–28; males have 7–9 precloacal pores, forming a narrow angular series; femoral pores and precloacal groove absent.

COLOUR Top of body grey or yellowish-brown, with 5–6 irregular dark brown, diamond-shaped paired paravertebral blotches; on flanks, interrupted dark brown stripe; nape has dark 'Y'- or 'V'-shaped mark; pale brown postocular stripe from eye to insertion of forearm; lips yellow-cream; upper surfaces of digits and limbs have dark blotches, and that of tail has dark bands; throat yellow-cream; rest of venter cream.

HABITAT AND BEHAVIOUR Inhabits riparian and swamp forests, in the lowlands and mid-hills, at up to 800m elevation. Nocturnal; active on leaves of saplings, as well as, more rarely, on stems and bases of trees. Diet presumably comprises arthropods. Eggs each 12 x 9mm; two eggs produced at a time.

DISTRIBUTION Endemic to Borneo. Found in Poring Hot Spring, Gunung Kinabalu Park and Deramakot in Sabah; Loagan Bunut National Park in northern Sarawak, and Ulu Temburong in Brunei Darussalam.

IUCN THREAT STATUS Least Concern.

KAPIT BENT-TOED GECKO *Cyrtodactylus kapitensis* Davis, Nashriq, Woytek, Wikramanayake, Bauer, Karin, Brennan, Iskandar & Das, 2024

(No vernacular names recorded)

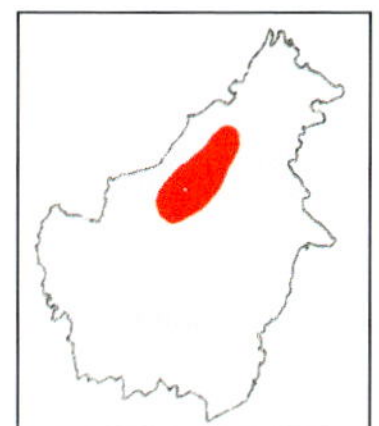

SIZE 112.4mm

IDENTIFICATION FEATURES Body fairly robust; tubercles present on occiput; paravertebral tubercles 43–54; longitudinal tubercle rows 16–20; ventrolateral dermal fold present; supralabials 9–12; infralabials 8–10; midventrals 47–63; precloacal groove absent but pit-like precloacal depression present; precloacal pores 4–10; femoral pores 0–16 in males; dorsal tubercles in longitudinal rows; subcaudals transversely widened; lamellae under toe IV 21–24.

COLOUR Top of body, forehead, limbs and tail dark brown; six pale, brown-edged, dark brown bands on body; bands jagged, at least two times wider than interspaces; indistinct vertebral stripe from nape to midbody; limbs dark brown banded; venter unpatterned pale grey.

HABITAT AND BEHAVIOUR Associated with lowland dipterocarp forests and mid-hills. Nocturnal, and found on old-growth trees. Diet and reproductive habits unknown.

DISTRIBUTION Endemic to Borneo. Known localities include Pelagus National Park, Ulu Baleh and Sungei Segaham in Sarawak.

IUCN THREAT STATUS Not Evaluated.

Five-banded Bent-toed Gecko *Cyrtodactylus limajalur* Davis, Bauer, Jackman, Nashriq & Das, 2019

(No vernacular names recorded)

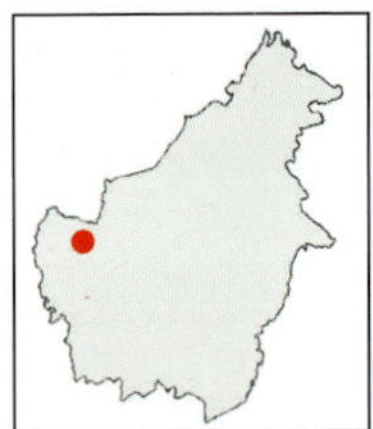

SIZE 94mm

IDENTIFICATION FEATURES Body moderately robust; supralabials 10–12; infralabials 9–11; dorsum has weak tubercules and no tubercles on ventral surfaces of forelimbs, gular region or ventrolateral folds; paravertebral tubercles 34–38; longitudinal tubercle rows 11–13; ventral scales 33–42; subdigital lamellae on toe IV 19–22; 5–6 enlarged femoral scales; no femoral pores; precloacal pores 7–8; precloacal pit present in males; enlarged median row of transverse scales.

COLOUR Top of body greyish-brown, with five dark brown bands; bands on dorsum edged anteriorly and posteriorly with narrow dark areas and are > twice the width of interspaces; 'V'-shaped line on rostrum; rostral chevron posterior to orbitals; no white edges of dorsal bands and nuchal loop; no scattered white tubercles on dorsum; forelimbs blotched with pale brown; hindlimbs have two dark bands; venter unpatterned cream; tail greyish-brown.

HABITAT AND BEHAVIOUR Associated with low limestone hills. Scansorial and associated with karst surfaces, such as boulders, and is nocturnal. Diet and reproductive habits unstudied.

DISTRIBUTION Endemic to Borneo. Restricted to the region near Kampung Tubih Mawang in Serian, in Sarawak.

IUCN THREAT STATUS Not Evaluated.

MALAYAN BENT-TOED GECKO *Cyrtodactylus malayanus* (de Rooij, 1915)

(No vernacular names recorded)

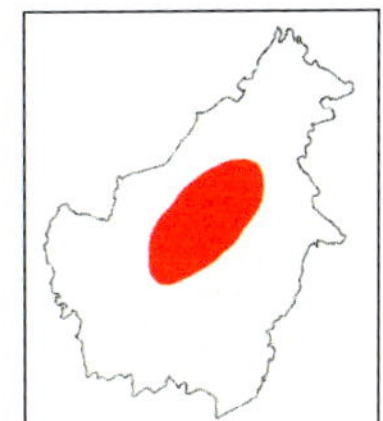

SIZE 93mm

IDENTIFICATION FEATURES Body relatively robust; head large; occiput tuberculate; dorsum with keeled tubercles in 18–20 rows; ventrals 58–62; tubercles present on upper arm; preanal groove absent; precloacal pores indistinct in males; precloacal groove absent; enlarged transverse median subcaudals.

COLOUR Top of body mid-brown; forehead has pale yellow network of reticulations; narrow tan-yellow cross-bars on dorsum of body; series of isolated dark spots along vertebral region; venter grey.

HABITAT AND BEHAVIOUR Inhabits lowland forests. Arboreal and nocturnal; found at at least 3m on large trees, seldom descending to base. Diet and reproductive habits unknown.

DISTRIBUTION Endemic to Borneo. Localities include the region of Kapit in central Sarawak; Ulu Temburong in Brunei Darussalam, and Gunung Pemantus, Bukit Baka Bukit Raya and Taman Nasional, Maruwai and Kutai Kartanegara in Kalimantan.

IUCN THREAT STATUS Least Concern.

Matsui's Bent-toed Gecko *Cyrtodactylus matsuii* Hikida, 1990

(No vernacular names recorded)

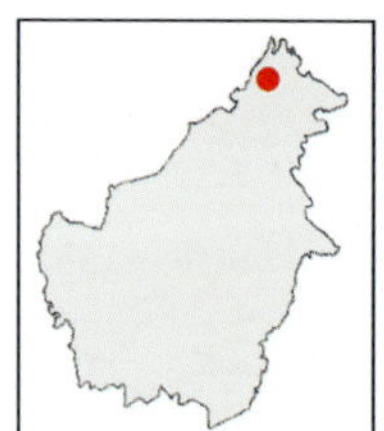

SIZE 105mm

IDENTIFICATION FEATURES Body relatively stout; supralabials 10–12; infralabials 10–11; dorsals have large, tuberculate scales arranged in 18 irregular rows; midventrals 48–51; males have 7–8 precloacal pores forming a wide angular series; precloacal groove absent; lamellae under toe IV 22; subdigital lamellae widened.

COLOUR Top of body yellowish-brown or pale brown, with irregular dark cross-bars; forehead has small, dark spots; dark band on interorbital region joining posterior edges of eyes; upper surface of limbs and tail dark banded; venter pale grey or brown, unpatterned.

HABITAT AND BEHAVIOUR Inhabits montane forests at elevations of 900–1,600m. Active at night on tree trunks and stems of saplings. Diet comprises insects and other arthropods. Reproductive habits unstudied.

DISTRIBUTION Endemic to northern Borneo. The best sites to see it are Gunung Kinabalu Park, Mendolong, Sipitang District, Gunung Lumaku, Gunung Emas and Crocker Range Park in Sabah.

IUCN THREAT STATUS Least Concern.

MIRI BENT-TOED GECKO *Cyrtodactylus miriensis* Davis, Das, Leaché, Karin, Brennan, Jackman, Nashriq, Chan & Bauer, 2021

(No vernacular names recorded)

SIZE 61mm

IDENTIFICATION FEATURES Body slender; tubercles present on occiput; paravertebral tubercles 47–57; longitudinal tubercle rows 12–20; ventrolateral dermal fold present; supralabials 10–15; infralabials 9–12; midventrals 36–45; precloacal groove present; precloacal pores absent; dorsal tubercles not in regular longitudinal rows; subcaudals not transversely widened; lamellae under toe IV 16–23.

COLOUR Top of body, forehead, limbs and tail greyish-brown; wide, pale-edged, dark brown nuchal loop extends to snout-tip; seven dark brown bands that are wider than their interspaces; limbs have pale brown-banded or blotched pattern; belly cream; tail has 9–10 dark bands.

HABITAT AND BEHAVIOUR Inhabits lowland dipterocarp forests and mid-hills. Nocturnal; active on low shrubs and tree branches, and man-made wooden walkways. Diet and reproductive habits unknown.

DISTRIBUTION Endemic to Borneo. Known localities include Gunung Mulu National Park, Niah National Park, Lawas and Lambir Hills National Park in Sarawak, and Ulu Temburong in Brunei Darussalam.

IUCN THREAT STATUS Not Evaluated.

MULU BENT-TOED GECKO *Cyrtodactylus muluensis* Davis, Bauer, Jackman, Nashriq & Das, 2019

(No vernacular names recorded)

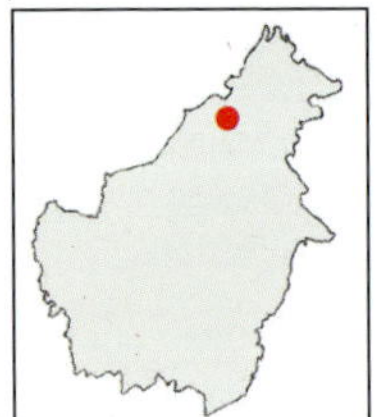

SIZE 88mm

IDENTIFICATION FEATURES Body moderately robust; supralabials 10–13; infralabials 8–11; weak tuberculation on body; no tubercles on ventral surface of forelimbs, gular region or ventrolateral folds; paravertebral tubercles 33–54; longitudinal tubercle rows 13–15; ventral scales 33–42; subdigital lamellae on toe IV 19–22; femoro-precloacal pores five; enlarged median row of transverse scales; shallow precloacal groove in males.

COLOUR Top of body greyish-brown, with 5–8 dark bands that are as wide as or slightly wider than interspaces; rostral chevron absent; dark body bands and dark brown nuchal loop not edged with white and lacking pale scattered tubercles; tail has nine dark caudal bands.

HABITAT AND BEHAVIOUR Encountered in lowland forests associated with Gunung Mulu's Melinau upland karst formation, specifically on karst surfaces, vines, branches and made-made structures, such as wooden bridges. Nocturnal and scansorial. Diet and reproductive habits unstudied.

DISTRIBUTION Endemic to Borneo. Known localities include Gunung Mulu National Park (Clearwater Cave, Long Cave and Lang Cave regions, and along major trails of the Park), in Sarawak.

IUCN THREAT STATUS Not Evaluated.

Close-up of head and forebody

GROOVED BENT-TOED GECKO *Cyrtodactylus pubisulcus* Inger, 1957

(No vernacular names recorded)

SIZE 74mm

IDENTIFICATION FEATURES Body slender; paravertebral tubercles 46–55; longitudinal tubercle rows 17–19; ventrolateral dermal fold present; supralabials 10–13; infralabials 9–12; midventrals 33–47; preanal groove present; preanal pores 7–8; dorsal tubercles not in regular longitudinal rows; subcaudals not transversely widened; lamellae under toe IV 20–23.

COLOUR Dorsal surface pale brown or grey, with dark brown blotches that are transversely enlarged on dorsal surface, sometimes forming cross-bars or arranged in a longitudinal series; forehead and limbs pale brown or grey, with dark brown variegations, tail dark banded; yellow ring around eyelids; dark brown postocular stripe extends to insertion of forelimbs; lips grey with pale yellow-grey spots; throat grey-cream; preanal sulcus area pale cream, rest of venter of body and under surfaces of limbs and tail unpatterned dark grey.

HABITAT AND BEHAVIOUR Lowland rainforests, and also peat swamps. Nocturnal and arboreal; active on stems and leaves of saplings and bases of tree trunks and buttresses. Diet includes arthropods. Clutches comprise two eggs, each 10.0–13.0mm.

DISTRIBUTION Endemic to Borneo. Verified records are from Kubah National Park, Semenggoh Wildlife Sanctuary, Gunung Gading National Park, Bako National Park, Ranchan Pool Forest, Serian and Tanjung Datu National Park in Sarawak.

IUCN THREAT STATUS Least Concern.

YOSHI'S BENT-TOED GECKO *Cyrtodactylus yoshii* Hikida, 1990

(No vernacular names recorded)

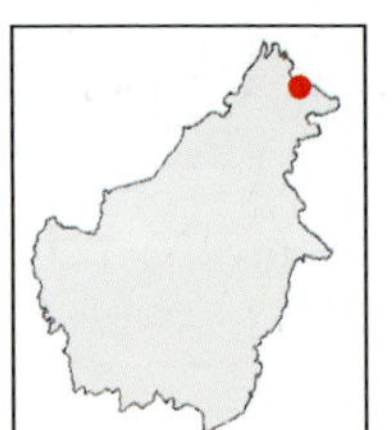

SIZE 96mm

IDENTIFICATION FEATURES Body relatively robust; supralabials 10–14; midventrals 50–58; males without femoral pores; precloacal pores 8–12, forming narrow angular series; precloacal groove absent; back has enlarged tubercles; lamellae under toe IV 25–30; subdigital lamellae not widened; tail ends in narrow point.

COLOUR Dorsal surface greyish-tan, with five dark brown, indistinct 'V'-shaped or irregular cross-bars between nape and inguinal region; dark olive-brown, 'V'-shaped stripe extends from posterior corner of orbit of eye, through top of ear opening, meeting on nape; labials yellowish-cream, with dark smudges; forehead brownish-yellow, with fine dark grey vermiculation; dorsal surfaces of limbs cream-yellow, with indistinct dark bands; under surfaces of limbs yellowish-cream; phalanges banded with grey and yellow-cream; venter yellow, with some grey smudges; tail has nine dark brownish-grey bands that are wider than the intervening pale areas; manus and pes grey.

HABITAT AND BEHAVIOUR Inhabits lowland forests, from sea level to about 500m. Nocturnal and arboreal; active on tree trunks and on walls of huts within forests. Diet and reproductive habits unknown.

DISTRIBUTION Endemic to northern Borneo. Restricted to eastern Sabah, including Danum Valley, Tawau Hills Park, Maliau Basin, Poring, Kipungit Satu, Sepilok Forest Reserve and Sepagaya Forest Reserve.

IUCN THREAT STATUS Least Concern.

Gehyra Four-clawed Geckos

Small (SVL to 64 mm) arboreal gekkonids; dorsal scales small, granular; digits free or webbed basally, strongly dilated, with undivided or mesially divided lamellae ventrally; terminal phalanges of outer four toes long, slender, clawed, free; inner digits have minute claw that is often concealed; males have precloacal and femoral pores; pupils vertical.

Common Four-clawed Gecko *Gehyra mutilata* (Wiegmann, 1834)

(No vernacular names recorded)

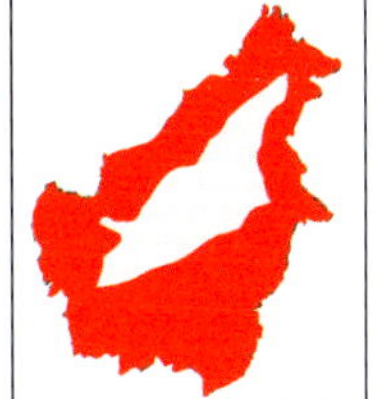

SIZE 64mm

IDENTIFICATION FEATURES Slender; head relatively large and oval; snout short; skin delicate, lacking tubercles; supralabials 8–11; midventrals 35–44; tail distinctly flattened dorsoventrally, widening at base, with sharp, somewhat denticulated edges; large, flat scales on tail and venter; limbs short; digits basally webbed; claw on inner digit absent; males have 25–41 preanofemoral pores; lamellae under toe IV 25; postcloacal tubercle absent.

COLOUR Dorsal surface pale translucent grey to pinkish-grey, usually with pale vertebral area, sometimes with dark and light spots; indistinct white band along face; juveniles have numerous pale spots on dorsum; venter pale pink or yellowish-cream.

HABITAT AND BEHAVIOUR Inhabits both human habitations and primary forests. Nocturnal, and occupies dark parts of walls. Call a series of 6–8 monosyllabic *tok*, that increases in intensity. Diet comprises insects and isopods. One to two eggs, each 7.1–8.5 x 10.5–11.2mm. Hatchlings 17–23mm.

DISTRIBUTION Northern and southwestern India, Andaman and Nicobar, Sri Lanka; to South-east Asia, including Myanmar, Thailand, Laos, Cambodia, Vietnam, Peninsular Malaysia and Singapore. In the Indo-Malayan archipelago, Sumatra, Java, Borneo, Flores, Sumba, Komodo, Sumbawa, Sulawesi, Timor, Halmahera, besides New Guinea and the Philippines. On Borneo, localities where it can be sighted include towns and cities, as well as forested area such as Kuching, Bako National Park, Lambir Hills National Park, Niah National Park, Loagan Bunut National Park in Sarawak; Ulu Temburong in Brunei Darussalam; Kota Kinabalu, Gunung Kinabalu, Sepilok Forest Reserve, Danum Valley Field Centre and Tawau Hills Park in Sabah, and Taman Nasional Kayan Mentarang, Gunung Niyut, Pontianak and Gunung Kenepai in Kalimantan.

IUCN THREAT STATUS Least Concern.

Gekko Giant Geckos

Medium-sized to large (to 180mm) arboreal gekkonids; digits free or partially webbed; moderately or strongly dilated, with undivided transverse lamellae ventrally; terminal phalanges of outer four digits slender, compressed, clawed; inner digits well developed, clawless; dorsal scales small, uniform or intermixed with larger scales; males have precloacal and femoral pores; pupils vertical. Includes 'gliding geckos' that were formerly allocated to the genera *Luperosaurus* and *Ptychozoon* (although one recent work has suggested that validity of *Luperosaurus*).

Key to Bornean species of *Gekko*

1a. Cutaneous folds absent on limbs **2**
1b. Cutaneous folds present on limbs **4**

2a. Rostral forming nostril border .... *G. monarchus* (p. 120)
2b. Rostral not forming nostril border **3**

3a. Head large; dorsal surface of body bluish-grey, spotted with red and white; juveniles not dark banded; scales between dorsal tubercles 3–5; precloacal pores 8–12 .... *G. gecko* (p. 117)
3b. Head not large; dorsal surface of body grey, spotted with black and white, juveniles dark banded; scales between dorsal tubercles 5–8; precloacal pores 5–8 .... *G. albofasciolatus* (opposite)

4a. Cutaneous folds absent on sides of body **7**
4b. Cutaneous folds present on sides of body **6**

5a. Anteriormost chin scales < posterior; dorsal scales flat or slightly convex; head and body elongated .... *G. browni* (p. 116)
5b. Anteriormost chin scales > posterior; dorsal scales conical, convex or spinose **11**

6a. Tail fringe with distinct serrations; head narrower than body; two intersupranasals .... *G. sorok* (p. 122)
6b. Tail fringe with irregular outline; head width subequal to body width; three intersupranasals .... *G. yasumai* (p. 123)

7a. Tail lacking lobes; no imbricate parachute support scales; digits ⅓–½ webbed .... *G. rhacophorus* (p. 121)
7b. Tail with lobes; imbricate parachute support scales present; digits almost entirely webbed **8**

8a. Tail without terminal flap; flanks lacking tubercles; males have 8–11 femoral pores .... *G. horsfieldii* (p. 118)
8b. Tail with broad terminal flap; flanks with two rows of tubercles; males lack femoral pores .... *G. kuhli* (p. 119)

White-banded Giant Gecko *Gekko albofasciolatus* (Günther, 1867)

(Bahasa Malaysia Kog-go)

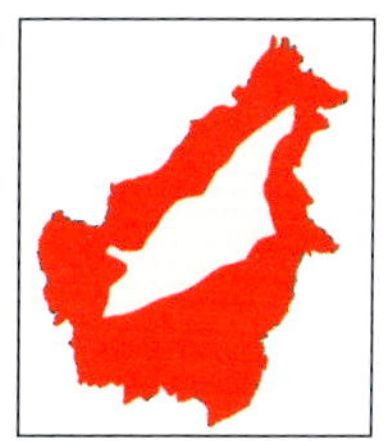

SIZE 180mm

IDENTIFICATION FEATURES Relatively robust; head large; body thick-set, with scattered tubercles on dorsal surface; males have 11–16 precloacal pores in a short angular series. Taxonomic status of this species remains unclear.

COLOUR Dorsal surface greyish-brown or pale greyish-yellow, with transverse series of white spots, composed of white-tipped tubercles and their surrounding areas; cream bands on sagittal and nuchal regions; tail has cream bands; dorsal surfaces of forelimbs and hindlimbs have pale grey bands; gular, pectoral and abdominal regions, as well as under surfaces of limbs and tail, yellowish-grey.

HABITAT AND BEHAVIOUR Inhabits forests in the lowlands and mid-hills, and enters human habitation. Call comprises 1–3 short barks, typically followed by a whirring sound. Diet comprises insects such as grasshoppers. Two eggs are laid.

DISTRIBUTION Borneo and associated islands to its north. The best places to see it are Gunung Matang, Niah National Park, Lambir Hills National Park, Loagan Bunut National Park in Sarawak; Ulu Temburong in Brunei Darussalam; Danum Valley Field Centre, Sandakan, Gunung Kinabalu Park, Taman Pulau Tiga, Pulau Gaya and Sepilok in Sabah, and Banjarmasin, Pulau Sibau and Sinkawang in Kalimantan.

IUCN THREAT STATUS Least Concern.

Forehead

Lateral view

Brown's Parachute Gecko *Gekko browni* (Russell, 1979)

(No vernacular names recorded)

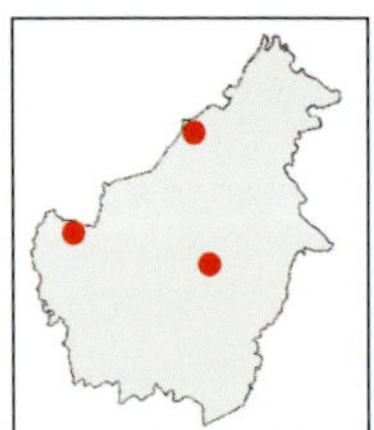

SIZE 67mm

IDENTIFICATION FEATURES Relatively slender; head and body elongated; supralabials 15; infralabials 13–14; anteriormost chin shields nearly as large as their successors; dorsal tubercles rounded, flattened or only slightly convex; ventrals flat, subimbricate; lamellae entire, numbering 16–19 under toe IV; preanofemoral pores 28–32; digits strongly dilated and half webbed; first digit unclawed; cutaneous folds on limbs; tail depressed, with lateral spines.

COLOUR Dorsal surface light grey, with minute black spots on head, body and limbs that form several broken oblique lines in postocular and temporal regions and on flanks; dark stripe in supralabial region; five dark, broken chevrons on middorsum of body; tail dark banded; venter white with several dark spots on tail.

HABITAT AND BEHAVIOUR Inhabits hilly rainforests, and associated with tree trunks. Diet unknown. Clutches comprise 1–2 egg, each 8.1–8.9 x 8.8–9.6mm; hatchlings 28.3–29.3mm.

DISTRIBUTION The Malay Peninsula, including Pulau Tekong in Singapore, and Borneo. Sites on the island include Kuching and Lambir Hills National Park in Sarawak, and Maruwai in Kalimantan.

IUCN THREAT STATUS Least Concern.

TOKAY GECKO *Gekko gecko* Linnaeus, 1758

(No vernacular names recorded)

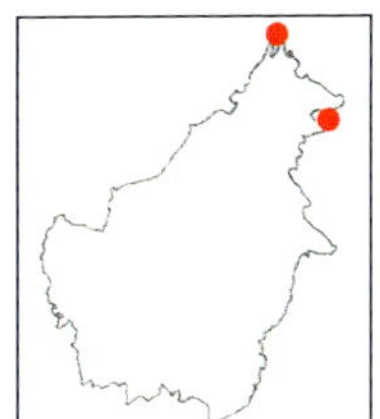

SIZE 176mm

IDENTIFICATION FEATURES Relatively robust; head relatively large, larger than snout length; body thick-set, with granular scales; supralabials and infralabials have weak keels; males have 10–24 precloacal pores; femoral pores present in population from islands off Borneo (and on Palawan), present in those from eastern India, Myanmar and Thailand; females in the area under study also lack perforations in the scales.

COLOUR Forehead as well as dorsal surface of body, limbs and basal region of tail slaty-grey to bluish-grey, with pale blue and red or orange spots; tail dark banded, with bands becoming more distinct posteriorly; venter cream, unpatterned or variegated with grey.

HABITAT AND BEHAVIOUR Associated with forested habitats on offshore islands. Diet comprises a variety of invertebrate prey, including coleopterans, orthopterans and lepidopterans, and also other geckos. Call heard on Pulau Balambangan can be syllabilized as *kog-go*, uttered 4–9 times in slow succession, and has been heard both by day and at night. Clutches comprise 1–2 eggs, each measuring 25mm; hatchlings 39.8–42.3 mm.

DISTRIBUTION Eastern India, Nepal, Bangladesh, through Myanmar, southern and eastern China, Thailand, Vietnam, offshore islands of the Malay Peninsula, Borneo, Sumatra, Java, Ambon, Sulawesi, Lombok, Komodo, Flores, Sumbawa, Sumba, Padar, Pulau Wetar, Pulau Babbar, East Timor, Rote, Aru and the Philippines; introduced into Florida and Hawaii in the USA, the French territory of Martinique in the West Indies, Taiwan and Madagascar. Records from Port Esslington, the Marianas and Pohnpei represent either erroneous localities or accidental introductions. Confirmed Bornean records are from offshore islands of northern Sabah, including Pulau Bohey Dulang, Balambangan and Banggi. Other records, such as Singkawang, Barabai, Buntut Bali, Pulaumajan and Sawah in Kalimantan, and a Sarawak locality from Sungei Rejang basin, need verification.

IUCN THREAT STATUS Least Concern.

Forehead

Lateral view

HORSFIELD'S PARACHUTE GECKO *Gekko horsfieldii* (Gray, 1827)

(No vernacular names recorded)

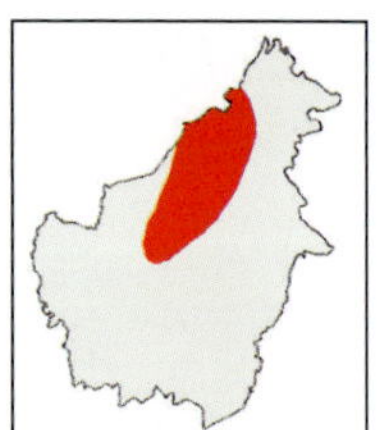

SIZE 80mm

IDENTIFICATION FEATURES Relatively robust; dorsal tubercles absent; supranasals not in contact; femoral pores 8–11; precloacal pores 10–11; femoral and precloacal pores in separated series; imbricate parachute support scales present; caudal lobes size reduction to terminus of tail gradual; 21–22 denticulate tail lobes.

COLOUR Dorsal surface grey to mid-brown, changeable to light grey under stress, peppered with black mottling, with dark brown wavy pattern; forehead and upper surfaces of limbs and digits grey with black mottling; broad dark band from posterior corners of eyes to beyond tympanum, followed by two large oval blotches that reach insertion of forelimbs; dark, butterfly-shaped mark on axilla; trunk has three other wavy bands; dorsal surface of patagium yellowish-brown, with five pale brown bands; chin yellow with a few scattered brown spots; lips yellow with thin bars between scales; six dark bands on tail between caudal constriction and tail-tip, which become darker posteriorly; pectoral and abdominal region dirty-yellow; under surfaces of forelimbs yellow; abdomen and under surfaces of limbs have dark spots; webs yellow with brown spots, both dorsally and ventrally.

HABITAT AND BEHAVIOUR Inhabits lowland rainforests, and found both in forests and on human-made structures. Diet presumably comprises small insects. Clutches comprise two eggs, each 13.7 x 11.9mm; hatchlings 34mm in SVL.

DISTRIBUTION Myanmar, Thailand, the Malay Peninsula, Sumatra and Borneo. On Borneo, localities where it may be encountered include Gunung Dulit, Niah National Park, Pelagus National Park and Lambir Hills National Park in Sarawak; Ulu Temburong in Brunei Darussalam; Mendolong, Sipitang District in Sabah, and Nanga Raun in Kalimantan.

IUCN THREAT STATUS Least Concern.

Head and forebody

Lateral view

KUHL'S PARACHUTE GECKO *Gekko kuhli* (Stejneger, 1902)

(No vernacular names recorded)

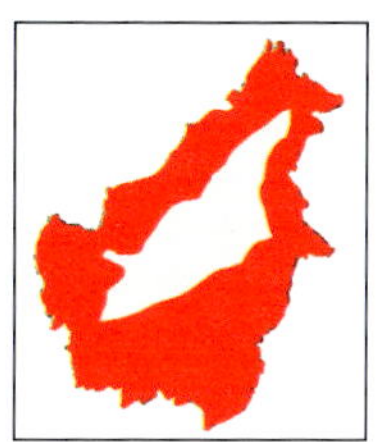

SIZE 108mm

IDENTIFICATION FEATURES Relatively robust; head rather large; elongated and widely expanded non-denticulate terminal tail-flap; large skin expansions on sides of body; imbricate parachute support scales present; scattered granules on dorsal surface of body that are strongly convex, raised or spinose; tail lappets set at right angles to tail; tail terminating in broad flap; feet webbed; femoral pores absent; precloacal pores 14–32, oriented in a curved line.

COLOUR Dorsal surface grey, greyish-yellow or reddish-brown, with 4–5 wavy transverse bands of dark brown on trunk; broad, dark brown postocular band from orbit to first dorsal band; labials cream with dark smudges; dorsal surfaces of limbs greyish-olive, unpatterned but darker than webbing or patagium; venter unpatterned yellow.

HABITAT AND BEHAVIOUR Inhabits large trees in lowland forests and occasionally walls of houses. Nocturnal, although basks on tree trunks by day. Diet includes small arthropods. Clutches comprise 1–2 eggs, each 9.5–15mm; hatchlings 34mm SVL.

DISTRIBUTION Southern Thailand, the Malay Peninsula, Sumatra, Enggano, the Mentawai Archipelago, Java and Borneo. On Borneo, it can be encountered in Bako National Park, Lambir Hills National Park, Gunung Dulit, Gunung Penrissen, Gunung Matang, Niah National Park in Sarawak; Sungei Rampayoh, Belait District and Ulu Temburong in Brunei Darussalam; Sepilok Forest Reserve, Danum Valley Field Centre, Gunung Rara, Purulon, Sukau, Kinabatangan District in Sabah, and Taman Nasional Kayan Mentarang, Pamabo Range and Pulau Sibau in Kalimantan.

IUCN THREAT STATUS Least Concern.

Type I

Hatchlings emerging from eggs

Type II

Warty Giant Gecko *Gekko monarchus* (Schlegel, in Duméril & Bibron, 1836)

(No vernacular names recorded)

SIZE 102mm

IDENTIFICATION FEATURES Body relatively robust and rugose; head oval; dorsals have large, tuberculate scales arranged in 16–17 longitudinal rows; scales on throat granular; midventrals 30–38; supralabials 10–11; infralabials 9–12; digits widened, basally webbed; weak webbing between toes IV and V; preanofemoral pores 23–42; median row of subcaudals widened; males larger than females.

COLOUR Forehead and dorsal surface of body, limbs and basal region of tail greyish-brown, with dark brown blotches arranged in 7–9 pairs on body; tubercles dark brown or yellow; hatchlings have bright yellow spots; forehead has several dark brown spots; venter yellowish-cream.

HABITAT AND BEHAVIOUR Can be encountered in both human-modified environments, such as cities and parks, and disturbed forests and forest edges in the lowlands and mid-hills up to 1,500m. Calls comprise more than 50 individual, low, *tock-tock* notes. Nocturnal; diet comprises insects and other invertebrates on walls and tree trunks. Clutches comprise two eggs, each 9.3–13.6 x 9.5–11.4mm. Incubation period 120 days; hatchlings 25–30mm.

DISTRIBUTION Southern Thailand, the Malay Peninsula, Sumatra, the Mentawai Archipelago, Pulau Simeulue, Java, Sulawesi, Ambon, Seram, the Philippines and Fakfak in Papua Propinsi, New Guinea. Introduced into Port Elizabeth, South Africa. Widespread on Borneo, with numerous records from Gunung Matang, Gunung Mulu National Park, Gunung Gading, Bako National Park, Gunung Santubong, Pulau Talang Talang and Niah National Park, as well as thoughout city centres of Sarawak; Bandar Seri Begawan, Kampung Baribi, Berakas, Brunei Muara District in Brunei Darussalam; Kota Kinabalu, Pulau Gaya, Taman Pulau Tiga, Sandakan, Gunung Kinabalu Park, Pulau Bohey Dulang and Tawau Hills Park in Sabah, and Banjarmasin, Pontianak, Putussibau, Sintang and Kutai in Kalimantan.

IUCN THREAT STATUS Least Concern.

KINABALU PARACHUTE GECKO *Gekko rhacophorus* (Boulenger, 1899)

(No vernacular names recorded)

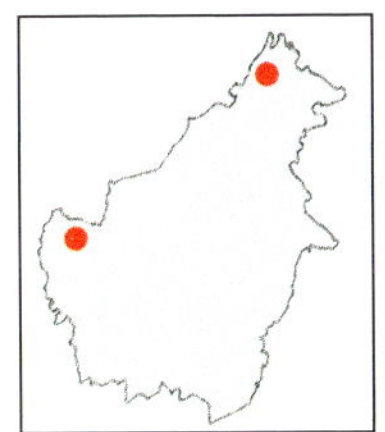

SIZE 75mm

IDENTIFICATION FEATURES Body relatively robust; dorsal surface has spinose or thorny tubercles that are widely scattered; cutaneous midbody expansions jagged and irregularly lobed; ventrals granular; supralabials nine; infralabials 10; supranasals widely separated; head lacks lateral skin fringes; sharply tapered tail that is short and without distinct lobes; terminal expansion of tail absent; reduced digital webbing; precloacal pores 17; femoral pores absent; postcloacal tubercles 2–3.

COLOUR Dorsal surface brownish-green, unpatterned or with dark mottling and indistinct wavy bands; venter light brown, speckled with dark brown.

HABITAT AND BEHAVIOUR Inhabits montane regions; seen on walls of buildings and on trunks of large trees, 600–1,600m tall. Nocturnal and insectivorous, being known to eat moths. Clutches comprise two eggs.

DISTRIBUTION Endemic to Borneo. Localities include upper reaches of Sungei Kedamaian, Gunung Kinabalu Park in Sabah, and Gunung Penrissen in Sarawak.

IUCN THREAT STATUS Data Deficient.

Hidden Parachute Gecko *Gekko sorok* (Das, Lakim & Kandaung, 2008)

(No vernacular names recorded)

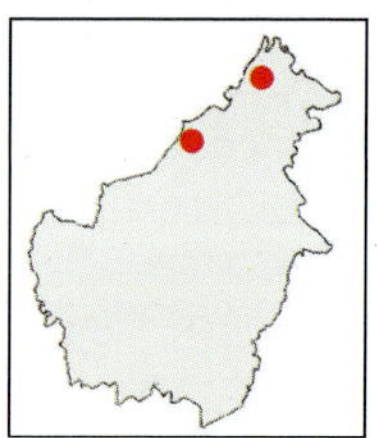

SIZE 35mm

IDENTIFICATION FEATURES Body robust; head narrower than body; auricular opening oval-squarish; subrictal tubercles present; two intersupranasals contact rostral; rostral enters nostrils; nine supralabials at midorbital position; 14–15 supralabials to posterior edge of orbit of eye; 11 infralabials; two scales separating supranasals; dorsal body scales rounded, convex and granular; 149 midbody scale rows; 1–3 anteriormost chin scales subequal, larger than those to their posterior; ventrolateral body tubercles spinose; three distal scansors deeply notched; tail fringe with distinct serrations.

COLOUR In preservative, dorsal surface pale grey, with dark grey double chevron marks, one on forehead, four on torso and one on pelvic region; six dark transverse bands on dorsum of tail; dark areas on back edged with black; dorsal surfaces of limbs pale grey with dark bands; lateral fringes of tail pale grey; chin pale, unpigmented; throat yellowish-cream, with darker speckles; rest of venter and under surfaces of limbs and tail yellowish-cream, with darker variegation; under surface of tail yellowish-cream, with darker smudges.

HABITAT AND BEHAVIOUR Inhabits hill dipterocarp forests. Arboreal and presumably nocturnal. Diet and reproductive habits unstudied.

DISTRIBUTION Endemic to Borneo. Known from Lambir Hills National Park in Sarawak, and Crocker Range Park in Sabah.

IUCN THREAT STATUS Data Deficient.

REMARKS A recent paper argues for reallocation of this species to *Luperosaurus*.

Head and neck

Holotype of *Gekko sorok* Das, Lakim & Kandaung, 2008 (SP 06618) from 'near bank of Sungai Bariawa, Crocker Range Park, Keningau District, Sabah, Malaysia (Borneo)'.

Yasuma's Parachute Gecko *Gekko yasumai* (Ota, Sengoku & Hikida, 1996)

(No vernacular names recorded)

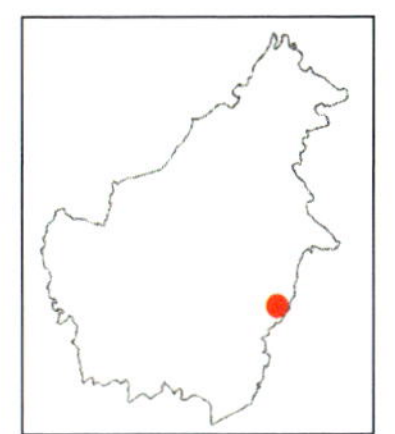

SIZE 39mm

IDENTIFICATION FEATURES Body slender; head depressed, as wide as body; snout strongly tapered; head and body not elongated; rounded at tip; supralabials 9–10; infralabials 10–11; anteriormost chin scales 2–3 times as large as their successors; dorsal tubercles conical or spinose; ventrals flat, subimbricate; lamellae entire, numbering seven under toe IV; tail strongly depressed.

COLOUR Dorsal surface of head and tail yellowish-brown; dorsal surface of body brownish-tan; numerous pale, cloudy markings on head, body and tail; two distinct, rounded, pale spots on middorsum, one before groin, the other above level of hindlimbs; venter light grey with several dark dots in gular region; eight indistinct dark broad bands on tail.

HABITAT AND BEHAVIOUR Inhabits lowland forests. Nocturnal and arboreal. Other aspects of biology unstudied.

DISTRIBUTION Endemic to Borneo. Known only from the Bukit Soeharto Experimental Forest, near Samarinda in Kalimantan.

IUCN THREAT STATUS Data Deficient.

Hemidactylus House Geckos

Small to medium-sized (SVL to 67mm in Borneo) arboreal gekkonids; digits free or partially webbed; rather strongly dilated, with divided series of lamellae ventrally; terminal phalanges long, slender, clawed, free; dorsal scales granular, uniform or intermixed with larger tubercles; pupils vertical; males have precloacal and sometimes also femoral pores.

Key to Bornean species of *Hemidactylus*

1a. Prominent cutaneous folds on lateral surfaces and posterior thigh; digits with prominent webbing **2**

1b. Cutaneous folds absent; digits without webs or only at base between toes III and IV **3**

2a. Body with narrow lateral fringe; fringe on sides of head and neck absent; tail has narrow denticulated fringe .... *H. platyurus* (p. 129)

2b. Body with broad lateral fringe; fringe on sides of head present; tail has broad denticulated fringe .... *H. craspedotus* (p. 126)

3a. Large dorsal tubercles numerous, strongly keeled, arranged in regular longitudinal series .... *H. brookii* (opposite)

3b. Large dorsal tubercles, if present, rounded, smooth, or feebly keeled, irregularly arranged **4**

4a. Tail rounded, with rows of enlarged tubercles; two pairs of large postmental chin shields contact infralabials .... *H. frenatus* (p. 127)

4b. Tail strongly depressed, with sharp denticulated, lateral edge and uniform small scales above; second pairs of large postmental chin shields excluded from infralabials by series of small scales .... *H. garnotii* (p. 128)

Brooke's House Gecko (opposite)

BROOKE'S HOUSE GECKO *Hemidactylus brookii* Gray, 1845

(No vernacular names recorded)

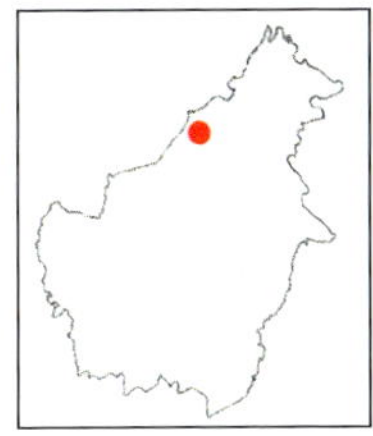

SIZE 63mm

IDENTIFICATION FEATURES Body relatively robust; head oval; head scales small; body flattened, with granular scales and rows of tubercles; tail plump, with spine-like tubercles on dorsal surface; males have 7–12 preanofemoral pores.

COLOUR Dorsal surface dark brown to light grey, with dark spots usually arranged in groups; two dark lines along nostrils and eyes; venter cream.

HABITAT AND BEHAVIOUR Little is known about the biology of the Bornean population, which has been seen on wooden structures at the edges of swamp forests. On the Asian mainland, it inhabits parks, gardens and houses, in addition to open forests. Largely terrestrial, although sometimes found climbing low walls. Call is a loud *chuck-chuck-chuck*. Diet comprises small insects. Two eggs, each 7 x 9mm; incubation period about 43 days.

DISTRIBUTION Widespread in the Old World tropics, from tropical Africa to southern China and South-east Asia, including Myanmar, Thailand, the Malay Peninsula, Borneo and the Philippines; also Kupang, southern Timor. The only Bornean locality is Loagan Bunut National Park in northern Sarawak. The nomenclature of these geckos remains uncertain.

IUCN THREAT STATUS Least Concern.

Frilly Gecko *Hemidactylus craspedotus* (Mocquard, 1890)

(No vernacular names recorded)

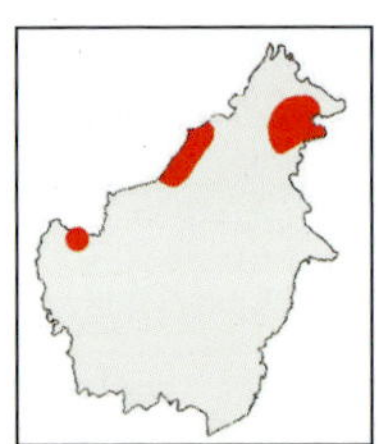

SIZE 62mm

IDENTIFICATION FEATURES Body relatively slender; body and tail depressed; dorsum has large, scattered tuberculate scales; skin frills on sides of body, tail, sides of throat and along lateral edges of limbs; digits nearly fully webbed.

COLOUR Dorsal surface of body greyish-brown, with two rows of dark, rectangular spots; broad, dark streak on sides of head; tail has dark bands on upper surface, lower surface reddish-orange basally, greyish-yellow distally, with pale grey bands; venter bright yellow, speckled with dark brown, on gular and pectoral regions; pale green on abdominal region; under surfaces of limbs grey-yellow.

HABITAT AND BEHAVIOUR Inhabits lowland rainforests. Active on tree trunks; insectivorous. Known to both parachute and glide for distances of up to 3m between trees. Diet and reproductive habits unknown.

DISTRIBUTION Peninsular Thailand, the Malay Peninsula and Borneo, and apparently also Java. Bornean records are from Bako National Park, Gunung Santubong, Lambir Hills National Park and Niah National Park in Sarawak, and Gunung Kinabalu, Danum Valley Field Centre and Tawau Hills Park in Sabah.

IUCN THREAT STATUS Least Concern.

ASIAN HOUSE GECKO *Hemidactylus frenatus* Duméril & Bibron, 1836

(No vernacular names recorded)

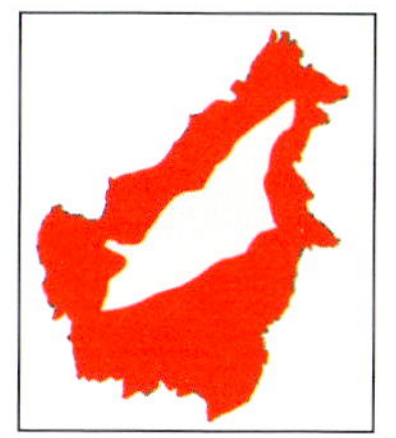

SIZE 67mm

IDENTIFICATION FEATURES Body slender and depressed; head large, dorsal scales smooth; lack of webbing on fingers and toes; sides of tail rounded, relatively long, with large tubercles; no flaps of skin along sides of body and at back of hindlimbs; supralabials 10–12; infralabials 7–10; preanofemoral pores in males 28–36; lamellae under toe IV 9–10.

COLOUR Dorsal yellowish-brown, greyish-brown to nearly grey-black, sometimes with darker markings; brown streak, light edged on top, runs alongside head, sometimes continuing on flanks; tail dark banded or unbanded, under surface yellowish-orange; venter unpatterned cream.

HABITAT AND BEHAVIOUR Inhabits man-made structures as well as forested areas to nearly 1,600m. Call a series of 4–5 loud, stacatto notes. Diet comprises arthropods, including coleoptera, isopods, blattarians, lepidopterans, orthopterans, dipterans, hemipterans, hymenopterans, zygopterans, isopterans and arachnids. Clutches comprise two eggs, each 8.2–10 x 7.2–8.4mm; hatchlings 19–22mm SVL.

DISTRIBUTION India, Sri Lanka, southern China, Myanmar, Thailand, Laos, Cambodia, Vietnam, the Malay Peninsula, Sumatra, Borneo, Java, Bali, Flores, Komodo, Padar, Lombok, Sulawesi, Ambon, East Timor, the Philippines, Saloman Atolls, Peros Banhos Atolls, Great Chagos Bank Atoll and Diego Garcia in the Chagos Archipelago. Introduced into Mexico, southern USA, South America, Madagascar, eastern and southern Africa, Mauritius, New Guinea, Polynesia and Australia. Widespread on Borneo, localities including Kuching, Gunung Santubong, Gunung Matang, Bako National Park, Samunsam Wildlife Sanctuary, Ranchan Pool Forest, Serian, Niah National Park, Lambir Hills National Park, Loagan Bunut National Park in Sarawak; Ulu Temburong and Bandar Seri Begawan in Brunei Darussalam; Ranau, Pulau Gaya, Sandakan, Sepilok Forest Reserve and Tawau Hills Park in Sabah, and Sintang, Pontianak, Sambas, Putussibau, Banjarmasin, Balikpapan in Kalimantan.

IUCN THREAT STATUS Least Concern.

Garnot's Gecko *Hemidactylus garnotii* Duméril & Bibron, 1836

(No vernacular names recorded)

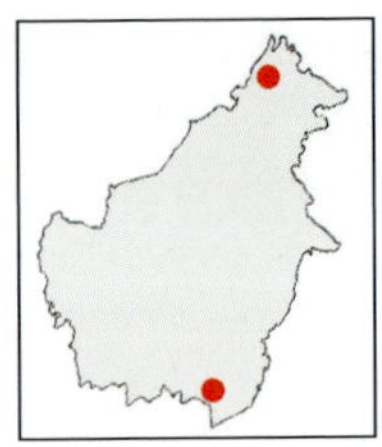

SIZE 65mm

IDENTIFICATION FEATURES Body slender and depressed; head large; head, body and tail covered with small ovoid scales; supralabials 12–13; infralabials 9–11; midventrals 29–34; original tail slender, depressed, with denticulate lateral edges; 14–19 enlarged femoral scales; lamellae under toe IV 11–14; rows of subdigital lamellae continue on to palms and soles.

COLOUR Dorsal surface brownish-grey, sometimes marbled with brown; cream spots sometimes present; venter unpatterned bright lemon-yellow; tail pale red, especially if regenerated, or yellow.

HABITAT AND BEHAVIOUR Inhabits primary forests as well as walls of buildings, to about 1,400m. Diet includes arthropods. An all-female, parenthogenetic gecko, producing two eggs, each 9 x 10mm; hatchlings 27–28mm.

DISTRIBUTION Eastern India, Myanmar, Thailand, southern China, the Malay Peninsula, Sumatra, Nias, Borneo, Java, Ambon, Flores, Sumbawa, the Philippines, New Caledonia, the Solomon Islands, Fiji and Tahiti; naturalized in Florida and Hawaii; introduced but not established in New Zealand. Bornean records are from Gunung Kinabalu in Sabah, and Banjarmasin in Kalimantan.

IUCN THREAT STATUS Least Concern.

FLAT-TAILED GECKO *Hemidactylus platyurus* (Schneider, 1792)

(No vernacular names recorded)

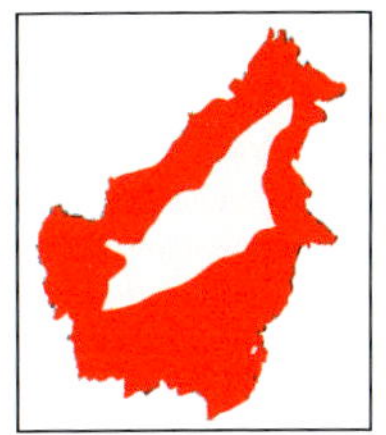

SIZE 69mm

IDENTIFICATION FEATURES Body slender; snout rather long; fingers and toes about half webbed; fringe of skin on sides of body and backs of hindlimbs; tail flattened, with serrated margin; body depressed, smooth, with tiny granules; lamellae under toe IV 7–9; males have 34–36 femoral pores.

COLOUR Dorsal surface light greyish-olive, lips light grey with darker smudges; short, medium-brown preocular stripe; postocular stripe of the same colour, widening on flanks of body and terminating a little beyond level of hindlimbs; four slightly darker areas on dorsal surface, forming bands, and best visible in stressed individuals; pupils black, vertical, with crenulated margins; dermal skin flaps, limbs and tail dorsum unpatterned grey; chin and gular region unpatterned cream; pectoral, abdominal and cloacal regions, and under surface of tail, bright yellow; under surface of limbs unpatterned cream; femoral pores dark brown, darker than adjacent scales.

HABITAT AND BEHAVIOUR Abundant in towns and cities; also known from open forests and trees along sea beaches. Diet comprises spiders and ants. Call is a continuous chuckling-whirr, with 4–5 notes uttered in quick succession. Clutches comprise two eggs, each 10–10.8 × 8.5–8.9mm; hatchlings 20.5–25mm SVL.

DISTRIBUTION Nepal and eastern India, Andaman and Nicobar Islands, Sri Lanka, Myanmar, Thailand, the Malay Peninsula to Sumatra, Borneo, Java, Bali, Komodo, Flores, Sulawesi, east to the Philippines and China. Introduced into Papua Propinsi in New Guinea, Indonesia. Localities where it can be sighted on Borneo include Kuching, Ranchan Pool Forest, Serian, Lambir Hills National Park and Gunung Santubong in Sarawak; Bandar Seri Begawan and Serasa beach in Brunei Darussalam; Kota Kinabalu, Sepilok Forest Reserve, Sandakan and Keningau in Sabah, and the upper reaches of Sungei Kapuas, Sawah, Penambo Range, Bulangan, Pontianak, Banjarmasin and Sintang in Kalimantan.

IUCN THREAT STATUS Least Concern.

Hemiphyllodactylus Worm Geckos

Small (SVL to 47mm for Bornean species) arboreal gekkonids; slender; vomers fused into single element; interclavicle dagger shaped, lacking laterally projecting cross-arm; series of precloacal pores slightly arched anteromedially; compressed distal portion of digit arises from within more dilated portion; preaxial-most digit of foot typically clawed, all others clawed; males in populations not parthenogenetic have precloacal and femoral pores.

Worm Gecko *Hemiphyllodactylus typus* Bleeker, 1860

(No vernacular names recorded)

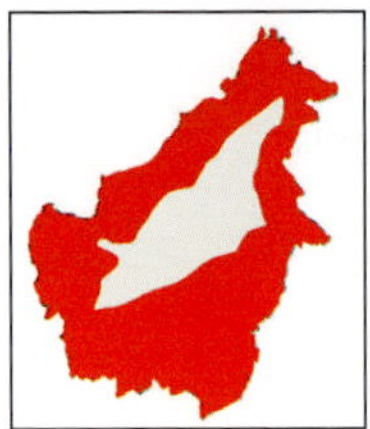

SIZE 47mm (on Borneo)

IDENTIFICATION FEATURES Build slender and depressed; head slightly distinct from neck; body dorsolaterally compressed and relatively long; limbs reduced; granular dorsal scales, but lacking tubercles; ventral scales smooth, rounded and imbricate; digits free; scansors divided, numbering 3–6; supralabials 8–12; infralabials 8–12; terminal phalange short, clawed; tail slender, rounded, prehensile, with uniform-sized scales that are larger than body scales; males have 10–12 angular series of precloacal pores, usually separated from 8–10 femoral pores.

COLOUR Dorsal surface dark brown to reddish-brown, with grey-brown network of narrow lines; dorsum, lips, flanks and upper surfaces of forelimbs and hindlimbs have pink spots and kidney-shaped marks; infralabials, postocular region and flanks of body have small, pinkish-yellow speckles; dark grey stripe from snout-tip, across nostrils to axilla; endolymphatic sacs indicated by yellow-grey areas; dark, chevron-shaped mark at tail-base, with broad tan posterior edge; tail has five pale oblong spots; gular region dark grey; pectoral and abdominal regions greyish-pink; under surfaces of limbs grey; under surface of tail brick-red with dark grey smudges.

HABITAT AND BEHAVIOUR Inhabits mangroves, beach forests, lowland dipterocarp forests and secondary forests, from sea level to 1,000m. Also enters human habitation. Noctural and arboreal; may be seen on walls of houses. Diet includes small insects. Unisexual and parenthenogenetic, producing two eggs. each 6.5–8 x 5-6mm; hatchlings 33–35mm TL.

DISTRIBUTION Mauritius, Nicobar Islands in India, Sri Lanka, Myanmar, China, Thailand, the Malay Peninsula, Sumatra, Borneo, Pulau Nias, Pulau Simeuleu, Java, Bali, Sumbawa, Komodo, the Philippines, including Palawan, New Guinea, Solomon Islands, New Caledonia, Polynesia, Hawaii and the Mascarene Islands; introduced into some of the islands of the Ryukyu Archipelago of Japan. Bornean records include Bako National Park, Kuching Division, Niah National Park, Lambir Hills National Park in Sarawak; Ulu Temburong and Tasek Lama, Bandar Seri Begawan in Brunei Darussalam; upper reaches of Sungei Lidon, Gunung Kinabalu Park and Sandakan in Sabah, and Sotek near Pemantus in Kalimantan.

IUCN THREAT STATUS Least Concern.

Dark morph

Pale morph

Lepidodactylus Mourning Geckos

Small (SVL to 48mm) arboreal gekkonids; relatively robust; vomers fused into single element; interclavicle dagger shaped, without laterally projecting cross-arms; series of precloacal pores slightly to moderately arched anteromedially; compressed distal portion of digit arises from anterior margin of dilated portion; preaxial-most digit of foot clawless; all other digits clawed; males in populations that are not parthenogenetic have precloacal and femoral pores.

Key to Bornean species of *Lepidodactylus*

1a. Terminal scansor entire; 35–37 large preanofemoral scales ....*L. ranauensis* (p. 132)
1b. Terminal scansors divided; 25–30 large preanofemoral scales........*L. lugubris* (below)

Common Mourning Gecko *Lepidodactylus lugubris* (Duméril & Bibron, 1836)

(No vernacular names recorded)

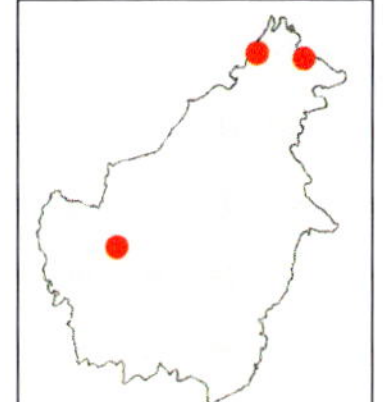

SIZE 49mm

IDENTIFICATION FEATURES Body relatively slender; head longer than broad; body elongated; tubercles on dorsal surface absent; preanofemoral scales 25–31; terminal scansors divided, moderately expanded; tail as long as or longer than head-body length, wide at base, transversely flattened.

COLOUR Dorsal surface cinnamon-brown, pale brown or greyish-brown; pale transverse sinuous pattern on dorsum, especially at pelvic region; yellow-orange canthal and postocular stripe extends to level of ears; lores darker than forehead; labials unpigmented pale yellow; black bar on each side of shoulder; small black blotch on each side of pelvic region; flanks of body pale yellow; throat unpigmented cream; pectoral and abdominal regions yellow; endolymphatic glands on sides of neck pale yellow; limbs unbanded; tail-base has two small black blotches; dorsum of tail rich brown with pale variegations; under surfaces of limbs pale yellow-cream; under surface of tail yellow-grey basally, medially and apically brown; digits pale brownish-yellow, unbarred; manus and pes grey.

HABITAT AND BEHAVIOUR Inhabits lowland forests, sea shores and especially mangroves; enters human habitation, and active on low vegetation such as branches and trunks at night. Diet comprises insects, including isopods, blattarians, lepidopterans, dipterans, hymenopterans and isopterans; also known to feed facultatively on fruits, nectar or other sweet plant derivatives. Can be both bisexual and unisexual and parthenogenetic, although Bornean population is unisexual and parthenogenetic, producing clutches of 1–2 eggs, each 7.0–8.7 x 9.5–10mm; hatchlings 17.2–17.8mm.

DISTRIBUTION Maldives, the Andaman and Nicobar Islands of India, Sri Lanka, Myanmar, Thailand, the Malay Peninsula, Borneo, the Riau Archipelago, Sulawesi, Halmahera, Lesser Sundas, Ternate, New Guinea, eastern China, Taiwan, the Ryukyu Archipelago of Japan, Diego Garcia, Salomon and Peros Banhos Atoll in the Chagos Archipelago, the Philippines, northern Australia, Micronesia, Fiji; introduced into Brazil, Colombia, Costa Rica, Ecuador, Galapagos Islands, Mexico, Nicaragua, Panama, Peru, Suriname and Hawaii. New Caledonian records suspected to be due to human introductions. Bornean records are from Sandakan and the Tanjung Aru Beach Resort, Kota Kinabalu in Sabah, and Sintang in Kalimantan.

IUCN THREAT STATUS Least Concern.

Ranau Mourning Gecko *Lepidodactylus ranauensis* Ota & Hikida, 1988

(No vernacular names recorded)

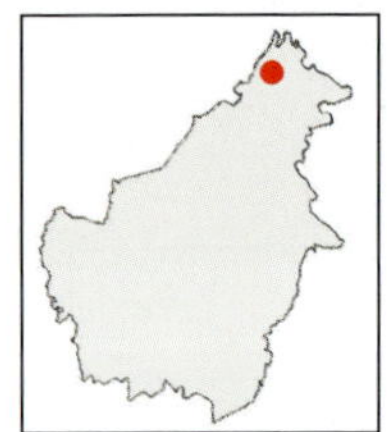

SIZE 48mm

IDENTIFICATION FEATURES Slender, and somewhat depressed; dorsals small, lacking large tubercles; supralabials nine; infralabials 9–10; midbody scale rows 108; digits long, relatively narrow; ventral surface with scansors on distal four-fifths; terminal lamellae entire; lamellae under toe IV 14–15; males have continuous series of 35–37 preanofemoral pores; femoral pores extend almost to distal edge of thighs; cloacal spur single; tail subcylindrical, lacking lateral ornamentation.

COLOUR Dorsal surface greyish-brown, with pale yellow, irregular series of oval spots on vertebral region; forehead greyish-brown or with slightly reddish-brown area; paired dark, triangular markings on dorsolateral part of tail-base; venter greyish-tan; dorsum of tail greyish-brown, with small, indistinct, dark-edged, pale yellow oval markings.

HABITAT AND BEHAVIOUR Inhabits lowlands and mid-hills, up to 1,600m. Only known from buildings but presumably also inhabits forests. Diet unstudied. Clutches comprise two eggs, measuring 4.1 × 3.7 and 4.5 × 3.6mm.

DISTRIBUTION Endemic to northern Borneo. Known only from the town of Ranau and the Park Headquarters of Gunung Kinabalu Park in Sabah.

IUCN THREAT STATUS Data Deficient.

Family Lacertidae Eurasian Lizards

Moderate to large lizards (SVL 40–260mm). Their characteristics include: skull has paired nasals, postorbitals and squamosals; dentition pleurodont; elongated species with conical head on distinct necks; long and robust trunk; long, moderately thick tail, and well-developed limbs; scalation uniform with large head scales, granular scales dorsally on neck and trunk, and large abutting scales on ventral surface; osteoderms absent on body scales and present on forehead scales; caudal autotomy present; egg-shells leathery.

Takydromus **Grass Lizards**

Medium-sized (SVL to 61mm) terrestrial and arboreal lacertids; body elongated; tail cylindrical, very long; longitudinal rows of large dorsal shields; continuous series of dorsal keels; ventrals have large, imbricate, plate-like scales, the outer ones keeled; lower eyelid scaly; temporal scales typically keeled; postmental shields reduced to 3–4 pairs; subdigital lamellae with tubercles; femoral pores present; tricuspid posterior teeth; hemipenis lacking a hemipenial sheath.

LONG-TAILED GRASS LIZARD *Takydromus sexlineatus* Daudin, 1802

(No vernacular names recorded)

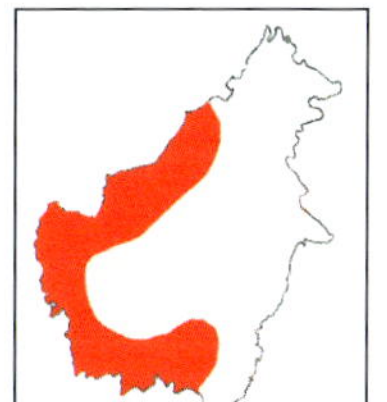

SIZE 61mm

IDENTIFICATION FEATURES Body slender; head rather long; dorsal surface has large, smooth, plate-like scales; three pairs of chin shields; sides of body have single row of large scales; tail 3–5 times as long as snout–vent length; transverse scale rows and plates 40; 1–2 femoral pores.

COLOUR Dorsal surface striped, including coffee-brown vertebral stripe from forehead, and yellow stripe from posterior corner of eyes to flanks of body; paravertebral stripe extends beyond tail-base, where it becomes indistinct, disappearing at around a fourth tail length; dorsolateral stripe dark brown, unpatterned; upper and lower lips pale green; brownish-black stripe on canthal region; scale

covering tympanum pale green; flanks greenish-yellow in anterior first third of length; each scale centrum in rest of flanks with large brown area; chin and throat cream; pectoral and abdominal regions cream with pale brown sheen; upper surfaces of limbs unbanded, pale brown; under surface of tail and lower parts of limbs pale brown; males brighter than females in colouration; juveniles resemble adult females.

HABITAT AND BEHAVIOUR Inhabits lowlands, in habitats such as grassland, marshes and swamps, up to 900m. Diurnal and terrestrial, as well as semi-arboreal. Diet includes insects and millipedes. Clutches comprise 2–5, eggs, each 10 x 5.5mm; hatchlings 81mm.

DISTRIBUTION A species composite, with range extending over Myanmar, Thailand, Vietnam, southern and eastern China, Peninsular Malaysia, Borneo, Sumatra, Java and smaller islands associated with those of the Greater Sundas. On Borneo, widespread and may be seen in Bukit Batu, Bau, Gunung Matang, Kota Samarahan and Lambir Hills National Park in Sarawak, and Sintang, Sinkawang, Pontianak, Gunung Niyut, Gunung Kenepai, Pulau Jambu and Banjarmasin in Kalimantan.

IUCN THREAT STATUS Least Concern.

Head and forebody

Juvenile

Family Lanthanotidae Earless Monitors

Lizards in this family measure 200mm SVL. Their characteristics include: pterygoid teeth present; nasals fused; supraoccipital has broad contact with parietal; frontals trapezoidal; prefrontal and postfrontal contact above orbit; dentition pleurodont; heterogenous scalation with reduced underlying osteoderms, not fused to skull; nine cervical vertebrae; upper temporal bar, parietal eye and ear opening absent.

Lanthanotus Bornean Earless Monitor

Small (SVL to 200mm) fossorial species; slender and elongated, with short limbs; heterogenous scalation with underlying osteoderms, concentrated on anterior of frontoparietal suture and in temporal regions; orbital rims form prominent bulges on sides of skull, projecting beyond level of snout and cheek; parietal eye and ear opening absent; palpebral reduced, roughly ovate and situated in dermis overlying upper eyelid; tooth characterized by development of plicidentine, indicated by striations on surface of tooth base; nine cervical vertebrae; lack of overlap of clavicles and interclavicles; sternum has two pairs of ribs; loss of a phalanx in fourth digit of manus and pes; egg-shells leathery.

Bornean Earless Monitor *Lanthanotus borneensis* Steindachner, 1877

(Iban Chenanum; Gana; Pangkor)

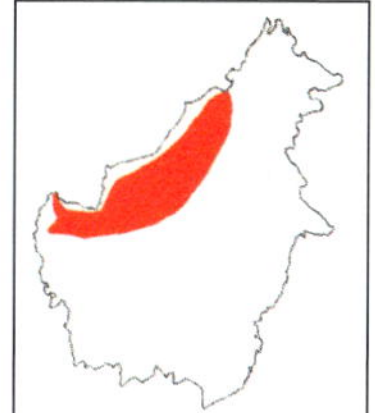

SIZE 200mm

IDENTIFICATION FEATURES Slender, elongated and cylindrical; head blunt; forehead flat, covered with small, granular, nodule-like scales; nostrils posterior of snout, situated on upper surface; eyes reduced, with movable lids; lower eyelid has transparent window; gular fold absent; tongue forked; males have blunt, rectangular jaws; females have relatively pointed jaws; six parallel rows of large tuberculate scales on dorsal surface from head to back of body; two central rows of large scales on tail; limbs short; claws recurved; tail not compressed.

COLOUR Dorsal surface unpatterned brownish-orange, or with dark vertebral stripes; venter yellow with brownish-orange and ochre mottling.

HABITAT AND BEHAVIOUR Inhabits lowland localities from lowland rainforests in the vicinity of streams to the mid-hills. Nocturnal, cryptozoic species, daylight hours being spent in a semi-

torpid condition in retreats such as burrows in the soil up to about 30cm, along riverbanks or under rocks and rotting logs. At night, forages on land and in water. Diet comprises earthworms, crabs and cockroaches. Skin shed in a single piece. Oviparous, producing 2–5 oval, leathery shelled eggs, each 20 x 30mm. These hatch 90 days later.

DISTRIBUTION Endemic to Borneo, western and central Sarawak and Kalimantan Barat. Protected within Baleh National Park, Niah National Park, Pelagus National Park and Samunsam Wildlife Sanctuary in Sarawak. Other Bornean locations include Ulu Temburong in Brunei Darussalam, and Landak and Embaloh Districts in Kalimantan.

IUCN THREAT STATUS Endangered.

Dorsal view

Dorsal view

Head and forebody

Head and forebody

Family Scincidae Skinks

Small to large lizards (SVL 27–35mm). Their characteristics include: skull has paired nasals and squamosals; single fused parietal; dentition pleurodont; pterygoid teeth present or absent; limbs range from well developed in many terrestrial and arboreal forms, to reduced or absent in some fossorial forms; scales smooth, shiny and cycloid, underlain by osteoderms, each composed of a mosaic of smaller bones; secondary palate ranges from partial to complete; tail autotomy; reproduction within the family comprises both oviparous and ovoviviparous modes; egg-shells of oviparous species leathery.

Key to Bornean genera of Scincidae

1a. Palatines separated in palate midline; limbs absent .... *Brachymeles*
1b. Palatines in contact in palate midline; limbs present **2**

2a. Supranasals absent **3**
2b. Supranasals present **7**

3a. Lower eyelid has transparent disc; light vertebral stripe .... *Lipinia*
3b. Lower eyelid scaly; no light vertebral stripe **4**

4a. External ear opening absent .... *Larutia*
4b. External ear opening present **5**

5a. Ear opening superficial; single large precloacal .... *Tropidophorus*
5b. Ear opening sunk; precloacals fragmented **6**

6a. Temporals larger than flank scales; digits not reduced .... *Sphenomorphus*
6b. Temporals subequal to flank scales; digits reduced .... *Tytthoscincus*

7a. Supranasals reduced, separated by rostro-frontonasal suture .... *Emoia*
7b. Supranasals large, forming median suture **8**

8a. Limbs reduced **9**
8b. Limbs well developed **10**

9a. Pterygoids rounded posteriorly .... *Subdoluseps*
9b. Pterygoids emarginate posteriorly .... *Lygosoma*

10a. Dorsal scales distinctly keeled; pterygoids separated .... *Eutropis*
10b. Dorsal scales weakly keeled; pterygoids in contact .... *Dasia*

Brachymeles Legless Lizard

Medium-sized (SVL to 131mm) terrestrial/fossorial live-bearing scincids; elongated; limbs reduced or absent; eyes small; lower eyelid scaly; external ear opening small or absent; pterygoid and palatine bones not in contact mesially; both lacking teeth; maxillary and mandibular teeth conical.

Kinabalu Legless Lizard *Brachymeles apus* Hikida, 1982

(No vernacular names recorded)

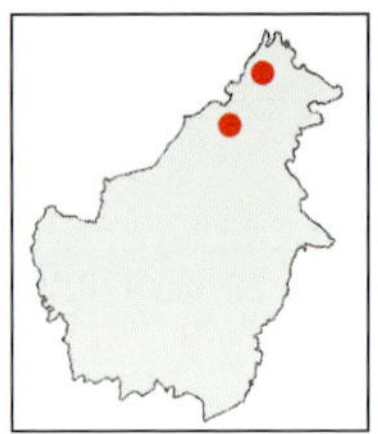

SIZE 131mm

IDENTIFICATION FEATURES Body slender, elongated and worm-like; lower eyelid has single large scale; ear opening absent; nostril between supranasal and supralabial I; postnasals absent; frontonasals wider than long, forming broad suture with rostral; prefrontals separated from each other; frontal longer than wide; parietals in broad contact behind interparietal; frontoparietals separated; nuchals absent or reduced to single pair; supraoculars five; supraciliaries two; supralabials six; infralabials five; single large postmental; dorsals and ventrals of equal size; midbody scale rows 22–24; ventrals 108–113; precloacals 5–8, large; limbs absent; tail thick, rounded, ending in blunt tip.

COLOUR Dorsal surface of body and forehead reddish-brown, darkening posteriorly; snout-tip lighter; tail-tip dark brown to reddish-black; ocular scales darker; venter unpatterned cream.

HABITAT AND BEHAVIOUR Inhabits montane areas at 1,300–1,520m, along Kinabalu-Crocker Range massif of Sabah, extending south-west to Sarawak's Gunung Murud. Diurnal, fossorial and probably a predator of small insects, their larvae and earthworms. Live bearer, producing clutches of four, each 42–43mm SVL.

DISTRIBUTION Endemic to Borneo. Gunung Murud in Sarawak; Kinabalu Park and Crocker Range Park in Sabah.

IUCN THREAT STATUS Least Concern.

Dasia Tree Skinks

Medium-sized to large (SVL to 130 mm) arboreal scincids; limbs and body robust; supranasals present; prefrontals large; frontoparietals paired; body scales smooth; paired nuchal scales; postorbital bones present or absent; palatal rami of pterygoids meet medially; pterygoid teeth present; juvenile colouration of dorsum comprises narrow bands.

Key to Bornean species of *Dasia*

1a. Parietals meet behind interparietal .... *D. vyneri* (p. 143)
1b. Interparietal completely separates parietals **2**

2a. Dorsal with numerous stripes .... *D. vittata* (p. 142)
2b. Dorsum banded or unicoloured **3**

3a. Ventrals ≥ 56; supranasals in broad contact **4**
3b. Ventrals ≥ 56; supranasals separated .... *D. olivacea* (p. 140)

4a. Adult have 5–8 broad, dark rings on body, between axilla and groin; prefrontals typically separated .... *D. semicincta* (p. 141)
4b. Adult has 8–14 narrow dark rings on body, between axilla and groin; prefrontals in broad contact .... *D. grisea* (below)

GREY TREE SKINK *Dasia grisea* (Gray, 1845)

(No vernacular names recorded)

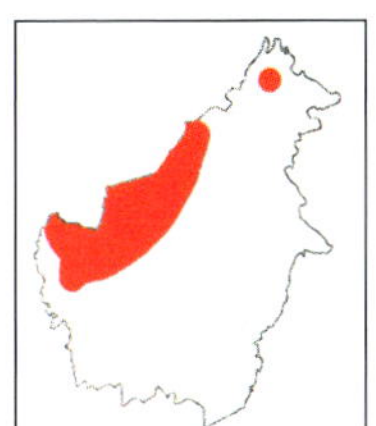

SIZE 130mm

IDENTIFICATION FEATURES Slender skink; paravertebrals not widened; ventrals 57–69; midbody scale rows 26–30; lamellae under toe IV 16–19; three strong keels on dorsal scales; prefrontals in broad contact; supranasals in broad contact; two large heel scales in males.

COLOUR Dorsal surface of adults light or dark brown, sometimes with green sheen, with 8–14 narrow dark rings, each 1–3 scales wide, between axilla and groin; forehead sometimes with faint dark spots; parietal spots present; eyelids distinctly yellow; dark caudal rings present; banded pattern more distinct in young, becoming obscure with growth; venter bright green.

HABITAT AND BEHAVIOUR Inhabits lowland forests, as well as forest edges and parks and gardens. Diet includes coleopterans, isopterans, formicids, molluscs and reportedly also fruits. Clutches comprise 2–6 eggs, each 22.5–24.0 x 14.5–15.5mm; hatchling 46–52mm SVL.

DISTRIBUTION Southern Malay Peninsula, Borneo, Sumatra and the Philippines. Widespread on Borneo, and most likely to be seen in Gunung Gading, Gunung Santubong and Niah National Park in Sarawak; Ranau in Sabah, and Sanggau in Kalimantan.

IUCN THREAT STATUS Least Concern

Olive Tree Skink *Dasia olivacea* Gray, 1839

(No vernacular names recorded)

SIZE 115mm

IDENTIFICATION FEATURES Body robust; snout pointed; scales under tail not large; ear opening small; paravertebrals not widened; ventrals 45–59; midbody scale rows 28–30; between 3 and 5 weak keels on dorsal scales; prefrontals separated; supranasals typically separated; paired nuchals; supraoculars four; supraciliaries 7–8; supralabials seven; infralabials eight; midbody scale rows 28–30; paravertebrals 41–46; lamellae under toe IV 17–21; postorbital bone present; two large heel scales in males.

COLOUR Dorsal surface of adults: forehead without stripes; head scales have black margins; dark caudal rings absent; surface of body olive to greenish-brown, sometimes black spotted; venter unpatterned green; occasionally dorsal surface more brown with darker speckling; juvenile colouration: dorsum golden-yellow, with 13–16 dark transverse bands, each three scales wide, broader than the orange-yellow bands, each a single scale wide; some dark bands, especially at posterior of body, incomplete; dorsal surface of tail yellowish-pink, unbanded; dark stripe from snout-tip to postocular region; lips unbarred, except the posterior ones; upper surfaces of limbs, especially forelimbs, have narrow yellow bands; chin, throat and under surface of tail unpigmented yellow; abdominal region and under surfaces of limbs cream-yellow; scales on manus and pes grey.

HABITAT AND BEHAVIOUR Inhabits relatively open forests, from the plains to 1,200m, and on offshore islands. Diurnal, utilizing tree trunks and branches, and ascending to level of canopy; associated with large trees, especially at edges of clearings, sheltering under peeling bark. Diet comprises insects, including bees, beetles, ants and flies, and other arthropods, as well as lizards. Clutches comprise 5–14 eggs, each 18.0–19.5 x 10.0–12.0mm; hatchlings 27–38mm SVL.

DISTRIBUTION Great Nicobar, Myanmar, Thailand, Cambodia, Peninsular Malaysia, Singapore, Sumatra, Borneo, Java, Natuna Islands, Laos, Cambodia and Vietnam. Widespread on Borneo, and most likely to be seen at Bako National Park, Pulau Satang, Niah National Park and Lambir Hills National Park in Sarawak; lower reaches of Gunung Kinabalu in Sabah, and Semitau and Pulau Sibau in Kalimantan.

IUCN THREAT STATUS Least Concern.

Head and forebody

Dorso-lateral view

HALF-BANDED TREE SKINK *Dasia semicincta* (Peters, 1867)

(No vernacular names recorded)

SIZE 130mm

IDENTIFICATION FEATURES Body relatively robust; head somewhat large; postorbital bone absent; paravertebrals not enlarged, numbering 59–63; midbody scale rows 28–30; lamellae under toe IV 17–22; three weak keels on dorsal scales; prefrontals typically separated, but occasionally in contact; supranasals in broad contact; two large heel scales in males.

COLOUR Adults have 5–8 broad dark rings between axilla and groin, the dark bands broader than light bands in juveniles, in which growth stage body can be glossy coal-black, with series of orange-yellow bars from snout-tip to end of tail; limbs, including digits, barred with yellow; transverse bands on head; dark caudal rings complete ventrally.

HABITAT AND BEHAVIOUR Arboreal species. The sole Bornean record with a precise locality is from Bario, from the submontane (> 1,000m asl) region of the Kelabit Highlands.

DISTRIBUTION Mindanao in the southern Philippines, and on Borneo only known from Bario, in the Kelabit Highlands of Sarawak.

IUCN THREAT STATUS Data Deficient.

Dasia semicincta (Peters, 1867) (FMNH 67340) from 'Bario, Kelabit Highlands, Miri Division, Sarawak.

Bornean Striped Tree Skink *Dasia vittata* (Edeling, 1864)

(No vernacular names recorded)

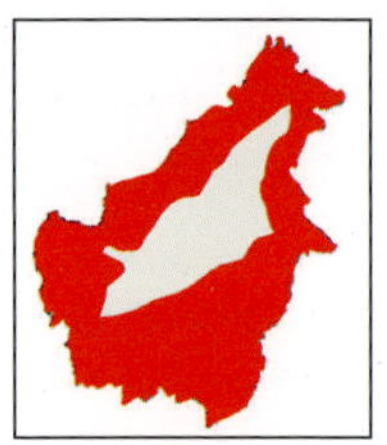

SIZE 96mm

IDENTIFICATION FEATURES Robust tree skink; dorsals keeled; lower eyelid has small scales; nostril situated within single nasal; frontonasal wider than long, forming suture with rostral and frontal; prefrontals separated from each other; frontal larger than frontoparietal and interparietal; nuchals generally absent, and when present, only a single pair; supraoculars 5; supraciliaries 8–9; supralabials 7–8; tympanum present; midbody scale rows 30; scales under toe IV 16–22; two large heel scales in males; tail rounded in cross-section, tapering to fine point.

COLOUR Forehead and forebody, up to behind scapula, glossy black; rest of dorsum brownish-grey or tan-brown, with less contrasting pattern; flanks have yellow spots edged with dark brown; supralabials and especially infralabials dark barred; temporal region has scattered yellowish-green spots; body has dark and light spots; pale greenish-yellow stripe from snout-tip to occiput; greenish-yellow dorsolateral stripe starts from snout-tip, crosses above eyes and continues up to back of scapular region; another stripe of same colour runs from snout-tip to insertion of forearm; tail and upper surfaces of limbs have dark and light mottled pattern; gular, pectoral, abdominal and under surfaces of tail and limbs bright green; scales on manus and pes grey.

HABITAT AND BEHAVIOUR Inhabits lowlands, from the sea coast up to mid-hills and offshore islands. Active on tree trunks and branches, from near ground level up to c. 37m near canopy. Diet comprises small insects such as ants, hymenopterans, coleopterans, blattarians, isopterans, orthopterans, dermapterans, hemipterans, thysanopterans, dipterans, arachnids, isopods and larvae of lepidopterans, in addition to geckos. Clutches of 2–4 produced.

DISTRIBUTION Endemic to Borneo. Known from across the island, including Kuching, Gunung Matang, Gunung Santubong, Bako National Park, Lambir Hills National Park, Niah National Park, and Gunung Mulu National Park in Sarawak; Bandar Seri Begawan, Ulu Temburong, Pulau Selirong, Tasek Merembun in Brunei Darussalam; Deramakot, Kota Kinabalu, Danum Valley Field Centre, Sepilok Forest Reserve, Gunung Kinabalu Parkmand Tawau Hills Park in Sabah, and Banjermasin and Barito Ulu, Taman Nasional Bentuang Karimun in Kalimantan.

IUCN THREAT STATUS Least Concern.

VYNER'S TREE SKINK *Dasia vyneri* (Shelford, 1905)

(No vernacular names recorded)

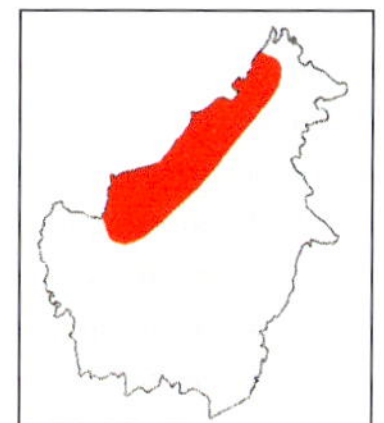

SIZE 72mm

IDENTIFICATION FEATURES Body slender; snout obtusely pointed; lower eyelid scaly; supranasals present, failing to contact each other; frontonasal as broad as long, in contact with rostral, but not with frontal; supraoculars five; supraciliaries eight; one pair of nuchals; tympanum small; midbody scale rows 21–26; precloacals slightly larger than preceding scales; body scales smooth; lamellae under toe IV 20; tail relatively short, equal in length to head and body.

COLOUR Head olive-grey or metallic brownish-green, some scales edged with black; dorsal surface of body has four longitudinal stripes, consisting of series of black dorsal scales that show central quadrate olive-grey spots; flanks of body and neck have black-edged brown scales; tail greyish-olive; venter pale green.

HABITAT AND BEHAVIOUR Inhabits forested mid-hills to 915m. Arboreal; active on trees. Diet and reproductive habits unstudied.

DISTRIBUTION Endemic to Borneo. Isolated records from Niah National Park; Gunung Balingean and Bukit Balian, Kapit Division in Sarawak; Kiudang Meriuk Farmstay in Tutong, Brunei Darussalam; Kiau in Kinabalu Park, Sandakan Bay and Sepilok Forest Reserve in Sabah, and Long Blu and Pulau Sibau in Kalimantan.

IUCN THREAT STATUS Least Concern.

Emoia Coastal Skinks

Small to medium-sized (SVL to 98mm), slender and supple, terrestrial scincids; limbs well developed; supranasals reduced, not forming median suture; interparietal typically fused with single frontoparietal; tympanum sunk; lower eyelids have undivided transparent discs; limbs well developed and pentadactyle.

Key to Bornean species of *Emoia*

1a. Interparietal fused with frontoparietals.......................................*E. cyanura* (p. 146)
1b. Interparietal not fused with frontoparietals...............................**2**

2a. Under surface of tail pink...*E. atrocostata* (below)
2b. Under surface of tail blue...*E. caeruleocauda* (opposite)

Mangrove Skink *Emoia atrocostata* (Lesson, 1830)

(No vernacular names recorded)

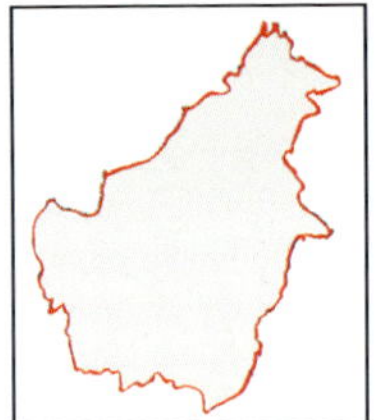

SIZE 98mm

IDENTIFICATION FEATURES Body slender; limbs and tail well developed; snout tapered; prefrontals narrowly in contact to separated; interparietal distinct, narrow; auricular lobules present; supralabials 6–8; infralabials 6–7; lamellae under toe IV 30–42.

COLOUR Top of body greyish-olive, flecked with dark brownish-grey; lower lateral parts greyish-tan; dark postocular stripe, with some pale flecks, extends beyond inguinal region and becomes broadest above axilla; venter bluish-grey to cream, with dark pigmentation under throat; juveniles have white venter; limbs greyish-olive with pale grey flecks; oral cavity pink.

HABITAT AND BEHAVIOUR Found in sandy and rocky beaches and mangrove swamps. Diet includes insects, isopods, decapods and – unusually for a skink – also fish and geckos. Clutches comprise 1–3 oval, smooth-shelled eggs, each 8 x 20mm, deposited in piles of driftwood and tree holes.

DISTRIBUTION The Malay Peninsula, Sumatra, Borneo, Java, the Lesser Sundas, Sulawesi, the Philippines, Ryukyu Archipelago in Japan, Taiwan, China, New Guinea, the Republic of Belau, Solomon Islands and northern Australia. Places to see it on Borneo include Gunung Santubong, Samunsam Wildlife Sanctuary, Bako National Park in Sarawak; Pulau Chermin, Serasa beach, Pulau Selirong in Brunei Darussalam; Pulau Banggi, Pulau Balambangan, Pulau Bohey Dulang, Pulau Gaya, Sandakan, Taman Pulau Tiga in Sabah, and Pulau Maratua and Tabanio in Kalimantan.

IUCN THREAT STATUS Least Concern.

Common Blue-tailed Skink *Emoia caeruleocauda* (De Vis, 1892)

(No vernacular names recorded)

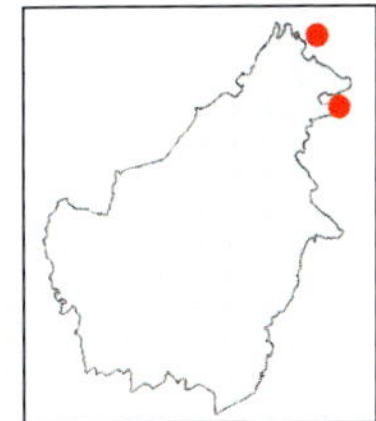

SIZE 65mm (on Borneo)

IDENTIFICATION FEATURES Body relatively robust; snout short, tapering; frontonasal broader than long, in contact with rostral and with frontal; parietal in contact with interparietal; interparietal distinct; one pair of large nuchals; supraoculars four; supralabials eight, the last reduced in size; tympanum smaller than eye, with three anterior auricular lobules; midbody scale rows 16, smooth; supraoculars four; two pairs of large precloacals.

COLOUR Back of forehead has dark smudges; dorsal surface of body yellowish-olive, with dark olive-brown vertebral stripe in males, yellow in females, which extends from rostral to beyond tail-base; second stripe from back of head, joining at tail-base; venter shiny buff; limbs dark olive-brown, unbanded; posterior half of tail pale blue.

HABITAT AND BEHAVIOUR Associated with island beaches. Terrestrial. Diet unstudied. Clutches comprise two eggs; hatchling 22.4mm SVL.

DISTRIBUTION Sulawesi, New Guinea, Vanuatu, Seram, New Britain, Fiji and the Solomon Islands. On Borneo, only known from Pulau Bakkungan Kecil and Pulau Sipadan, Semporna District of Sabah.

IUCN THREAT STATUS Least Concern.

Adult

Juvenile

Blue-tailed Skink *Emoia cyanura* (Lesson, 1830)

(No vernacular names recorded)

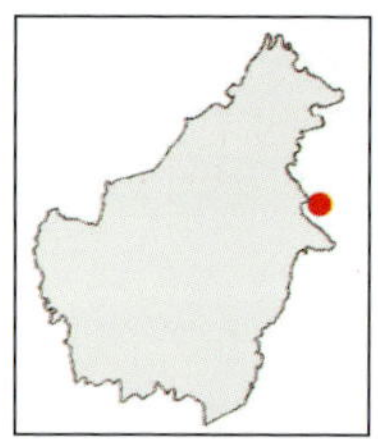

SIZE 61mm

IDENTIFICATION FEATURES Body robust; snout tapering; prefrontals separated; interparietal fused with frontoparietals; supralabials 6–7; infralabials 6–7; paravertebral scale rows 52–64; midventral scale rows commonly 25–32, 28; toe IV lamellae 59–86.

COLOUR Dorsal surface brownish-black or dark brown, with distinct pale vertebral and dorsolateral stripes, ground colour fading posteriorly to blue or brown; pale midlateral stripe; venter bluish-cream to cream, or greyish-tan; tail green, blue or tan, lacking prominent black spots or bars.

HABITAT AND BEHAVIOUR Inhabits coastal areas and edges of closed forests. Diurnal and generally terrestrial, active on leaf litter, palm trash and plantation rubble; also reported to climb low vegetation. Diet comprises insects, spiders, woodlice and earthworms. Eggs each 10–13 × 6–9mm; hatchlings 22mm SVL.

DISTRIBUTION Sula Archipelago and Halmahera of eastern Indonesia, Oceania, from the archipelago of Hawaii, Clipperton Island, Easter Island, west through Polynesia and Melanesia, to the Admiralty and Bismarck Islands, Sulawesi, Seram and the west coast of Central or South America. Sole Bornean record is from Pulau Derawan in Kalimantan.

IUCN THREAT STATUS Least Concern.

Eutropis Ground Skinks

Medium-sized to large (SVL to 137mm for Bornean species) terrestrial scincids; build robust; snout obtusely pointed; palatines in contact; nostril situated in single nasal; eyelids movable; lower eyelid of all Bornean species scaly; supranasals present; dorsal scales keeled; limbs well developed, pentadactyle; tympanum deeply sunk; pterygoids separated, pterygoid teeth present; presacral vertebrae 26; precloacal scales not large; both oviparity and viviparity as reproductive mode shown by its members.

Key to Bornean species of *Eutropis*

1a. Dorsal scales have 5–9 strong keels in adults (3 in juveniles)........**2**
1b. Dorsal scales have 3 (rarely 4 or 5) weak to strong keels in adults........**3**

2a. Head scales at posterior smooth; midbody scale rows 32–34; prefrontals in broad contact; lateral band dark brown or orange........*E. multifasciata* (p. 148)
2b. Head scales at posterior rugose; midbody scale rows 28–30; prefrontals not in contact; lateral band black, edged on top and below with white stripes........*E. rudis* (p. 149)

3a. Dorsal surface has 5–7 pale longitudinal lines........*E. rugifera* (p. 150)
3b. Dorsal surface lacks pale longitudinal lines........*E. indeprensa* (below)

Philippine Ground Skink *Eutropis indeprensa* (Brown & Alcala, 1980)

(No vernacular names recorded)

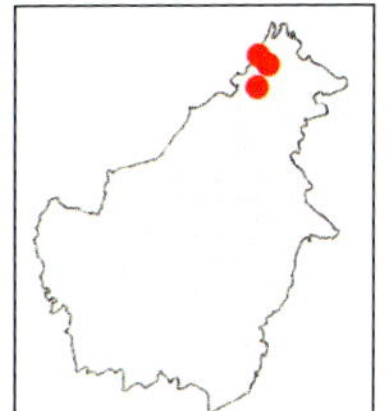

SIZE 67mm

IDENTIFICATION FEATURES Body relatively robust; tympanum reduced; nostril situated within nasal; supranasals separated; frontonasal long and narrow, forming suture with rostral and frontal; prefrontals in contact or separated; parietals typically divided by large interparietal; supraoculars four; supralabials 6–7; infralabials six; dorsal scales have three keels; midbody scale rows 27–34; paravertebrals 40–48; lamellae under toe IV 18–24; tail tapering.

COLOUR Dorsal surface brownish-tan to light brown, typically with 2–4 longitudinal rows of small dark brown blotches; flanks dark brown; light narrow stripe from labials to insertion of forelimbs; throat of breeding males orangish-red; venter greyish-blue.

HABITAT AND BEHAVIOUR Inhabits lowland and mid-hill forests and vicinity of human settlements. In the Philippines, recorded at altitudes of up to 1,200m. Terrestrial and insectivorous. Hatchlings 22.5–24.7mm SVL.

DISTRIBUTION The Philippines (Mindoro, Caluya, Negros, Cebu, Leyte, Camiguin, Sámar, Mindanao and Palawan). Northern Bornean records are from the upper reaches of Sungei Purulon and Keningau in the Crocker Range, Bukit Padang, Kota Kinabalu Kiau, Kinokok and Lobang in Gunung Kinabalu Park, in Sabah.

IUCN THREAT STATUS Least Concern.

Common Sun Skink *Eutropis multifasciata* (Kuhl, 1820)

(Bahasa Indonesia: Kadal Kebun; Bahasa Malaysia; Bengkarong Matahari; Bruneian Malay: Balinkarong; Kelabit: Ale to'o)

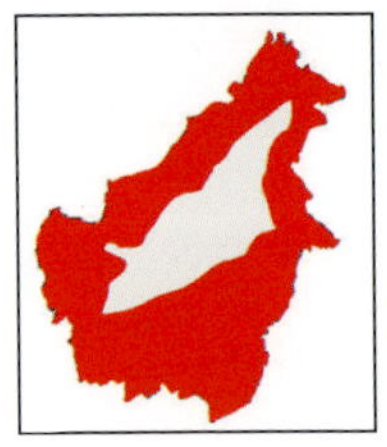

SIZE 137mm

IDENTIFICATION FEATURES Body robust; lower eyelid scaly; dorsal scales have three, rarely five, keels; tympanum rather large, rounded; postnasal present; supranasals separated; nostril situated within single nasal; frontonasal twice as wide as long; prefrontals typically in contact; frontoparietals larger than interparietal; parietals divided by interparietal; pair of nuchals; temporals smooth; supraoculars four; supraciliaries 5–6; supralabials 6–7; infralabials 6–7; midbody scale rows 30–35; paravertebrals 42–48; lamellae under toe IV 16–20; dorsal surface of thighs smooth.

COLOUR Dorsal surface bronze-brown, usually with dark brown band, with series of white spots or streaks along sides of body; during the breeding season, both adult males and females have a bright orange or reddish-orange side band; bright yellow throat and red lips; pale dorsolateral line present; venter unpatterned cream.

HABITAT AND BEHAVIOUR Inhabits open habitats, including forest clearings and edges of riparian forests, to 1,800m. Diurnal and terrestrial. Diet insectivorous, and including termites, orthopterans, cockroaches, isopods, spiders, centipedes and snails; smaller lizards also ingested. Ovoviviparous; clutch size 1–10; neonates 33.1–43mm SVL.

DISTRIBUTION Northeastern India and the Nicobar Islands, Myanmar, southern China, Taiwan, Thailand, Vietnam, the Malay Peninsula, Sumatra, Borneo, Java, Bali, Komodo, Lombok, Sumbawa, Rintja, Sumba, Flores, Maluku, Sulawesi and Halmahera, the Sula and Togian Archipelagos, and the Philippines. Records from New Guinea, Darwin, Australia and Dade County, Florida, USA, are suspected to be through inadvertent human introductions. Can be found across Borneo, and some localities include Bako National Park, Gunung Matang, Gunung Santubong, Loagan Bunut National Park, Gunung Mulu National Park and Niah National Park in Sarawak; Pulau Labi, Tasek Merembun, Tutong District and Tasek Lama, Bandar Seru Begawan in Brunei Darussalam; Taman Pulau Tiga, Kota Kinabalu, Gunung Kinabalu, Danum Valley Field Centre, Sepilok Forest Reserve and Sandakan District in Sabah, and Sintang, Pulau Sibau, Banjarmasin, Samarinda and Taman Nasional Kayan Mentarang in Kalimantan.

IUCN THREAT STATUS Least Concern.

Black-banded Ground Skink *Eutropis rudis* (Boulenger, 1887)

(No vernacular names recorded)

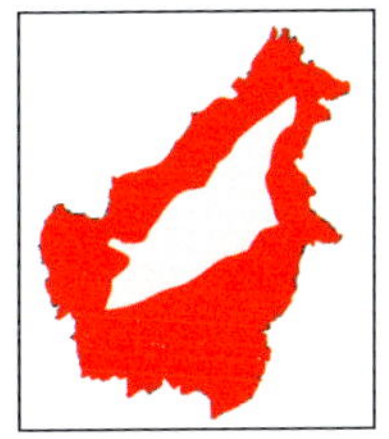

SIZE 120mm

IDENTIFICATION FEATURES Body robust; head nearly indistinct from neck; forehead scales at posterior distinctly rugose; dorsal scales have three strong keels; midbody scale rows 28–30; lamellae under toe IV 18–21.

COLOUR Top of body olive-brown, with a light-edged dark brown line along sides of head and body; sides white spotted; throat of adult males crimson, sometimes dark spotted; blue or yellow; that of females unpatterned cream; venter greenish-white.

HABITAT AND BEHAVIOUR Inhabits forested habitats in the lowlands and mid-hills up to about 1,300m asl. Diurnal and terrestrial; active on the leaf litter. Diet includes orthopterans, blattids, lepidopterans, dipterans and isopods. Clutches of 2–4 eggs are laid inside rotting logs.

DISTRIBUTION The Mentawai Archipelago, Sumatra, Borneo, the Nicobar Islands, the Sula Archipelago, Sulawesi and the Sulu Archipelago of the southern Philippines. Widespread on Borneo, localities including Gunung Gading, Bako National Park, Gunung Matang, Gunung Santubong, Lanjak-Entimau Wildlife Sanctuary, Subis Forest Reserve, Niah National Park, Lambir Hills National Park and Loagan Bunut National Park in Sarawak; Ulu Temburong in Brunei Darussalam; Kota Kinabalu, Sepagaya Forest Reserve, Sepilok Forest Reserve, Tawau Hills Park, Gunung Kinabalu Park, Deramakot, Kinabatangan District, Taman Pulau Tiga, Beaufort District, Gunung Trus Madi and Danum Valley Field Centre in Sabah, and Bagan Manggis, the upper reaches of Sungei Palung, Pulau Sibau, Pontianak, Sintang, Bulangan and Taman Nasional Kayan Mentarang in Kalimantan.

IUCN THREAT STATUS Least Concern.

REMARKS We relegate *E. lewisi* (Bartlett, 1895) to the synonymy of the current species, following our work in progress.

RED-THROATED SKINK *Eutropis rugifera* (Stoliczka, 1870)

(No vernacular names recorded)

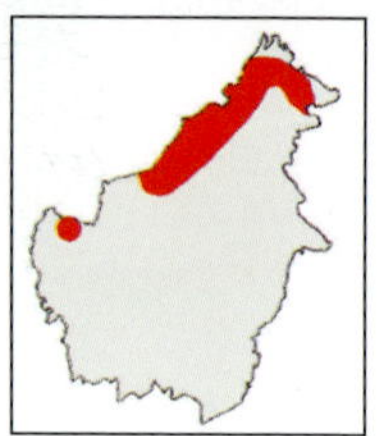

SIZE 65mm

IDENTIFICATION FEATURES Body relatively robust; head indistinct from neck; lower eyelids scaly; dorsal scales have five, rarely seven, distinct keels; ear opening reduced, horizontal to oval, edged with lobules; postnasals typically absent; supranasals narrow and widely separated; frontonasal wider than long, contacting rostral and frontal; prefrontals separate; parietals in contact by interparietal or separated; nuchals two although occasionally absent; supraoculars four; supraciliaries 5–6; supralabials nine; infralabials seven; midbody scale rows 24–28; lamellae under toe IV 18–26.

COLOUR Dorsal surface bronze-brown, lighter on flanks, sometimes with 5–7 greenish-cream longitudinal stripes, sometimes broken to form spots, giving it a grizzled appearance; forehead mid-brown, each scale bordered with dark brown; no dark postocular stripe; narrow, yellowish-grey postocular stripe extends to above level of axilla, from where it extends along flanks to about a third of tail; lower flanks pale brown; supralabials and infralabials lemon-yellow or cream-grey, with grey smudges; gular and pectoral regions lemon-yellow, unpatterned or dark spotted, turning bright red in breeding season; upper surfaces of limbs and tail unpatterned bronze-brown; phalanges dark brown; abdominal and cloacal regions cream or greenish-cream; under surface of tail cream, with dark brown smudges laterally and medially; under surfaces of forelimbs pale yellow, those of hindlimbs pinkish-cream; superciliaries bright yellow.

HABITAT AND BEHAVIOUR Inhabits mid-hills of dipterocarp forests and peat swamps. Terrestrial with some arboreal activity. Diet includes insects. Reproductive habits unknown.

DISTRIBUTION The Nicobar Islands of India, Peninsular Thailand, the Malay Peninsula, Sumatra, Borneo, Java and Bali. Widespread on Borneo and known from Gunung Matang, Bako National Park, Gunung Gading, Lambir Hills National Park and Niah National Park in Sarawak; Gadong, Bandar Seri Begawan and Tasek Merimbun in Brunei Darussalam, and Sepilok Forest Reserve, Taman Pulau Tiga, Tawau Hills Park, Danum Valley Field Centre and Crocker Range Park in Sabah.

IUCN THREAT STATUS Least Concern.

Female

Female

Male

Larutia Limbless (or Reduced Limbed) Skinks

Medium-sized (SVL to 175 mm) fossorial scincids; body elongated; limbs reduced; forelimbs and hindlimbs have two digits each; eyes small; parietal eye-spot absent; supranasals absent; single presubocular; large posterior chin shield separated from infralabials by smaller scales; midbody scale rows 22; median pair of precloacals large; external ear opening absent; postorbital bone present; presacral vertebrae 50–59; dorsal colour pattern with longitudinal stripes or cross-bars

Key to Bornean species of *Larutia*

1a. Eyelid coved by large, plate-like scales; midbody scale rows 22 .....*L. kecil* (below)
1b. Eyelid coved by small scales; midbody scale rows 23 ....*L. puehensis* (p. 152)

Penrissen Small-limbed Skink *Larutia kecil* Fukuyama, Hikida, Hossman & Nishikawa, 2019

(No vernacular names recorded)

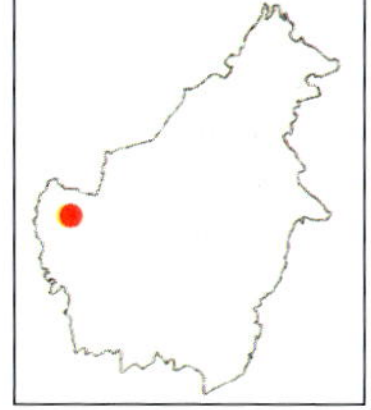

SIZE 84mm

IDENTIFICATION FEATURES Body slender; head elongated, triangular; lower eyelid coved by small, plate-like scales; two irregular frontoparietals; left frontoparietal overlaps right; supraoculars six; supralabials six; infralabials six; body elongated, covered with smooth, cycloid imbricate scales; midbody scale rows 22; ventrals large and rounded; limbs small, degenerate, covered by small, round imbricate scales; two clawed digits, two subdigital lamellae on toe II; subcaudal scales subequal to dorsal caudals.

COLOUR Dorsum dark brown, lacking yellow or pale bands, stripes or spots; lateral margins of dorsal and dorsolateral scales light brown; certain head scales, such as rostral, supralabial I, nasal, frontonasal, mental and infralabial I opaque; throat pinkish-grey, mottled with pale brown; rest of venter yellowish-brown; under surface of tail yellowish-white with dark brown mottling.

HABITAT AND BEHAVIOUR Inhabits edges of submontane forests at around 1,000m asl. Subfossorial, with at least some terrestrial activity, and apparently crepuscular, active on grass. Diet and reproductive habits unstudied.

DISTRIBUTION Endemic to Borneo. Known only from Gunung Penrissen in western Sarawak.

IUCN THREAT STATUS Not Evaluated.

Gunung Pueh Small-limbed Skink *Larutia puehensis*

Grismer, Leong & Yaakob, 2003

(No vernacular names recorded)

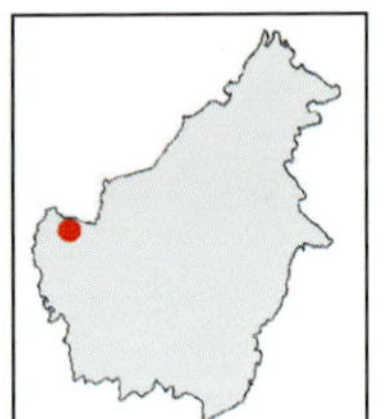

SIZE 141mm

IDENTIFICATION FEATURES Body slender; head elongated, triangular; lower eyelid coved by small, plate-like scales; two kidney-shaped frontoparietals; right frontoparietal overlaps left; supraoculars four; supralabials six; infralabials six; body elongated, covered with smooth, cycloid imbricate scales; midbody scale rows 23; ventrals large and rounded; limbs small, degenerate, covered by small, round imbricate scales; digits III and IV clawed, lamellae absent beneath digits; subcaudal scales subequal to dorsal caudals.

COLOUR Dorsal surface orangish-brown with scattered dark brown flecks covering supraoculars and sides of head; nuchal region uniformly dark brown with light radiating linear pattern; dark, linearly arranged spots resembling stripes extend from nuchal region to tail-tip; venter uniformly orangish-brown.

HABITAT AND BEHAVIOUR Inhabits hill dipterocarp forest at 300m. Probably a fossorial species, inhabiting the leaf litter and feeding on small invertebrates. Reproductive habits unstudied.

DISTRIBUTION Endemic to Borneo. Known only from Gunung Berumput in western Sarawak.

IUCN THREAT STATUS Data Deficient.

Lipinia Striped Skinks

Small (SVL to 56mm) arboreal or terrestrial scincids; body slender; premaxillary teeth nine; vomers fused; pterygoid teeth absent; nuchals oriented obliquely to parietals; postorbital absent; prefrontal large, forming a median suture; frontoparietals distinct or fused; scales on vertebral rows typically wider than those on lateral rows; median precloacals large; lower eyelid has clear window (in Bornean species); auricular lobules absent; body scales smooth; longitudinal scale rows at midbody < 28; basal subdigital lamellae typically expanded; dorsal colour pattern comprising pale (rarely dark) middorsal stripe at least anteriorly.

Key to Bornean species of *Lipinia*

1a. External ear opening absent **2**
1b. External ear opening present **3**

2a. Midbody scale rows 22 .... *L. nitens* (p. 154)
2b. Midbody scale rows 18–20 .... *L. inexpectata* (below)

3a. Midbody scale rows 28; toe IV lamellae 25 .... *L. vittigera* (p. 155)
3b. Midbody scale rows 24; toe IV lamellae 21 .... *L. miangensis* (p. 154)

Bornean Striped Skink *Lipinia inexpectata* Das & Austin, 2007

(No vernacular names recorded)

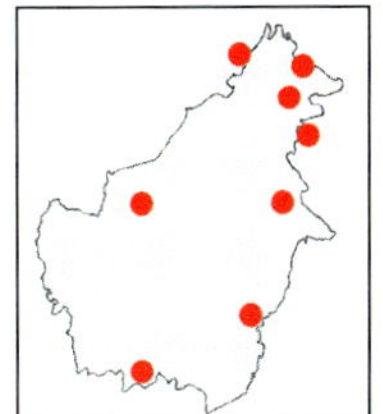

SIZE 41mm

IDENTIFICATION FEATURES Body slender; snout acute; external ear opening absent; lower eyelid has a clear spectacle; midbody scale rows 20; longitudinal scale rows between parietals and base of tail 46–50; lamellae under toe IV 16–17; supralabials six; infralabials 6–7; subcaudals 68–74.

COLOUR Dorsal surface tan-brown with series of dark grey-brown stripes. Vertebral stripe commences from interorbital region, continues across nape and midbody, and on to tail up to tip, narrowing on midbody; paired paravertebral stripes commence from posterior edge of supraoculars, and continue on to tail-tip; series of dark spots arranged in longitudinal series bilaterally on each outer side of paravertebral stripe; paired dorsal stripes start from temporals and terminate at inguinal region; snout-tip dark grey; labials unbarred; venter yellowish-cream with pale grey smudges, forming longitudinal stripe, especially along pectoral region; limbs unbanded; scales on under surface of limbs light grey; under surface of tail yellowish-cream.

HABITAT AND BEHAVIOUR Inhabits lowland forests, including offshore islands. Diet unknown. Clutches comprise two eggs.

DISTRIBUTION Endemic to Borneo. Localities include Nanga Tekalit on Sungei Mengiong, Kapit Division in Sarawak; Pulau Manukan, Sepagaya Forest Reserve, Sepilok, Danum Valley Field Centre and Dewhurst Bay in Sabah, and Pulau Nunukan and Bulangan in Kalimantan.

IUCN THREAT STATUS Least Concern.

Pulau Miang Striped Skink *Lipinia miangensis* (Werner, 1910)

(No vernacular names recorded)

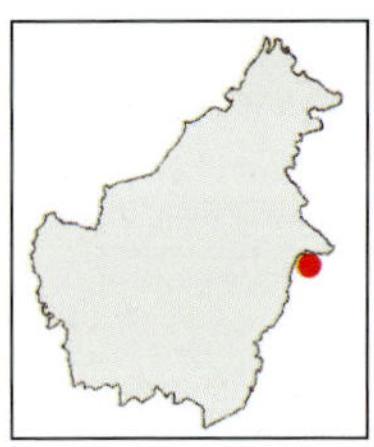

SIZE 39mm

IDENTIFICATION FEATURES Body slender; tail long; snout pointed; two vertebral series of scales large; parietals in contact with interparietal; supraoculars four; midbody scale rows 24; nuchals three pairs; toe IV has 21 smooth lamellae.

COLOUR Dorsal surface golden with two dark brown longitudinal stripes from snout-tip, over eyes, to around middle of tail; dark line from eye to base of forelimb; limbs light brown, with indistinct yellow dots; tail yellow dorsally; venter unpatterned greenish-white.

HABITAT AND BEHAVIOUR Presumably an inhabitant of lowland rainforests. Ecology unstudied, and no museum specimens are extant.

DISTRIBUTION Endemic to Borneo. Known only from Pulau Miang in Kalimantan.

IUCN THREAT STATUS Data Deficient.

Sarawak Striped Skink *Lipinia nitens* (Peters, 1871)

(No vernacular names recorded)

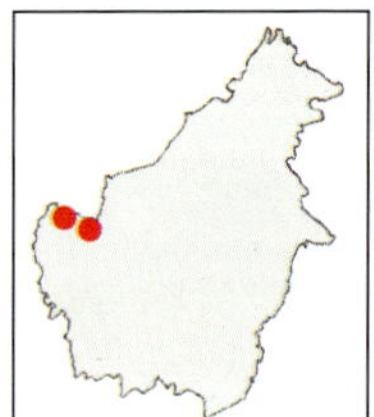

SIZE 34mm

IDENTIFICATION FEATURES Body slender; snout pointed; two paravertebral scale rows large; dorsal and lateral scales smooth; frontoparietals fused; external ear opening absent, replaced by scaly dimple; outer precloacals overlap inner; limbs reduced; midbody scale rows 20–22; lamellae under toe IV 16; supralabials six; supralabial IV in subocular position; fourth in suborbital position; median precloacal scales large; subcaudal scales enlarged.

COLOUR Dorsal surface metallic green; sides of body spotted with black and green; pale yellow vertebral stripe, with jagged edged black lines, one on each side from supraorbital to tail-base; colour pattern extends along length of body; supralabials and infralabials dark barred; limbs grey with irregular yellow blotches; belly greyish-yellow with scattered dark spots, especially on cloacal and abdominal regions; tail yellow with isolated black speckling, especially on sides.

HABITAT AND BEHAVIOUR Inhabits lowland and Kerangas forests. Terrestrial and diurnal. Diet comprises ants. Reproductive biology unstudied.

DISTRIBUTION Endemic to Borneo. Localities include Gunung Pueh and Gunung Matang in western Sarawak.

IUCN THREAT STATUS Data Deficient.

Common Striped Skink *Lipinia vittigera* (Boulenger, 1894)

(No vernacular names recorded)

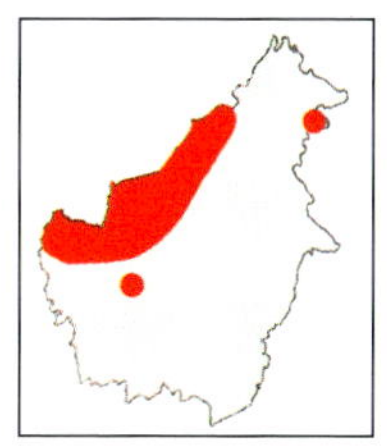

SIZE 42mm

IDENTIFICATION FEATURES Body slender; tail long, slender, tapering; snout elongated and acute; lower eyelid has transparent disc; ear opening small, lacking lobules; parietals in contact with interparietal; supraoculars four; supraciliaries seven; nuchals 3–4 pairs; fifth supralabials large and in midorbital position; midbody scale rows 28, smooth; toe IV lamellae 25.

COLOUR Dorsal surface brownish-black, with bright yellow vertebral stripe, commencing from snout-tip; flanks have dark and pale spots; bright yellow ocular ring; dark cantal and postocular stripe extends beyond tympanum; limbs black spotted; venter green.

HABITAT AND BEHAVIOUR Inhabits primary and lightly disturbed lowland rainforests, as well as peat-swamp forests. Arboreal; generally found on trunks of large and medium-sized trees with peeling bark. Diet unstudied. Clutches comprise 2–4 eggs.

DISTRIBUTION Thailand, the Malay Peninsula, the Mentawai Archipelago, Sumatra and Borneo. Known across Borneo, including Bukit Meraja, Gunung Matang, Pulau Talang Talang Besar Ranchan Pool, Serian in Sarawak; Tasek Merimbun Heritage Park in Brunei Darussalam; Danum Valley Field Centre and Tawau Hills Park in Sabah, and also in Kalimantan.

IUCN THREAT STATUS Least Concern.

Lygosoma Giant Skinks

Medium-sized to large skinks (SVL 49–168mm); head and limbs short; body and tail elongated; digits short; lower eyelid scaly; prefrontals not in contact; parietals in contact behind interparietal; pyerygoids emarginated posteriorly; palatines with posteriormedially projecting processes.

Key to Bornean species of *Lygosoma*

1a. Nuchals four........*L. kinabatangense* (p. 157)
1b. Nuchals seven........*L. bampfyldei* (below)

Bampfylde's Giant Skink *Lygosoma bampfyldei* Bartlett, 1895

(No vernacular names recorded)

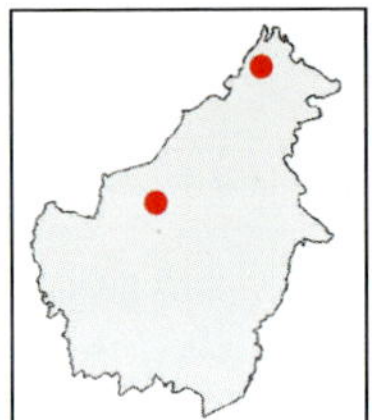

SIZE 142mm

IDENTIFICATION FEATURES Body robust; snout obtuse; lower eyelids scaly; three large auricular lobules; limbs short; supranasals contact behind rostral; prefrontals small; supraoculars five; supraciliaries eight; supralabials seven; infralabials six; parietals form suture behind interparietal; nuchals seven; precloacals 10; midbody scale rows 36–40, smooth; digits short, somewhat compressed; lamellae under toe IV 14; deep groove from posterior corner of nostrils to below orbit; tail slightly narrower than body, tip acute.

COLOUR Dorsal surface pale grey-brown up to lower flanks, unpatterned; brownish-black band across back of head, 5–6 scales wide, over nuchals and extending ventrally to slightly below tympanum; cream band anteriorly from back of eyes; area around eyes has dark patches; grey vertical bar across rostrum; flanks lack bands; gular, pectoral and abdominal regions unpatterned cream; upper surfaces of limbs dark brownish-grey; venter cream; tail dark brownish-grey dorsally, pale grey ventrally, unbanded; scales on manus and pes grey.

Head and forebody, lateral view

HABITAT AND BEHAVIOUR Known from isolated localities in the lowlands and mid-hills. Terrestrial and probably also semi-fossorial. Ecology unstudied.

DISTRIBUTION Endemic to Borneo. Known localities include the Sungei Rejang catchment in central Sarawak, and Fernarium, Crocker Range Park in Sabah.

IUCN THREAT STATUS Data Deficient.

Head and forebody, dorsal view

Lygosoma bampfyldei Bartlett, 1895 (SP 06841) from the Fernarium, Crocker Range National Park, Keningau District, Sabah.

Kinabatangan Giant Skink *Lygosoma kinabatangense*

Grismer, Quah, Dzulkafly & Yambun, 2018
(No vernacular names recorded)

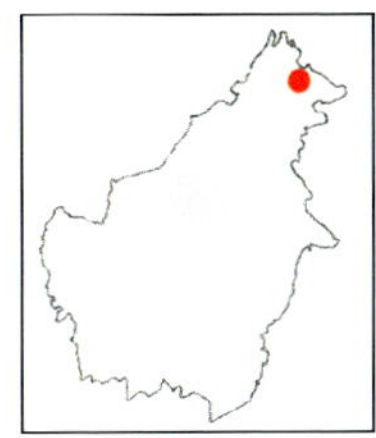

SIZE 119mm

IDENTIFICATION FEATURES Body slender; snout obtuse; lower eyelids scaly; three large auricular lobules; limbs short; supranasals contact behind rostral; prefrontals small; supraoculars five; supraciliaries eight; supralabials seven; infralabials 6–7; parietals form suture behind interparietal; nuchals four; preanals nine; midbody scale rows 42, smooth; digits short, somewhat compressed; lamellae under toe IV 15; deep groove from posterior corners of nostrils to below orbit; tail slightly narrower than body, tip acute.

COLOUR Dorsal surface of body mid-brown; forehead brown, with light band across parietal region and dark band connecting supralabial regions; neck and flanks paler, scales margined with pale brown; venter dark brown.

HABITAT AND BEHAVIOUR Known from lowland forests and diurnal in activity. Diet and reproductive habits unstudied.

DISTRIBUTION Deramakot, Kinabatangan District.

IUCN THREAT STATUS Not Evaluated.

Sphenomorphus Litter Skinks

Medium-sized to large (SVL to 82 mm) terrestrial or semi-arboreal scincids; body slender to relatively robust; prefrontals large, often forming a median suture; frontoparietals paired; precloacals large; supranasals absent; eyelids well developed; ear opening present; limbs large and pentadactyle.

Key to Bornean species of *Sphenomorphus*

1a.	Supraoculars four	**2**
1b.	Supraoculars ≥ five	**6**
2a.	Midbody scale rows > 30 rows	**3**
2b.	Midbody scale rows in < 30 rows	**4**
3a.	Midbody scale rows 32	*S. crassus* (p. 161)
3b.	Midbody scale rows 34	*S. shelfordi* (p. 169)
4a.	Nuchal present	*S. stellatus* (p. 170)
4b.	Nuchal absent	**5**
5a.	Midbody scale rows 26; toe IV lamellae 24	*S. tenuiculus* (p. 172)
5b.	Midbody scale rows 24; toe IV lamellae 21–23	*S. buettikoferi* (p. 160)
6a.	Eight supraoculars; midbody scale rows 36	*S. maculicollus* (p. 165)
6b.	Five to seven supraoculars; midbody scale rows 30–49	**7**
7a.	Midbody scale rows < 36	**8**
7b.	Midbody scale rows > 36	**9**
8a.	Supraoculars 7; parietals not in contact with supraoculars; midbody scale rows 32–35	*S. kinabaluensis* (p. 164)
8b.	Supraoculars 6; parietals in contact with supraoculars; midbody scale rows 30–32	*S. murudensis* (p. 166)
9a.	Dorsal surface with pale patches	*S. haasi* (p. 163)
9b.	Dorsal surface without pale patches	**10**
10a.	Dark dorsolateral band	**11**
10b.	No dark dorsolateral band	**12**
11a.	Five supraoculars	*S. tanahtinggi* (p. 171)
11b.	Six supraoculars	*S. cyanolaemus* (p. 162)
12a.	Dark labial bars absent, or when present narrower than light bars	*S. multisquamatus* (p. 166)
12b.	Labials strongly barred, dark bars wider than light ones	*S. sabanus* (p. 168)

ALFRED'S FOREST SKINK *Sphenomorphus alfredi* (Boulenger, 1898)

(No vernacular names recorded)

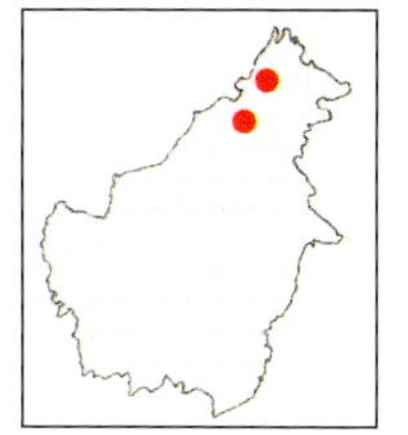

SIZE 33mm

IDENTIFICATION FEATURES Body slender, elongated; snout short; lower eyelid scaly; supraoculars four; supralabials IV–VI subocular; auricular opening large, subequal to orbit; enlarged nuchals absent; precloacal scales not enlarged; limbs short; tail thick basally, tapering to a point; prefrontals large, in contact; toe IV lamellae 7–12; midbody scale rows 28–30.

COLOUR Top of body reddish-brown; dark brown spots on nape; dark lateral stripe along sides of head crosses flanks; gular region has brown spots; belly unpatterned cream.

HABITAT AND BEHAVIOUR Inhabits primary forests in the low hills. Diet and reproductive habits unstudied.

DISTRIBUTION Endemic to Borneo. Known from Keningau in Sabah, and Sawah, Banjaran Penambo in Sarawak.

IUCN THREAT STATUS Data Deficient.

REMARKS Not included in key to *Sphenomorphus*; suspected to be a member of the genus *Tytthoscincus*.

After Boulenger (1898).

Büttikofer's Forest Skink *Sphenomorphus buettikoferi*

(van Lidth de Jeude, 1905)

(No vernacular names recorded)

SIZE 35mm

IDENTIFICATION FEATURES Body slender; snout short, obtusely shaped; lower eyelid scaly; frontonasal broader than long; supraoculars four; supraciliaries nine; large nuchals absent; ear opening rounded, lacking lobules; supralabials III–V in midorbital position; dorsal scales smooth; midbody scale rows 24; limbs well developed; lamellae under toe IV 22; tail thick.

COLOUR Dorsal surface reddish-brown, with four longitudinal rows of darker spots, two in the paravertebral region, one on each flank; venter unpatterned grey.

HABITAT AND BEHAVIOUR Appears restricted to hill dipterocarp forest. Biology unstudied.

DISTRIBUTION Endemic to Borneo. Restricted to Bukit Liang Kubung in Kalimantan.

IUCN THREAT STATUS Data Deficient.

Syntypes of *Sphenomorphus buettikoferi* (van Lidth de Jeude, 1905) (RMNH 4471), from 'Liang Koeboeng' (= Bukit Liang Kubung), on Sungei Kapuas, Kalimantan.

THICK FOREST SKINK *Sphenomorphus crassus* Inger, Tan, Lakim & Yambun, 2001

(No vernacular names recorded)

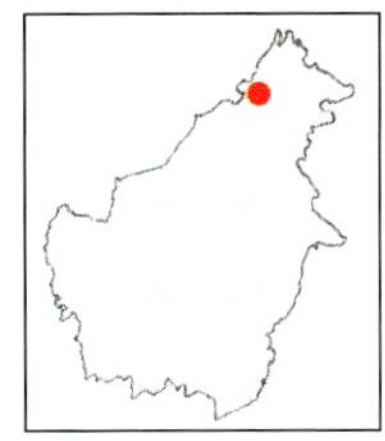

SIZE 82mm

IDENTIFICATION FEATURES Body robust; head deep; lower eyelid scaly; tympanum deeply sunk; ear lobules absent; limbs relatively short, pentadactyle; tail rounded in cross-section; supraoculars four; prefrontals widely separated; frontoparietals distinct; supraciliaries eight; supralabials seven; infralabials seven; lamellae under toe IV 18; midbody scale rows 32; postmental larger than mental; two pairs of large chin shields, the first pair in contact medially; median subcaudals enlarged; median precloacals distinctly enlarged.

COLOUR Dorsal surface and flanks medium brown, with faint dark lines formed by dark centres of most dorsal scales; trunk spotted with yellow, especially on flanks and tail; dark lateral band absent; venter unpatterned cream.

HABITAT AND BEHAVIOUR Inhabits forested mid-hills at around 670m. Terrestrial and associated with the leaf litter. Ecology unstudied.

DISTRIBUTION Endemic to Borneo. Known only from Mendolong, Sipitang District in Sabah.

IUCN THREAT STATUS Data Deficient.

Holotype of *Sphenomorphus crassus* Inger, Tan, Lakim & Yambun, 2002 (FMNH 243839), from 'Mendolong, Sipitang District, Sabah'.

Blue-throated Forest Skink *Sphenomorphus cyanolaemus*

Inger & Hosmer, 1965

(No vernacular names recorded)

SIZE 60mm

IDENTIFICATION FEATURES Body slender; limbs relatively long; tail long, tapering, slightly exceeding head and body length; snout obtuse; ear opening lacks lobules; prefrontals in broad contact; anterior loreals two, superimposed; nuchals absent; supraoculars six; supraciliaries 12–15; supralabials 6–8; infralabials six; mental as wide as rostral, followed by azygous postmental and three pairs of chin shields; midbody scale rows 37–42, smooth; lamellae under toe IV 16–19; two large precloacals.

COLOUR Head, throat and pectoral regions of adult males deep indigo blue, in females, light blue; lower part of abdomen pale blue in males, yellowish-orange in females; lips pale blue with dark brown bars; dorsal surface of torso bronze-brown to olive-brown, with longitudinal series of yellowish-brown blotches, extending from postocular region, along flanks, to mid-tail region; other less distinct rows of blotches on dorsum; dark grey-brown stripe from posterior of orbit, along flanks, to beyond insertion of hindlimbs, continuing along tail; sides of neck and flanks have brown-grey circles that enclose light blue areas in males; upper surfaces of forelimbs and hindlimbs dark brown or brownish-grey, with yellowish-brown blotches; under surfaces of forelimbs unpigmented yellow-cream, of hindlimbs unpigmented pale yellow; cloacal scales yellowish-pink; postcloacal region pale yellow (in females) or salmon-pink (in males); rest of tail venter pale blue or unpatterned cream.

HABITAT AND BEHAVIOUR Inhabits lowland dipterocarp forests and edges of peat swamps. Diurnal and terrestrial/semi-arboreal, climbing trees up to at least 5m. Clutches of two eggs, each 6 x 11mm, deposited in ant heaps within buttresses of trees; hatchlings 25–26mm SVL.

DISTRIBUTION Endemic to Borneo. The best places to see it include Subis Forest Reserve, Lambir Hills National Park, Gunung Mulu National Park, Niah National Park, Tinbarap Palm Oil Estate and Loagan Bunut National Park in Sarawak; Ulu Temburong and the middle or upper reaches of Sungei Ingei, Belait District in Brunei Darussalam, and Gunung Kinabalu Park, Malutut, Hutan Simpan, Mandamai, Tawau Hills Park, Sinsuron and Crocker Range Park in Sabah.

IUCN THREAT STATUS Least Concern.

Haas's Forest Skink *Sphenomorphus haasi* Inger & Hosmer, 1965

(No vernacular names recorded)

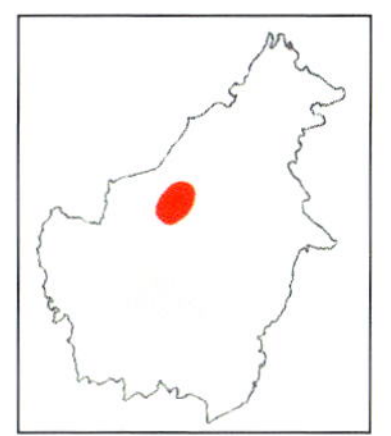

SIZE 57mm

IDENTIFICATION FEATURES Body slender; tail thick basally, tapering to a point; snout obtuse; ear opening lacking lobules; prefrontals narrowly separated or in contact; supraoculars six; anterior loreals two; nuchals absent; supraciliaries 13–14; supralabials seven; infralabials six; large pairs of chin shields 2–3; midbody scale rows 41–42, smooth; lamellae under toe IV 16–18.

COLOUR Dorsal surface greyish-brown with pale olive blotches, producing mottled appearance; throat and abdomen unpigmented cream; upper and lower lips, temporal region and sides of neck orange; black spots on upper and lower labials; dark dorsolateral band of flanks absent.

HABITAT AND BEHAVIOUR Inhabits lowland dipterocarp forests. Terrestrial, associated with the leaf litter. Ecology unstudied.

DISTRIBUTION Endemic to Borneo. Localities include Nanga Tekalit on Sungei Mengiong and Sungei Segaham, Kapit Division in Sarawak.

IUCN THREAT STATUS Data Deficient.

Holotype of *Sphenomorphus haasi* Inger & Hosmer, 1965 (FMNH 148565), from 'Nanga Tekalit, Mengiong River, Kapit District, Third Division, Sarawak'.

GUNUNG KINABALU FOREST SKINK *Sphenomorphus kinabaluensis*

(Bartlett, 1895)

(No vernacular names recorded)

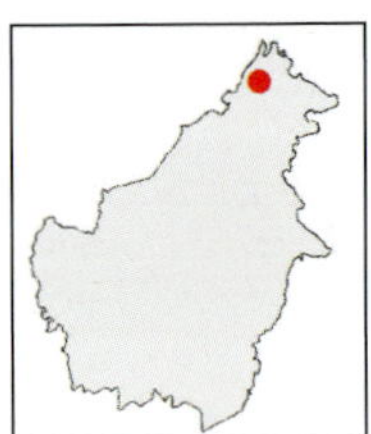

SIZE 58mm

IDENTIFICATION FEATURES Body slender; limbs long; prefrontals in wide contact with each other; supraoculars five; parietals not in contact with supraoculars; supralabials seven; infralabials seven; nuchals absent; midbody scale rows 32–38; lamellae under toe IV 15–17; tail thick.

COLOUR Dorsal surface light to dark brown, with several longitudinal rows of dark brown to yellow spots, and occasionally also dark brown speckles; black dorsolateral stripe with small yellowish flecks; throat sometimes speckled with dark brown; venter cream, yellow or grey; under surface of tail pink, bright red or grey.

HABITAT AND BEHAVIOUR Inhabits montane forests at 1,600–2,200m. Diurnal and insectivorous, inhabiting leaf litter. Clutches comprise 1–2 eggs, each 9.8–13.8 x 7.9–9.4mm, laid communally in nests of ponerine ants, in fallen trees of moss forests. Hatchlings 20.1–21.5mm SVL.

DISTRIBUTION Endemic to Borneo. Known from Gunung Kinabalu and a few other adjacent ranges in Sabah.

IUCN THREAT STATUS Least Concern.

WHITE-THROATED FOREST SKINK *Sphenomorphus maculicollus* Bacon, 1967

(No vernacular names recorded)

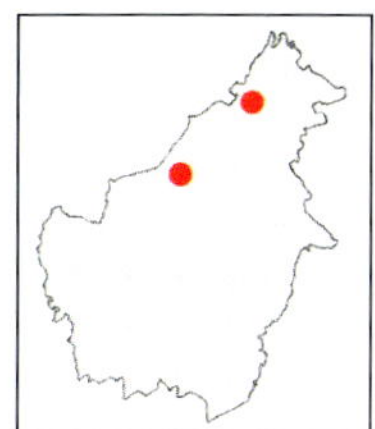

SIZE 49mm

IDENTIFICATION FEATURES Body slender; forehead scales smooth, iridescent; frontonasal twice as broad as long, not contacting frontal; prefrontals separated by an azygous scale; anterior loreals two; posterior loreals two; supraciliaries 15–16; supraoculars eight; supralabials five; infralabials six; a single preocular; nuchals absent; auricular lobules absent; tympanum slightly sunken; midbody scale rows 36; scales between mental and vent 90; two large precloacal scales.

COLOUR Head light brown, speckled with minute dark brown spots; scales with dark spot on each side of neck; dorsal surfaces of limbs light brown with irregular white spots; chin, throat, venter of body and under surface of tail-base unpatterned white; posterior two-thirds of tail has dark mottling.

HABITAT AND BEHAVIOUR Inhabits lowland and mid-hills of northern Borneo, up to submontane limits. Ecology unstudied.

DISTRIBUTION Endemic to Borneo. Known only from northern Sarawak (vicinity of Sungei Pesu, Miri Division) and southern Sabah (Mendolong, Sipitang District).

IUCN THREAT STATUS Data Deficient.

Holotype of *Sphenomorphus maculicollus* Bacon, 1967 (FMNH 161484), from 'near Sungei Pesu, Bintulu District, Sarawak'.

Many-scaled Forest Skink *Sphenomorphus multisquamatus* Inger, 1958

(No vernacular names recorded)

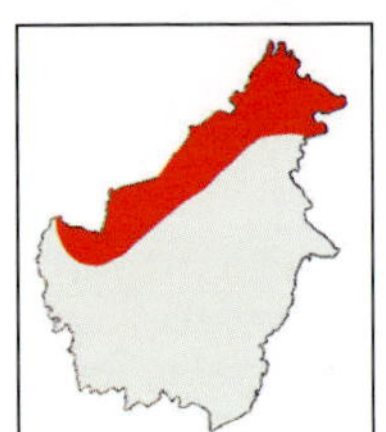

SIZE 69mm

IDENTIFICATION FEATURES Body robust; snout very short; body longer than tail; auricular lobules lacking; prefrontals in broad contact; nuchals absent; supraoculars 6–7; supraciliaries 14–15; supralabials 6–7; infralabials five; midbody scale rows 40–49; lamellae under toe IV 18–23.

COLOUR Forehead and dorsal surface of body dark greyish-brown, with 2–4 rows of squarish black spots, with or without dark dorsolateral bands; upper eyelids grey-brown; no black cervical spot; lips without black bars; males usually have blue throat and sides of neck; upper surfaces of limbs have dark brown variegation; digits without cross-bars; venter unpatterned cream; bright yellow ring around eye; under surfaces of limbs pinkish-yellow; under surface of tail pinkish-yellow; manus and pes grey.

HABITAT AND BEHAVIOUR Inhabits lowland rainforests and peat swamps; often found along banks of rocky hill streams, in tree buttresses as well as on leaf litter. Diet comprises small insects. Reproductive biology unstudied.

DISTRIBUTION Endemic to Borneo. The best places to see it are Gunung Pueh, Subis Forest Reserve, Niah National Park, Lambir Hills National Park and Gunung Mulu National Park in Sarawak, and Gunung Kinabalu Park, Danum Valley Field Centre, Mendolong in Sipitang District and Tawau Hills Park in Sabah.

IUCN THREAT STATUS Least Concern.

Gunung Murud Forest Skink *Sphenomorphus murudensis* Smith, 1925

(No vernacular names recorded)

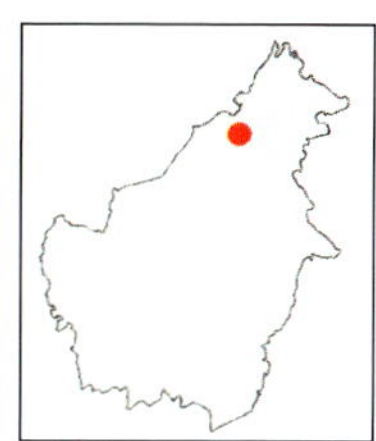

SIZE 49.3mm

IDENTIFICATION FEATURES Body slender; snout rounded; lower eyelid scaly; tympanum two-thirds of eye diameter; ear lobules absent; supranasals absent; nostrils in single nasal; prefrontals in contact; supraoculars six; parietals in contact with supraoculars; supraciliaries eight; loreals 2 + 2; supralabials six, the fourth in midorbital position; temporals large; nuchals absent; midbody scale rows 30–32, scales smooth; ventrals larger than dorsals; precloacals enlarged; toe IV lamellae 16.

COLOUR Dark brown dorsally, with black spots; dark band on sides, comprising dark spots on pale background; tail unpatterned brown above, cream below; hatchlings and juveniles have reddish-brown tails.

HABITAT AND BEHAVIOUR Inhabits montane forests at 1,500–2,450m, hiding under stones and fallen tree trunks. Clutches comprise two eggs, measuring 17.3–18.0 x 10.5–10.7 and 18.0 x 10.7mm.

DISTRIBUTION Endemic to Borneo. Restricted to Gunung Murud, Kelabit Highlands in Sarawak.

IUCN THREAT STATUS Data Deficient.

SABAH FOREST SKINK *Sphenomorphus sabanus* Inger, 1958

(No vernacular names recorded)

SIZE 58mm

IDENTIFICATION FEATURES Body relatively robust; auricular lobules lacking; prefrontals in broad contact or separated; parietals in contact with supraoculars; supraoculars typically six; supraciliaries 14–17; supralabials seven; infralabials 5–7; midbody scale rows 38–42; lamellae under toe IV 18–22.

COLOUR Forehead orange-brown, with dark grey areas in each scale; upper lips bright yellow, with large black spots; dark grey preocular stripe; forehead scales, especially supraoculars, iridescent blue; Dorsal surface of body olive-yellow, with pale grey broken paravertebral rows of elongated marks; dorsal surfaces of limbs, digits and tail olive-yellow with dark grey variegation; gular and pectoral regions unpatterned cream; abdominal region and under surface of original tails unpatterned pale yellow; of regenerated tails, unpatterned grey; manus and pes pale grey.

HABITAT AND BEHAVIOUR Inhabits lowlands to submontane forests. Associated with tree trunks and buttresses, and can ascend low trees. Diet comprises coleopterans, arachnids, formicids, blattids, orthopterans, lepidopterans, dipterans and isopods. Lays 2–3 eggs, each 10–12mm.

DISTRIBUTION Endemic to northern Borneo, including Sandakan Forest Reserve, Sepilok Forest Reserve, Tawau Hills Park, Poring Hot Spring, Lembah Danum, Dewhurst Bay, Mendolong, Maliau Basin in Sabah. The Bulangan record from Kalimantan needs confirmation.

IUCN THREAT STATUS Least Concern.

Shelford's Forest Skink *Sphenomorphus shelfordi* (Boulenger, 1900)

(No vernacular names recorded)

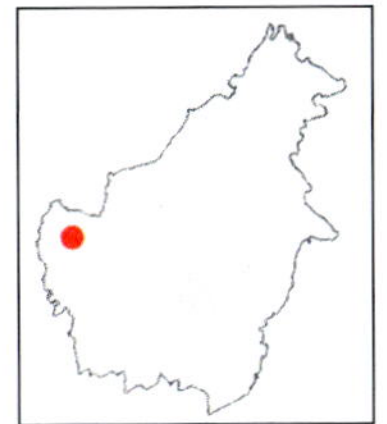

SIZE 67mm

IDENTIFICATION FEATURES Body slender; snout short and obtusely oriented; lower eyelids scaly; nostril in a single nasal; supranasals absent; rostral as long as broad, in contact with frontonasal; frontal narrow, as long as frontoparietals and interparietal; supraoculars four; supraciliaries seven; nuchals one pair; ear opening oval; auricular lobules absent; midbody scale rows 30–34; lamellae under toe IV 27–29; pair of large cloacals.

COLOUR Dorsal surface olive-brown, irregularly spotted with black; black lateral stripe from snout to inguinal region, which is broken up into spots on flanks of body; venter grey.

HABITAT AND BEHAVIOUR Inhabits submontane forests at altitude of more than 1,200m. Ecology unstudied.

DISTRIBUTION Endemic to Borneo. Known only from the upper reaches of Gunung Penrissen in Sarawak.

IUCN THREAT STATUS Data Deficient.

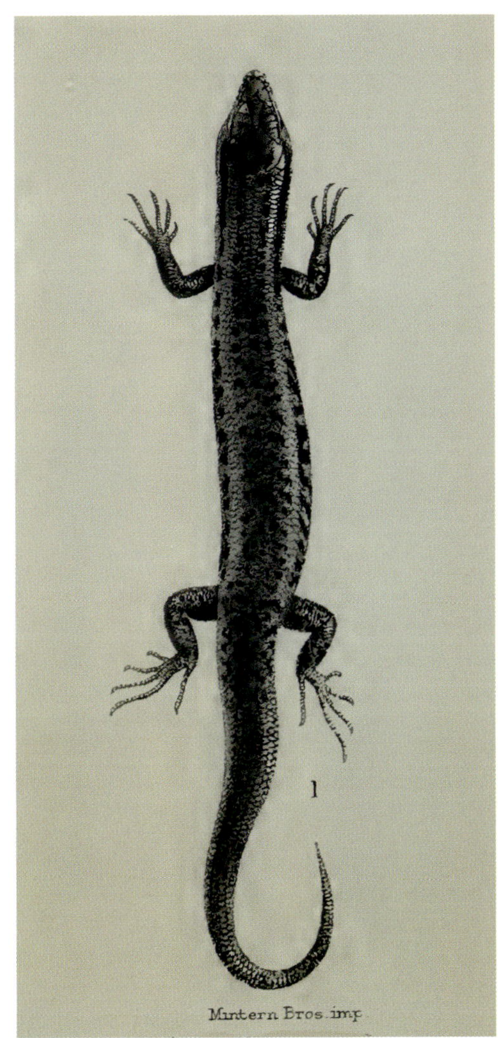

After Boulenger (1900).

Starred Forest Skink *Sphenomorphus stellatus* (Boulenger, 1900)

(No vernacular names recorded)

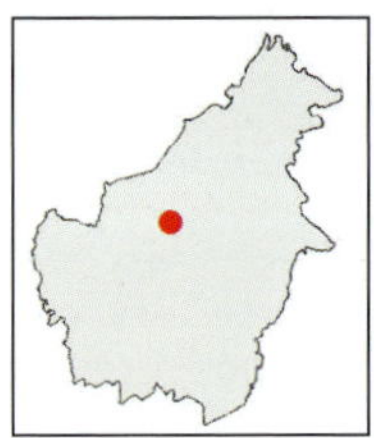

SIZE 80mm

IDENTIFICATION FEATURES Body slender; snout pointed; head distinct from neck; scales smooth; tail rounded; frontonasal width less than twice length; prefrontals large, juxtaposed; frontal longer than parietals and interparietal; supraoculars four; loreals two; presuboculars three; supraciliaries eight; supralabials 5–6; infralabials six; nuchals numbering six in double transverse rows; midbody scale rows 22–24; paravertebral scales large; upper posterior temporals large; lamellae under toe IV 18–23; two enlarged precloacal scales; median subcaudals large.

COLOUR Top of body greenish- or bronze-brown, with small, star-like white areas; dorsolateral dark stripe absent; labials dark; limbs bronze with white flecks; throat has 7–8 distinct longitudinal dark bands between scale rows; abdominal region has traces of dark pigmentation; tail has white spots laterally.

HABITAT AND BEHAVIOUR Natural history unstudied.

DISTRIBUTION Solitary record that needs verification from Nanga Tekalit Camp on Sungei Mengiong, Kapit Division of Sarawak. The species was described from Gunung Larut, at 1,341m in Perak State, Peninsular Malaysia, making the Sarawak record doubtful.

IUCN THREAT STATUS Least Concern.

Specimen from Peninsular Malaysia.

MONTANE FOREST SKINK *Sphenomorphus tanahtinggi* Inger, Tan, Lakim & Yambun, 2001

(No vernacular names recorded)

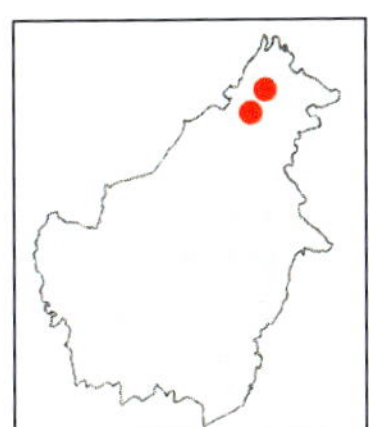

SIZE 63.5mm

IDENTIFICATION FEATURES Body robust; supraoculars five; three small superimposed anterior loreal scales; prefrontals narrowly in contact or separated; ear lobules absent; no large nuchals; midbody scale rows 40–42; adpressed limbs widely overlap.

COLOUR Top of body olive-brown, darker over tail, without markings or with widely scattered, small dark spots on centres of a few scales; sides of head and flanks dark brown; light line along dorsal margin of dark lateral band; 10 scale rows between dark bands; ventral margin of lateral band gradually fades into venter colour; small light spots in lower portion of band; belly pale olive-grey, unpigmented.

HABITAT AND BEHAVIOUR Known from leaf litter of selectively logged forests at altitudes of 850–1,180m. Diet and reproductive habits unstudied.

DISTRIBUTION Endemic to Borneo. Known only from Gunung Lumaku, Upper Sungai Pangas and Gunung Trus Madi in Sabah.

IUCN THREAT STATUS Data Deficient.

Holotype of *Sphenomorphus tanahtinggi* Inger, Tan, Lakim & Yambun, 2001 (FMNH 239855), from 'Mt Lumaku, Sipitang District, Sabah'.

Narrow-necked Forest Skink *Sphenomorphus tenuiculus*

(Mocquard, 1890)
(No vernacular names recorded)

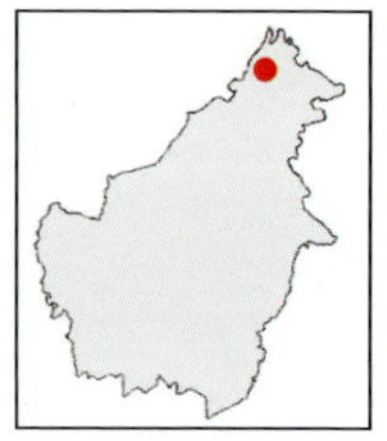

SIZE 46mm

IDENTIFICATION FEATURES Body slender; limbs relatively long; parietals in contact with one another; supraoculars four; supralabials seven; infralabials six; midbody scale rows 26; lamellae under toe IV 21–24; tail thick.

COLOUR Dorsal surface brown, spots on dorsum not fusing to form longitudinal lines; flanks dark, with white spots; venter unpatterned yellow.

HABITAT AND BEHAVIOUR Inhabits submontane forests at 1,225m asl. Ecology unstudied.

DISTRIBUTION Endemic to Borneo. Known only from Bundu Tuhan, in the mid-hills of Gunung Kinabalu in Sabah.

IUCN THREAT STATUS Data Deficient.

Holotype of *Sphenomorphus tenuiculus* (Mocquard, 1890) (MNHN 1889.186) from 'Kina Balu' (= Gunung Kinabalu, Sabah).

Subdoluseps Asian Supple Skinks

Small skinks (SVL 35–70mm); head and limbs short; body and tail elongated; prefrontals not in contact; parietals in contact behind interparietal; pyerygoids rounded posteriorly; palatines with posteriormedially projecting processes.

Key to Bornean species of *Subdoluseps*

1a. Robust species, SVL to 70 mm; enlarged nuchals absent; flanks brown, fading to white .... *S. samajaya* (p. 174)

1b. Slender species, SVL to 58 mm; enlarged nuchals typically present; flanks red and yellow, fading to white .... *S. bowringii* (below)

Bowring's Supple Skink *Subdoluseps bowringii* (Günther, 1864)

(No vernacular names recorded)

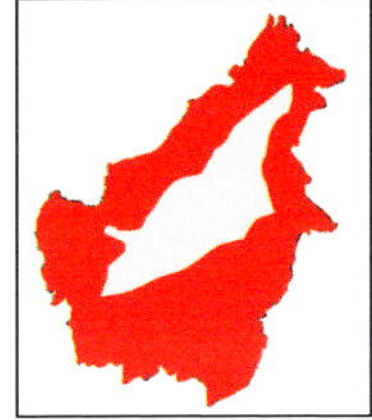

SIZE 58mm

IDENTIFICATION FEATURES Body slender, elongated; head scarcely distinct from neck; lower eyelid scaly; ear opening small, rounded, lacking lobules; limbs reduced, not meeting when adpressed; scales smooth or weakly keeled; supranasals large; prefrontals reduced, widely separated; paired nuchals; supraoculars four; supraciliaries seven; supralabials 6–7; infralabials 6–7; midbody scale rows 24–28; paravertebrals 52–62; lamellae under toe IV 10–13; tail slightly longer than SVL, rather thick, rounded, tapering to narrow point.

COLOUR Dorsal surface chocolate-brown or bronze-brown; golden-yellow stripe from postocular region to tail-base; black stripe from postnarial region to beginning of tail; infralabials and supralabials cream, scales edged with dark grey; paravertebral stripe golden-yellow, two scales wide; lateral stripe black, two scales wide, with cream spots; scales on flanks cream with black edges; throat cream with yellow patches; pectoral and abdominal regions greyish-cream, turning unpatterned yellow in males during breeding season; upper surfaces of limbs grey-brown with pale spots; lower surfaces of limbs unpatterned grey; tail of juveniles bright red, becoming grey or brown with growth.

HABITAT AND BEHAVIOUR Inhabits open areas, and most commonly seen in forest clearings and within human habitation, for example in parks and gardens. Semi-fossorial and terrestrial; diet comprises small insects. Clutches include 2–4 eggs, each 7.3 x 12.4mm; hatchlings 22–23mm SVL.

DISTRIBUTION Andaman Islands of India, Myanmar, Thailand, the Malay Peninsula, Borneo, Java, Sulawesi, possibly Sumatra, Buton Island, Vietnam, Cambodia, Laos and the Sulu Archipelago of the southern Philippines. Introduced through human agencies to Christmas Island, Indian Ocean. Widespread on Borneo, and recorded in Kuching, Bintulu and Kota Samarahan in Sarawak; Bandar Seri Begawan in Brunei Darussalam; Labuan, Kota Kinabalu, Bukit Padang and Pulau Gaya in Sabah, and west Kalimantan.

IUCN THREAT STATUS Least Concern.

SAMA JAYA SKINK *Subduloseps samajaya* (Karin, Freitas, Shonleben, Grismer, Bauer & Das, 2018)

(No vernacular names recorded)

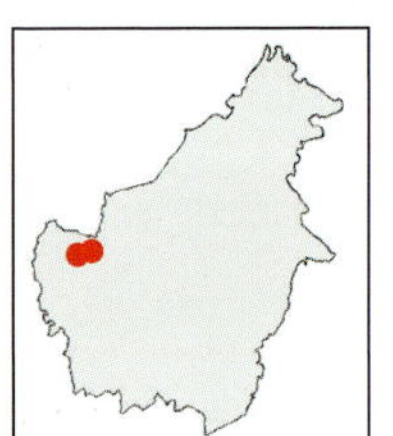

SIZE 70mm

IDENTIFICATION FEATURES Body robust, slightly compressed dorsoventrally; head slightly distinct from neck; frontoparietals paired; no enlarged nuchals; supranasals in contact; glossy, quinquecarinate dorsal and lateral scales; subdigital lamellae on fourth toe 13–14; subdigital lamellae on fourth finger; paravertebral scale rows 61; midbody scale rows 28–30; tail shorter than body.

COLOUR Dorsal surface of head, body, limbs and tail uniform mid-brown; ventral surfaces uniform cream or pale yellow; dark brown lateral stripe along flanks, from nostrils to midbody, bordered above by pale, narrow line above eye to beginning of forelimb; underneath of tail darkens posteriorly.

HABITAT AND BEHAVIOUR Inhabits a lowland forest that includes dipterocarp and Kerangas patches. Terrestrial and associated with the leaf litter. Diet and reproductive habits unstudied.

DISTRIBUTION Endemic to Borneo. Known from Sama Jaya Nature Reserve, Libiki Bamboo Resort in Bau and base of Gunung Gumbang in western Sarawak.

IUCN THREAT STATUS Not Evaluated.

Head and forebody

Tropidophorus Water Skinks

Medium-sized to large (SVL to 101 mm) semi-aquatic scincids; body relatively slender in juveniles, more robust in adults, especially males; palatines and pterygoids in contact; no palatine teeth; limbs and eyelids well developed and movable; lower eyelid scaly; supranasals absent; prefrontals well developed; frontoparietal distinct from interparietal; tympanum large, superficial; and two or three large precloacals.

Key to Bornean species of *Tropidophorus*

1a. Dorsal scales smooth..**2**
1b. Dorsal scales keeled..**3**

2a. Midbody scale rows 28–30; dark dorsolateral stripe along flanks..*T. beccarii* (p. 176)
2b. Midbody scale rows 33–34; no dark dorsolateral stripe along flanks..**6**

3a. Four supraoculars..*T. micropus* (p. 179)
3b. Five supraoculars..**4**

4a. Dorsal scales bicarinate, grooved..*T. iniquus* (p. 178)
4b. Dorsal scales unicarinate; not grooved..**5**

5a. Supraciliaries 14; midbody scale rows 32. ....*T. brookei* (p. 177)
5b. Supraciliaries 5–6; midbody scale rows 30..*T. perplexus* (p. 181)

6a. Four supraoculars..*T. sebi* (p. 182)
6b. Five supraoculars..*T. mocquardii* (p. 180)

Beccari's Water Skink *Tropidophorus beccarii* Peters, 1871

(No vernacular names recorded)

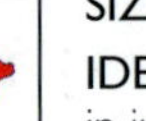

SIZE 98mm

IDENTIFICATION FEATURES Body relatively robust in adults, slender in juveniles; body scales smooth; forehead scales smooth; prefrontals in broad contact or separated; tympanum smaller than orbit of eye; supraoculars 4–5; supraciliaries 6–7; supralabials 7–8; infralabials 4–5; midbody scale rows 28–30; lamellae under toe IV 19–21; single large precloacal plate.

COLOUR Dorsal surface rich reddish-brown, with scattered dark grey blotches and cross-bars, up to about anterior half of tail; sides of head and flanks have light spots; lips and throat cream, with dark brown bars; thin dark grey stripe from snout-tip through nostril to eye; broad postocular stripe of same colour extending to inguinal region, where it is broadest; along flanks, dark lateral band with vertical light bars or wedge-shaped marks; dark infralabial stripes extend to around middle of throat; dark stripe runs diagonally across from below tympanum to axilla; limbs pale brown with dark brown variegation; venter cream or yellowish-cream, with incomplete dark pattern of longitudinal lines; ventral surface of tail grey with darker variegation.

HABITAT AND BEHAVIOUR Inhabits edges of rocky streams in lowlands and mid-hills to 1,000m, within dipterocarp forests. Diet unstudied. Ovoviviparous, producing four young measuring 30mm SVL.

DISTRIBUTION Endemic to Borneo. The best localities for it include Gunung Santubong, Gunung Matang, Gunung Penrissen, Pelagus National Park and Loagan Bunut National Park in Sarawak; the middle or upper reaches of Sungei Ingei, Belait District and Ulu Temburong in Brunei Darussalam; Tawau Hills Park, Sepilok Forest Reserve, Sepagaya Forest Reserve and Maliau Basin in Sabah, and Maruwai in Kalimantan.

IUCN THREAT STATUS Least Concern.

Brooke's Water Skink *Tropidophorus brookei* (Gray, 1845)

(No vernacular names recorded)

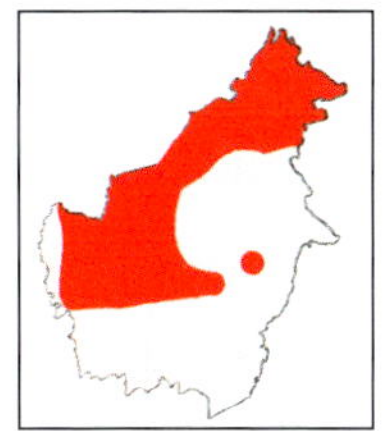

SIZE 101mm

IDENTIFICATION FEATURES Body robust in adults, slender in juveniles; head long; tympanum smaller than orbit of eye; supraciliaries 14; supraoculars five; supralabials eight; infralabials five; single postmental; midbody scale rows 32, keeled, forming eight longitudinal ridges on dorsum and oblique ones on flanks; large precloacal plate.

COLOUR Dorsal surface dark greyish-brown with dark grey transverse band-like blotches on body and tail; lips orange-yellow; tympanum with dark-edged, light cream patch, black spot on sides of neck; sides of neck and flanks peppered, comprising scales with dark grey and cream scales; large, dark grey patch in axilla; venter of body, tail, tail-base and under surface of limbs unpatterned cream; rest of under surface of tail grey with darker variegation, darkening posteriorly; supraciliaries bright yellow.

HABITAT AND BEHAVIOUR Inhabits lowland dipterocarp forests, up to mid-hills. Diurnal and semi-aquatic, at edges of small rocky hill streams. Diet comprises small arthropods. Produces 1–5 live young.

DISTRIBUTION Endemic to Borneo. Widespread on the island; localities where it can be sighted include Gunung Penrissen, Gunung Matang, Bako National Park, Gunung Pueh, Lanjak-Entimau Wildlife Sanctuary, Gunung Santubong, Ranchan Pool Forest, Serian, Gunung Mulu National Park, Loagan Bunut National Park and Gunung Dulit in Sarawak; Ulu Temburong, Bukit Patoi, Sungei Ingei, Belait District and Tasek Merimbun in Brunei Darussalam; Deramakot, Tawau Hills Park, Gunung Rara, Sandakan, Sepilok Forest Reserve, Danum Valley Field Centre and Mendolong in Sabah, and Taman Nasional Bentuang Karimun and Bulangan in Kalimantan.

IUCN THREAT STATUS Least Concern.

Sungei Kajan Skink *Tropidophorus iniquus* van Lidth de Jeude, 1905

(No vernacular names recorded)

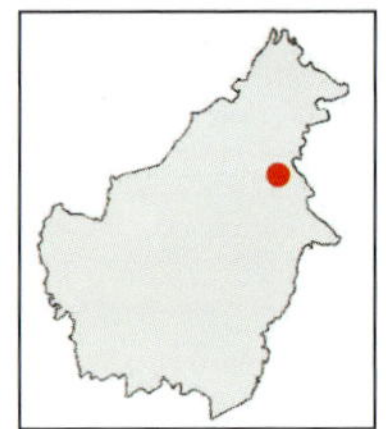

SIZE 96mm

IDENTIFICATION FEATURES Body slender; snout acute; forehead shields rugose; tympanum smaller than orbit of eye; supraciliaries six; supraoculars five, the first longest; fifth reduced; fifth supralabial largest, contacting orbit of eye; supralabials eight; infralabials five; large azygous postmental, followed by two pairs of large postmentals that are in contact; three swollen bands on nuchals, separated by grooves; dorsals bicarinate, grooved; midbody scale rows 34; single large precloacal scale; subdigital lamellae smooth.

COLOUR Dorsal surface of the only known specimen described as 'dark grey' (probably an artefact of preservation); venter unpatterned white; under surfaces of tail and feet dark.

HABITAT AND BEHAVIOUR Inhabits hill dipterocarp forests. Ecology unstudied, and suspected to be associated with hill streams and, like its congeners, to be an insectivore and ovoviviparous.

DISTRIBUTION Endemic to Borneo, and only recorded from the upper reaches of Sungei Kajan in Kalimantan.

IUCN THREAT STATUS Data Deficient.

Holotype of *Tropidophorus iniquus* van Lidth de Jeude, 1905 (RMNH 4451), from '...upper part of the Kajan river..' (Kalimantan Barat).

Small-scaled Water Skink *Tropidophorus micropus* van Lidth de Jeude, 1905

(No vernacular names recorded)

SIZE 40mm

IDENTIFICATION FEATURES Body relatively slender; snout long, acute; scales on forehead distinctly striated; frontonasal as long as broad; supraoculars four; supraciliaries seven; scales on flanks relatively small; ventrals smooth, larger than laterals; midbody scale rows 26–34; subdigital lamellae smooth.

COLOUR Dorsal surface dark brown, with eight paired pale brown blotches (including a pair on neck, a pair at level of axilla and five on torso, between forelimbs and hindlimbs), each edged with black; on inguinal region, broad, paired band of the same colour; forehead dark brown; supralabials and infralabials black with cream spots; flanks greyish-brown, with large, pale-edged dark oblique bars; dorsal surfaces of forelimbs and hindlimbs dark brown; dorsal surface of tail dark brown, bearing darker bands; gular region up to level of forearms black with bluish-cream spots; pectoral and abdominal regions salmon-pink; pectoral region has darker variegation; abdominal region unpatterned; under surfaces of limbs pale pink with darker variegation; manus and pes grey; ventral surface of tail dark grey with large, pale areas.

HABITAT AND BEHAVIOUR Inhabits lowland forests and frequents edges of rocky streams, inside rock crevices, as well as in adjacent leaf litter. Diurnal and insectivorous. Two live young produced at a time.

DISTRIBUTION Endemic to Borneo. Only known localities include Upper Baleh in Sarawak and Long Blu, on upper reaches of Sungei Mahakam in Kalimantan.

IUCN THREAT STATUS Data Deficient.

Mocquard's Water Skink *Tropidophorus mocquardii* Boulenger, 1894

(No vernacular names recorded)

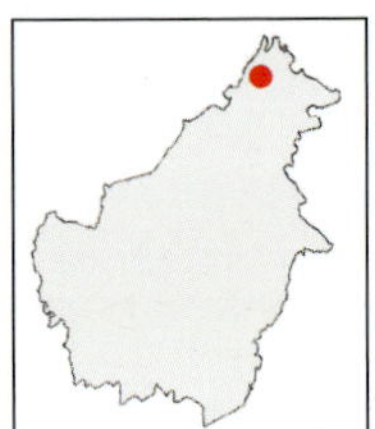

SIZE 95mm

IDENTIFICATION FEATURES Body slender; head large, wide at temples; forehead scales smooth; tympanum smaller than orbit of eye; supraoculars five; supralabials seven; infralabials five; midbody scale rows 34, smooth; single large precloacal scale; digits short with smooth lamellae.

COLOUR Dorsal surface brown, with dark transverse bands; flanks have white spots; venter cream.

HABITAT AND BEHAVIOUR Inhabits rocky hill streams in hill dipterocarp forests, at elevations above 1,000m. Diet unstudied. Ovoviviparous.

DISTRIBUTION Endemic to Borneo. Known only from the mid-hills of Gunung Kinabalu and Crocker Range Park in Sabah.

IUCN THREAT STATUS Least Concern.

PERPLEXING WATER SKINK *Tropidophorus perplexus* Barbour, 1921

(No vernacular names recorded)

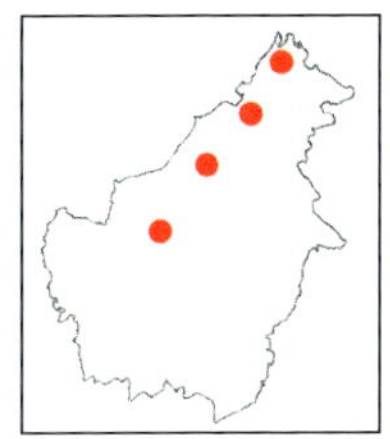

SIZE 73mm

IDENTIFICATION FEATURES Body slender; dorsal and lateral scales large and strongly keeled; scales on forehead rugose; frontonasal divided, as broad as long; supraoculars five; supraciliaries 5–6; supralabials six; tympanum as large as eye; midbody scale rows 30; gulars keeled; ventrals large, cycloid, imbricate and smooth; lamellae under digits feebly keeled; tail long, slightly compressed, with keeled scales, upper rows spinose.

COLOUR Dorsal surface of body rich brown, with paler narrow cross-bars; venter yellow.

HABITAT AND BEHAVIOUR Inhabits mid-hills of northern and central-northern Borneo, in hill dipterocarp forests, being associated with forest litter. Diet and reproductive habits unstudied.

DISTRIBUTION Endemic to Borneo. Records are from Long Lobang, Sungei Tinjar in Sarawak, and Mendolong, Sipitang District and Marak Parak, Kota Marudu District in Sabah.

IUCN THREAT STATUS Least Concern.

Holotype of *Tropidophorus perplexus* Barbour, 1921 (MCZ 14632), from '...near the Fort at Long Loba, Tinjar River, Sarawak' (= vicinity of Fort Hose, Marudi, Miri Division).

Baleh Water Skink *Tropidophorus sebi* Pui, Karin, Bauer & Das, 2017

(No vernacular names recorded)

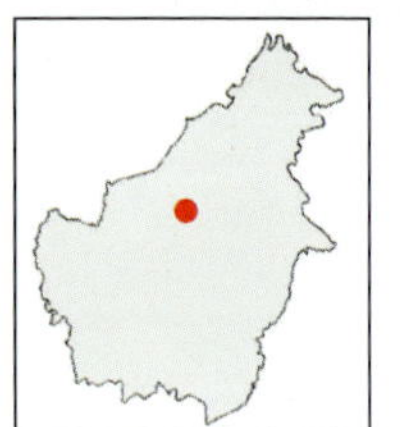

SIZE 86mm

IDENTIFICATION FEATURES Body robust; upper head shields, dorsal and lateral scales smooth; parietal scales in two pairs; supraciliaries eight, first largest; supraoculars four, fourth contacting orbit; supralabials seven; infralabials four, second longest; postmental undivided; longitudinal scale rows 58; ventrals 53; transverse scale rows at midbody 34; subcaudals 98; precloacals enlarged, single; subdigital lamellae on toe IV 19.

COLOUR Dorsal surface chocolate-brown, with dark greyish-brown transverse bars on trunk and tail; bars on nape form collar-like pattern; blackish-grey postocular band from posterior corner of orbit to axillary region, narrowing on flanks; forehead unpatterned chocolate-brown; temporal region grey with cream-coloured flecks; axilla and lower flanks have scattered yellow flecks; labials brown with greyish-black bars; dorsal surface of tail greyish-brown with darker areas forming incomplete bands; dorsal surfaces of limbs chocolate-brown, with darker variegation; digits brown with darker bands; gular region pale grey, with dark grey variegation, forming about 10 lines that extend to the pectoral region; upper two-thirds of pectoral region pale yellow with indistinct grey stripes; lower third orange, each scale edged with yellow; cloacal region deep orange; lowest edge of cloacal scale dark grey; tail-base grey with large, orange-coloured scales; rest of tail dark grey; lower parts of forearm unpatterned pale grey; hindlimbs unpatterned pale orange.

HABITAT AND BEHAVIOUR Inhabits hill dipterocarp forests. Associated with rocky hill streams. Diet comprises small arthropods. Ovoviviparous, producing two young, measuring 64mm TL.

DISTRIBUTION Endemic to Borneo. Restricted to headwaters of Sungei Baleh.

IUCN THREAT STATUS Data Deficient.

Tytthoscincus Litter Skinks

Diminutive (SVL < 45 mm) skinks with smooth scales; slender body; temporal scales undifferentiated from lateral body scales in size or shape; and reduced digits. On account of its morphological similarities to *Sphenomorphus* (and possible other genera), it is likely that several members of this newly described genus are currently allocated to other genera.

Key to Bornean species of *Tytthoscincus*

1a. Parietals not in contact with supraocular .... *T. hallieri* (p. 185)
1b. Parietals in contact with supraocular 2

2a. Precloacal scales not enlarged .... *T. aesculeticola* (below)
2b. Precloacal scales enlarged 3

3a. Prefrontals in contact .... *T. batupanggah* (p. 184)
3b. Prefrontals not in contact .... *T. leproauricularis* (p. 186)

Oak Forest Litter Skink *Tytthoscincus aesculeticola* (Inger, Tan, Lakim & Yambun, 2001)

(No vernacular names recorded)

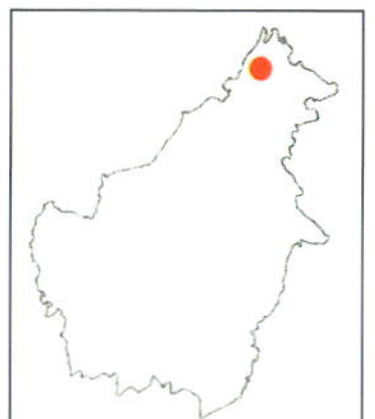

SIZE 42mm

IDENTIFICATION FEATURES Body slender; head not wider than neck or trunk; limbs relatively short; prefrontals separated; supraoculars four; frontonasal absent; supraciliaries 8–10; midbody scale rows 28–30; lamellae under toe IV 6–10.

COLOUR Forehead and dorsum mid-brown to dark brown or reddish-brown in life, many scales dark spotted, forming series of dark lines or checkered pattern; sides of snout and lips dark brown; labial scales have light centres; dark lateral band from posterior corner of eyes, across tympanum, across trunk, as wide as 2–3 scales, interrupted with small white spots; dorsal surfaces of limbs spotted with black and brown; and venter unpatterned cream, except under surface of tail, where scales have small, dark centres.

HABITAT AND BEHAVIOUR Inhabits submontane forests and associated with the leaf litter and rocks within oak forests at 1,350–1,650m. Diet unstudied. Clutches comprise two eggs; hatchling 15mm SVL.

DISTRIBUTION Endemic to Borneo. Known only from Sabah, including Gunung Lumaku, Mesilau, Gunung Kinabalu Park, Crocker Range Park and Gunung Trus Madi.

IUCN THREAT STATUS Least Concern.

Holotype of *Sphenomorphus aesculeticolus* Inger, Tan, Lakim & Yambun, 2001 (FMNH 239852), from 'Mount Lumaku, Sipitang District, Sabah'.

BATU PANGGAH LITTER SKINK *Tytthoscincus batupanggah* Karin, Das & Bauer, 2016

(No vernacular names recorded)

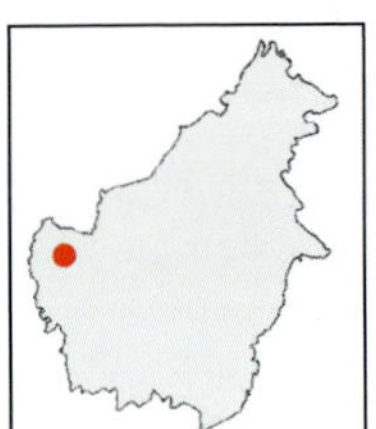

SIZE 33.2mm

IDENTIFICATION FEATURES Body slender; midbody scale rows 30–31; paravertebral scale rows 62–66; vertebral scale rows 55–61; four enlarged supraoculars; parietals in contact with last supraocular; prefrontals in contact with each other; two loreals; six supralabials; five infralabials; lamellae on toe IV 9; smooth texture of subdigital lamellae; enlarged precloacal scales.

COLOUR Dorsum and flanks dark brown with cream-yellow spots along body; cream-coloured dorsolateral stripe from orbit to midbody; another lateral stripe of the same colour from orbit to auricular opening; belly cream coloured.

HABITAT AND BEHAVIOUR Inhabits submontane forests at 1,050m. Subfossorial, and associated with the leaf litter, and possibly nocturnal or crepuscular. Diet and reproductive habits unstudied.

DISTRIBUTION Batu Panggah on Gunung Penrissen, Sarawak. Endemic to Borneo.

IUCN THREAT STATUS Data Deficient.

HALLIER'S LITTER SKINK *Tytthoscincus hallieri* (van Lidth de Jeude, 1905)

(No vernacular names recorded)

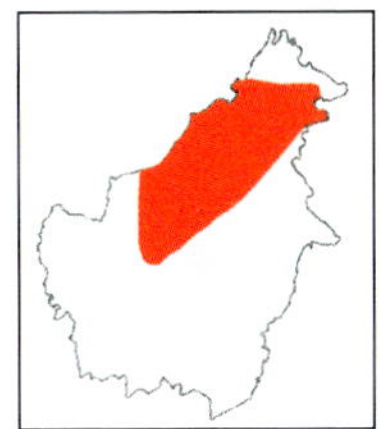

SIZE 48mm

IDENTIFICATION FEATURES Body slender; large upper temporal wedged between last supraocular and parietal; parietals reduced; no large nuchals or precloacals; loreals two; tympanum slightly sunk, lacking auricular lobules; supraoculars 5–6; supraciliaries 9–10; infralabials five; preoculars 3–4; midbody scale rows 36–39; lamellae under toe IV 10–15.

COLOUR Dorsal surface brown, often with red or rust-coloured dorsolateral streak from eye to behind shoulder that is broken up into spots or bars on trunk; flanks light brown or olive with numerous small yellow or greenish spots; throat of males light blue with dense black spots, sides of head nearly black; females have pink or yellow throats.

HABITAT AND BEHAVIOUR Inhabits forested lowlands and mid-hills. Associated with leaf litter. Ecology unstudied.

DISTRIBUTION Endemic to Borneo. Isolated records from Baleh and interior of Kapit Division in Sarawak; the middle or upper reaches of Sungei Ingei, Belait District and Batang Duri, Temburong District in Brunei Darussalam; Purulon, Tenom District, Crocker Range Park, Danum Valley Field Centre, Mendolong and Tawau Hills Park in Sabah, and Putussibau in Kalimantan.

IUCN THREAT STATUS Least Concern.

Scaly-eared Litter Skink *Tytthoscincus leproauricularis*

Karin, Das & Bauer, 2016

(No vernacular names recorded)

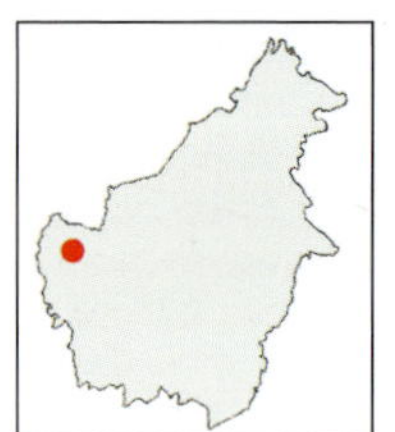

SIZE 34mm

IDENTIFICATION FEATURES Body slender; midbody scale rows 35–37; paravertebral scale rows 68–74; vertebral scale rows 80–81; four enlarged supraoculars; parietals in contact with last supraocular; prefrontals separated; two loreals; six or seven supralabials; five infralabials; lamellae on toe IV 10; smooth texture of subdigital lamellae; enlarged precloacal scales

COLOUR Dorsum dark; lacking dark dorsolateral stripe; no light postorbital stripe or pale dorsolateral stripe.

HABITAT AND BEHAVIOUR Inhabits submontane forests. Subfossorial and possibly nocturnal or crepiuscular. Diet and reproductive habits unstudied.

DISTRIBUTION Endemic to Borneo. Gunung Penrissen, Sarawak.

IUCN THREAT STATUS Data Deficient.

Family Varanidae Monitor Lizards

Large (SVL to 3m) terrestrial, aquatic or arboreal lizards; body ranging from relatively gracile to robust, depending on species, ontogenetic stage and health; tail compressed; head elongated; tongue long, slender, smooth and bifid at tip; pupil rounded; cephalic scales small, juxtaposed; dorsal scales rounded or oval; ventral scales quadrangular, in transverse rows; precloacal pores present and eggs soft shelled.

The derivation of the generic name is from the latinization of 'Waran', the Arabic/Egyptian name for *Varanus niloticus* (Linnaeus, 1766). The commonly used English name for these lizards is 'monitor', derived from the Latin 'monere', meaning 'one who admonishes or gives advice or warning to another as to his conduct'. Although the name 'monitor', in contemporary folklore, is thought to derive from the lizards' behaviour of warning humans of the presence of crocodiles, the name 'Varanus' (used by Merrem in 1820), may have originated from the Arabic name for varanids, 'waran'.

Key to Bornean species of *Varanus*

1a. Nostrils oblique slits; body without transverse yellow bars speckled with black.......... 2
1b. Nostrils rounded or oval; body without transverse yellow bars speckled with black.......... *V. salvator* (p. 190)

2a. Nuchal scales smooth or posteriorly feebly keeled; snout moderately long and broad; neck with 'U'-shaped black line enclosing black spot; tail reddish-brown, with indistinct bands........ *V. dumerilii* (p. 188)
2b. Nuchal scales strongly keeled; snout very long and narrow; neck with three black stripes; tail with black and yellow bands...... *V. rudicollis* (p. 189)

Duméril's Monitor Lizard (p. 188)

DUMÉRIL'S MONITOR LIZARD *Varanus dumerilii* (Schlegel, 1839)

(No vernacular names recorded)

SIZE 1.5m

IDENTIFICATION FEATURES Body relatively robust; head rather small, flattened; snout short and broad; nostrils elongated, closer to eyes than to snout; tympanum large, rounded; nuchal scales oval, flat, smooth or posteriorly feebly keeled; abdominal scales weakly keeled, in 37–41 rows; tail laterally compressed.

COLOUR Forehead orange or yellowish-orange, especially in juveniles, showing less contrast with growth; dorsal surface brownish-yellow or tan, with dark temporal streak from eye to ear, which may be confluent with 'U'-shaped mark on neck and enclosing a black spot; dark vertical bars on lips; throat yellow with 6–8 orange stripes; venter yellow with dark transverse bars; tail reddish-brown with indistinct dark bands; neonates brightly coloured; bright orange patch on forehead that fades with growth.

HABITAT AND BEHAVIOUR Inhabits lowland forests and mid-hills, including mangrove swamps. Diet comprises crabs, ants, scorpions, beetle larvae, spiders, fish, other lizards, eggs and rodents. Clutches comprise 12–16, more rarely up to 23; hatchlings 81–83.5mm SVL.

DISTRIBUTION Southern Myanmar, Thailand, the Malay Peninsula, Sumatra, Borneo, Pulau Bangka and Pulau Belitung. Some accessible localities for the species on Borneo include Buntal, Gunung Penrissen, Gunung Dulit and Niah National Park in Sarawak; Jalan Labi, Belait District in Brunei Darussalam; Mendolong, Deramakot, the upper reaches of Sungei Tibas, Kalabakan and Dewhurst Bay in Sabah, and Tanjung, Nanga Raun, Anjungan, Sungei Pengkaran, Sinkawang, Bongan and Banjarmasin in Kalimantan.

IUCN THREAT STATUS Data Deficient.

Head and forebody of juvenile. In adults, forehead is olive.

Rough-necked Monitor Lizard *Varanus rudicollis* Gray, 1845

(No vernacular names recorded)

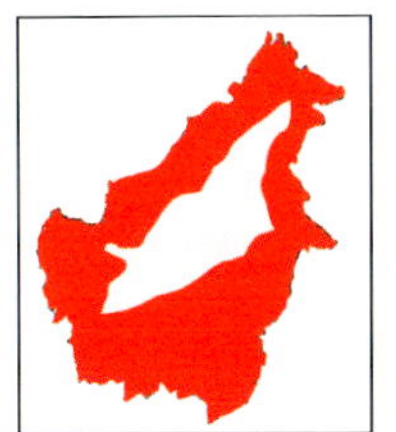

SIZE 1.46m

IDENTIFICATION FEATURES Body relatively slender; snout relatively long; nostrils elongated, closer to eyes than to nostrils; nuchal scales narrow and strongly keeled, with large, pointed scales, especially in adults; abdominal scales keeled, in 79–90 transverse rows; body, neck and limbs relatively slender.

COLOUR Dorsal surface dark, almost black in adults, with yellow tinge on neck and foreparts of body in juveniles; neck has three black stripes; flanks have yellow ocelli; hatchlings have yellow and black horizontal bands on venter of body; limbs yellow spotted, forming bands, and tail has dark and yellow bands that become obscure with growth.

HABITAT AND BEHAVIOUR Inhabits mid-hills of primary and secondary forests, including mangroves. Diet comprises ants and termites as well as large invertebrates, including phasmids, cockroaches, grasshoppers, spiders, scorpions, small mammals, frogs, fish and crabs. Clutches comprise 13–14 eggs; hatchlings 23.8–26cm.

DISTRIBUTION Southern Myanmar, Thailand, the Malay Peninsula, Sumatra, Pulau Bangka, Borneo and the Philippines. Localities where it can be seen on Borneo include Bako National Park, Gunung Penrissen, Ulu Baleh National Park and Niah National Park in Sarawak; Ranau, Tawau Hills Park, Sandakan and Sepagaya Forest Reserve in Sabah, and Bongan, Anjungan and Samarinda in Kalimantan.

IUCN THREAT STATUS Data Deficient.

Ex-situ hatchling

Wild hatchling

Half-grown individual

WATER MONITOR LIZARD *Varanus salvator* (Laurenti, 1768)

(Bahasa Malaysia Biawak)

SIZE 3.2m

IDENTIFICATION FEATURES Body relatively robust in adults, slender in juveniles; snout depressed; nostril rounded or oval, twice as far from orbit as from snout-tip; nuchal scale strongly keeled; crown scales large, flat, smooth, larger than nuchal scales; supraoculars well differentiated; midventral scales feebly keeled, numbering 148–153; tail strongly compressed, with a double-toothed crest above; caudal scales keeled dorsally and ventrally.

COLOUR Juveniles brighter than adults, with black forehead, some scales olive margined; canthal region olive; eyelids yellow-cream; black postocular stripe extends to level of tympanum; dorsal surface of body black-olive, with six rows of transversely arranged spots that may form ocelli; flanks have a black-olive, net-like design, forming ocelli-like marking in parts towards axilla; upper surface of forelimbs and hindlimbs blackish-grey, with uniformly scattered yellow scales; dorsal surface of tail blackish-olive with lemon-yellow bands, more distinct towards tip and indistinct at tail-base; upper surfaces of digits blackish-olive, with large yellow scales arranged to form bands; chin region yellow, with greyish-olive transverse bars; gular region lemon-yellow with incomplete black bars; pectoral and abdominal region lemon-yellow, with indistinct bars; cloacal region and under surface of tail lemon-yellow with blackish-olive brown; under surface of forelimbs and hindlimbs yellow, lacking bars.

HABITAT AND BEHAVIOUR Commonly seen in towns and cities across Borneo, beside forests from the lowlands to the mid-hills, and in mangrove swamps and other coastal areas. Typically associated with water, inhabiting edges of marshes, estuaries, rivers and canals, within mangroves and dipterocarp forests, but reported from islands lacking fresh water. Diet includes

Adult

arthropods such as insects and spiders, fish, snakes, crabs, freshwater turtles, eggs and adult water birds, crocodiles, small mammals and sea turtles, as well as other varanids and rodents. Clutches comprise 5–30 eggs, each 32.3–42.9 x 64.0–82.6mm; hatchlings 18–30mm.

DISTRIBUTION Sri Lanka and eastern and northeastern India, the Andaman and Nicobar Islands, Myanmar, Thailand, Vietnam, southern and eastern China and Hong Kong, the Malay Peninsula, Sumatra, Borneo, the Riau Archipelago, Pulau Bangka, Pulau Billiton, Java, Sulawesi, the Lesser Sundas and the Philippines. On Borneo, Bako National Park in Sarawak is arguably the best place to see it. Other localities include Pulau Talang Talang Besar, Pulau Satang, Gunung Matang, Gunung Santubong, Lambir Hills National Park, Niah National Park, Loagan Bunut National Park, all in Sarawak; also Bukit Patoi and Ulu Temburong in Brunei Darussalam; Deramakot, Sukau, Taman Pulau Tiga, Pulau Gaya and Sepagaya Forest Reserve in Sabah, and Pulau Sibau, Sinkawang, Banjarmasin, Samarinda, Taman Nasional Kayan Mentarang, Pulau Bilangbilangan, Bulangan, Labuan Kelambu in Kalimantan.

IUCN THREAT STATUS Least Concern.

Head and neck of hatchling, lateral view

Juvenile

SNAKES • Order SQUAMATA SERPENTES

Key to Bornean families of snakes

1a. Ventrals similar in size to dorsals..2
1b. Ventrals dissimilar in size to dorsals..9

2a. Eyes located under scales..Typhlopidae
2b. Eyes visible..3

3a. Scales tiny, granular; midbody scales in > 80 rows..Acrochordidae
3b. Scales non-granular; midbody scales in < 30 rows..7

4a. Mental groove present..Cylindrophiidae
4b. Mental groove absent..Anomochilidae

5a. Loreal grooved; front teeth hinged..Viperidae
5b. Loreal teeth lacking grooves; front teeth not hinged..11

6a. Supralabials with pits..Pythonidae
6b. Supralabials lack pits..13

7a. Interparietals present..Xenopeltidae
7b. Interparietal absent..15

8a. Loreal absent; anterior teeth of maxilla enlarged..Elapidae
8b. Loreal present; anterior teeth of maxilla not enlarged..17

9a. Prefrontals enlarged; internasals fused with nasals..Xenophidiidae
9b. Prefrontals not enlarged; internasals not fused with nasals..19

10a. Nostrils oriented upwards; ≥ 23 dorsal scale rows..Homalopsidae
10b. Nostrils oriented sideways; if upwards, dorsal scale rows 19..21

11a. Subcaudals single..Xenodermatidae
11b. Subcaudals paired..23

12a. Scales on nuchal region overlapping; three median rows of dorsal scales keeled..Pseudoxenodontidae
12b. Scales on nuchal region non-overlapping; median rows of dorsal scales not as above..25

13a. Median chin groove absent..Pareidae
13b. Median chin groove present..27

14a. Hypapophyses project backwards..Natricidae
14b. Hypapophyses do not project backwards..Colubridae

Family Acrochordidae Wart Snakes

This family includes three species worldwide, two of them occurring on Borneo. They are recognizable by their heavy bodies; loose, folded skin with rough, granular scales and bristle-tipped tubercles; valvular nostrils; eyes positioned on top of the head; and flap for closing the lingual opening of the mouth – adaptations for a highly aquatic mode of life. They inhabit fresh waters and sea coasts, and are nocturnal, secreting themselves beneath fallen logs and other debris underwater, and emerging to hunt crabs, fish and other snakes at night. Large-growing species are harvested for their durable skins and also for their flesh, while at other localities they are killed by fishermen on account of their diet, which largely comprises fish (and therefore, presumably, makes them competitors).

Acrochordus Wart Snakes

Medium-sized to large aquatic snakes of rivers, swamps and sea coasts; head and body dorsum covered with granular or tuberculate scales on loose skin; ventral or subcaudal scutes not transversely enlarged; very high midbody scale count (130–150); centre of scale has bony denticle (osteoderm); nostrils valvular, dorsally placed.

Key to Bornean species of *Acrochordus*

1a. Dorsal surface has pale grey bands.............................*A. granulatus* (below)
1b. Dorsal surface unbanded.............................*A. javanicus* (p. 194)

WART SNAKE *Acrochordus granulatus* (Schneider, 1799)

(Bahasa Malaysia Ular Kadut; Bahasa Indonesia Ular Air Tawar Kecil; Iban Ular Paiie)

SIZE 1m

IDENTIFICATION FEATURES Body stout, compressed; head indistinct from neck, and covered with small, juxtaposed scales; eyes tiny with vertical pupils; midbody scale largest on vertebral region; tail short and prehensile; distinct fold of skin along mid-belly; supralabials 8–11; row of small scales separates supralabials from mouth; infralabials 12–18; rostral absent; chin shields absent; midbody scale rows 100.

COLOUR Top of body olive, blue or blackish-grey, marked with distinct transverse cream-coloured bands, especially in juveniles.

HABITAT AND BEHAVIOUR Inhabits coastal areas, including estuaries, mangroves and sea coasts. Diet includes crabs, eels, burrowing gobies and other snakes. Ovoviviparous, producing 6–12 neonates; 360–400mm.

DISTRIBUTION India, Sri Lanka, Bangladesh, Myanmar, Thailand, Peninsular Malaysia, Borneo, Sumatra, Java, New Guinea and Australia. Throughout coastal Borneo, can be encountered off the Santubong coast of western Sarawak, with other records from the Buntal areas nearby; Kampung Pelambayan in Brunei Darussalam; Sabah records include Kimanis and Marudu Bays and Sandakan. Expected to occur in Kalimantan, from where there are currently no records.

IUCN THREAT STATUS Least Concern.

Elephant Trunk Snake *Acrochordus javanicus* Hornstedt, 1787

(Bahasa Malaysia/Indonesia: Ular Belalai Gajah. Hakka: Nai She. Iban: Ular Pai)

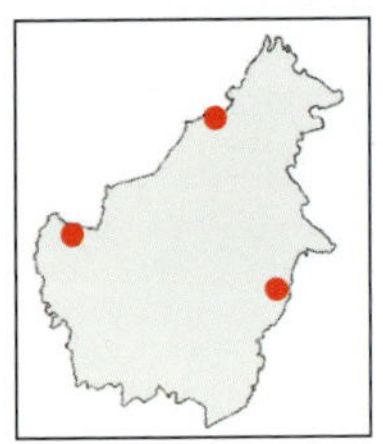

SIZE 2m

IDENTIFICATION FEATURES Body very stout and slightly compressed; head indistinct from neck; forehead scales small and rough; eyes small with vertical pupils; dorsals keeled; midbody scale rows largest around vertebral region; tail short, prehensile; supralabials 22–36; infralabials 31–34; midbody scale rows 120–150.

COLOUR Top of body greyish-black, the head with darker lines; two diffuse longitudinal stripes and elongated dark blotches on flanks; belly cream.

HABITAT AND BEHAVIOUR Freshwater wetlands, including peat swamps and black-water rivers, plus ditches and canals. Diet comprises fish, including eels and catfish. Ovoviviparous, producing 6–48 neonates; 290–460mm. Known to produce embryos without mating.

DISTRIBUTION Thailand, Peninsular Malaysia, Singapore, Borneo, Sumatra, Java, Cambodia and Vietnam. Known from lowlands of Borneo; the species common around the drainage of Sungei Sadong at Serian and Balai Ringin, near Kuching, Sarawak State. Kampung Dato Gandhi in Brunei Darussalam. Records from Kalimantan include lower reaches of Sungei Mahakam and Tayan Hilir.

IUCN THREAT STATUS Least Concern.

Close-up of head

Full view of body

Head and forebody

Family Anomochilidae Giant Blind Snakes

This family of burrowing snakes is restricted to Sundaland, and is represented by three species, with subcylindrical bodies, lack of chin groove, and teeth on the pterygoid and palatine bones. They inhabit lowland and montane forests.

Anomochilus Giant Blind Snakes

Small family of burrowing snakes with cylindrical bodies; supralabials four; no mental (chin) groove.

Key to Bornean species of *Anomochilus*

1a. Parietofrontal scales paired; flanks pale striped................................*A. weberi* (p. 197)
1b. Parietofrontal scale single; lacking pale stripes **2**

2a. Large, pale spots in paravertebral region; midventrals 214–252...*A. leonardi* (below)
2b. No large pale spots in paravertebral region; midventrals 258–261 ....*A. monticola* (p. 196)

Malayan Giant Blind Snake *Anomochilus leonardi* Smith, 1940

(No vernacular names recorded)

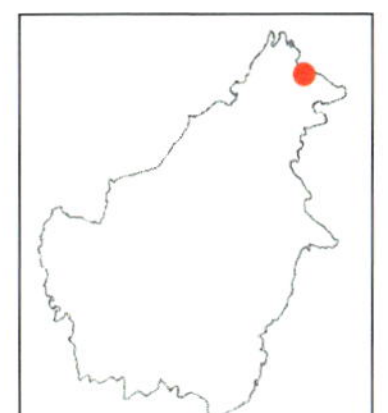

SIZE 228mm

IDENTIFICATION FEATURES Body stout, rounded in cross-section; head small, indistinct from neck; forehead covered with large scales; azygous parietofrontal; nostril in single nasal, contacts with supralabial II; loreal and preocular absent; single postocular; eyes small; mental groove absent; tail short and conical; dorsals smooth, slightly larger than ventrals at same level; midbody scale rows 17 or 19; ventral scales 214–252; subcaudals 6–7.

COLOUR Top of body glossy black or purplish-brown, with oval yellow spots; yellow bar covers most of frontal and part of preoculars; venter black; subcaudals red.

HABITAT AND BEHAVIOUR Inhabits plains ascending mid-hills, to c. 250m asl. Nocturnal and subfossorial. Diet and reproductive habits unstudied.

DISTRIBUTION PENINSULAR Malaysia and Borneo. The sole Bornean record is from Sepilok in Sabah State.

IUCN THREAT STATUS Least Concern.

Kinabalu Giant Blind Snake *Anomochilus monticola* Das, Lakim, Lim & Tan, 2008

(Bahasa Malaysia/Bahasa Indonesia: Ular Tanah)

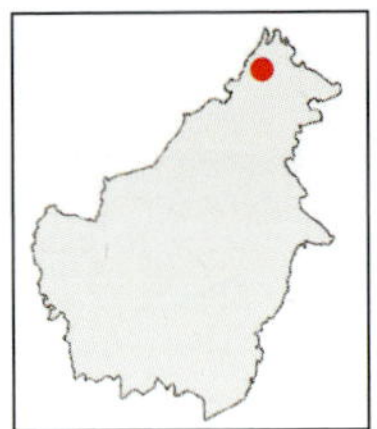

SIZE 520mm

IDENTIFICATION FEATURES Body stout, rounded in cross-section; head small, indistinct from neck; forehead covered with large scales; eyes small; tail short and conical; dorsals smooth, slightly larger than ventrals at same level; midbody scale rows 19; ventrals 258–261; subcaudals 7–8.

COLOUR Top of body blue-black, and lacks pale lateral lines and large pale blotches on either side of vertebral region; transverse yellow bar across snout; series of isolated pale yellow scales on flanks; belly dark brown.

HABITAT AND BEHAVIOUR Inhabits submontane forests at 1,450m asl, at edges of waterbodies and in human-modified areas. Diet includes arthropods. Reproductive habits unknown.

DISTRIBUTION Endemic to Borneo. The sole locality is the Park Headquarters area of Gunung Kinabalu Park in Sabah.

IUCN THREAT STATUS Data Deficient.

Sumatran Giant Blind Snake *Anomochilus weberi*

(van Lidth de Jeude, 1890)

(No vernacular names recorded)

SIZE 230mm

IDENTIFICATION FEATURES Body stout, rounded in cross-section; head small, indistinct from neck; forehead covered with large scales; paired parietofrontals; nostril set in single nasal, contacting supralabial II; loreal and preocular absent; single postocular; eyes small; mental groove absent; tail short and conical; dorsals smooth, slightly larger than ventrals at same level; midbody scale rows 19; ventrals 242–248; subcaudals 6–8.

COLOUR Top of body black with pale stripe along flanks and large pale blotches on either side of vertebral; venter black.

HABITAT AND BEHAVIOUR Inhabits montane forests. Subfossorial. Diet unstudied. Oviparous, producing clutches of four eggs.

DISTRIBUTION Borneo and Sumatra. The sole Bornean record is from Kutai in Kalimantan.

IUCN THREAT STATUS Data Deficient.

After van Lidth de Jeude (1890).

Family Cylindrophiidae Pipe Snakes

Snakes with blunt heads, subcylindrical bodies that can be dorsoventrally depressed, short tails, and belly with black and white checkered pattern. They are associated with swamps, and may also be terrestrial and subfossorial, and specialize in eating elongated prey, including other snakes and eels. Pipe snakes are widespread in South-east Asia, and one species occurs in Sri Lanka.

Cylindrophis Pipe Snakes

Small, burrowing snakes with small heads that are nearly indistinct from neck; body robust, cylindrical; forehead scales large; eyes small; nostril located within nasal scute; supralabials 5–6; scales smooth; ventrals slightly enlarged; mental groove present and tail short.

Key to Bornean species of *Cylindrophis*

1a. Midbody scale rows 17 .... *C. engkariensis* (below)
1b. Midbody scale rows ≥ 19 2

2a. Paired red stripe along body; tail under surface white .... *C. lineatus* (opposite)
2b. No paired red stripe along body; tail under surface red .... *C. ruffus* (p. 200)

Engkari Pipe Snake *Cylindrophis engkariensis* Stuebing, 1994

(No vernacular names recorded)

SIZE 485mm

IDENTIFICATION FEATURES Body robust, and flattened when displaying; head long, indistinct from neck; Supralabial III contacts orbit; tail tapers to narrow point; dorsals smooth; midbody scale rows 17; ventrals 234; subcaudals 6.

COLOUR Top of body black; short, light postocular streak; irregular rows of paravertebral spots; tail black with black and white mottling; belly has black and white cross-bars, divided at midline.

HABITAT AND BEHAVIOUR Inhabits hilly dipterocarp forests in the lowlands. Subfossorial; associated with leaf litter. Diet and reproductive habits unstudied.

DISTRIBUTION Lanjak Entimau area of western Sarawak. Endemic to Borneo.

IUCN THREAT STATUS Data Deficient.

Holotype of *Cylindrophis engkariensis* Stuebing, 1994 (ZRC 2.3398) from 'headwaters of the Engkari River, Lanjak-Entimau Wildlife Sanctuary'. Dorsal (left) and ventral (right) views.

LINED PIPE SNAKE *Cylindrophis lineatus* Dennys, 1880

(No vernacular names recorded)

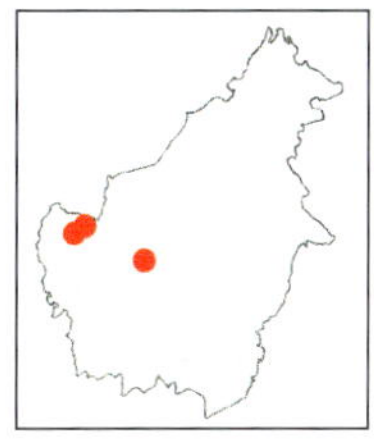

SIZE 982mm

IDENTIFICATION FEATURES Body robust, elongated, and flattened when displaying; mental groove present; head long, blunt and indistinct from neck; eyes reduced; tail tapers to narrow point; dorsal scales smooth; midbody scale rows 21; ventrals 210–215; subcaudals 9–10; cloacal scute divided.

COLOUR Top of body bright reddish-pink; series of longitudinal red or yellow stripes from back of head to base of tail; forehead has scattered dark spots.

HABITAT AND BEHAVIOUR Associated with lowland forests and low hills at up to 400m asl. Subfossorial. Diet and reproductive habits unknown.

DISTRIBUTION Endemic to Borneo. Gunung Penrissen, Gunung Matang, Bungo Range and the Bau region of western Sarawak.

IUCN THREAT STATUS Data Deficient.

COMMON PIPE SNAKE *Cylindrophis ruffus* (Laurenti, 1768)

(Bahasa Malaysia: Ular Kepala Dua, Ular Tanah. Bahasa Indonesia: Ular Kepala Dua. Iban: Ular Bangkit, Ular Untup)

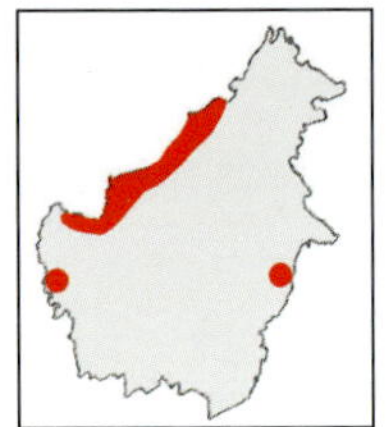

SIZE 900mm

IDENTIFICATION FEATURES Body robust, elongated, and flattened when displaying; mental groove present; head short, blunt and indistinct from neck; eyes reduced, with rounded or vertical pupils; tail tapers to narrow point; dorsals smooth; midbody scale rows 19 or 21; ventrals 185–245; subcaudals 5–10; cloacal scute divided.

COLOUR Top of body black, typically with pale collar; cream bands on dorsal surface in some populations; belly has black cross-bars.

HABITAT AND BEHAVIOUR Low, swampy areas within forested habitats, salt-water lagoons and agricultural fields at < 1,676m asl. Subfossorial. Diet comprises snakes and eels. Ovoviviparous, producing 5–13 neonates, measuring 205mm.

DISTRIBUTION Myanmar, Thailand, Peninsular Malaysia, Borneo, Sumatra, Pulau Bangka, Pulau Belitung, Riau Archipelago, Java, Laos, Cambodia, Vietnam, China, Sangihe and Sula archipelagos and Sulawesi. Widespread on Borneo, with localities including Kuching, the area around Sungei Baram, Sungei Labang and Sungei Pesu drainages, Bintulu District, Gunung Gading in Sarawak; Bandar Seri Begawan in Brunei Darussalam, and Pontianak District and Tenggarong in Kalimantan.

IUCN THREAT STATUS Least Concern.

Ventral view

Dorsal view

Family Pythonidae Pythons

This family includes the largest snakes in the region, with one species reaching 10m in total length. Famous in literature and legend, pythons are swallowers of small and medium-sized mammalian prey, with at least two species reaching sizes large enough to subdue and swallow adult humans, although such instances are rare. Pythons have teeth on their pre-maxilla, and also have supraorbital bones on the dorsal margin of the orbit. In addition, they have rows of heat-sensing labial organs and vestigial pelvic and hindlimb bones (visible as paired spurs on each side of the cloaca). All snakes in this family are egg-layers. Their global distribution covers the Old World tropics and subtropics, from Africa, through Asia and the archipelagos of Australasia, to Australia.

Key to Bornean species of Pythonidae

1a. Supralabials I–IV with thermal pits > 295 ventrals............................*Malayopython reticulatus* (below)
1b. Supralabials I–II with thermal pits < 280 ventrals............................*Python breitensteini* (p. 203)

Malayopython Giant Pythons

These are two gigantic species of South-east Asian python, characterized by infralabial pits more distinct than thermoreceptive pits; infralabial pits in longitudinal groove, defined ventrally by longitudinal fold; colour of suborbital supralabial region similar to rest of supralabials; elongated medial anterior process of ectopterygoid, extending further anteriorly than lateral anterior process; suborbital portion of maxilla without lateral flare or projection; mandibular foramen of compound bone lying below posterior of dentary tooth row; large, and medially divided frontal.

Reticulated Python *Malayopython reticulatus* (Schneider, 1801)

(Mandarin: Mang Seh. Hakka: Kim She; Bahasa Brunei: Ular Penalan. Bahasa Malaysia: Ular Sawa Batik, Ular Sawa Cindai. Bahasa Indonesia: Ular Sanca Batik, Ular Saab. Iban: Ular Sawah)

SIZE 10m

IDENTIFICATION FEATURES Body relatively elongated and slender; head distinct from neck; some infralabials with heat-sensitive pits; eyes small with vertical pupils; cloacal spurs in both sexes; midbody scale rows 69–79; ventrals 297–330; subcaudals 78–102, paired; cloacal scute entire.

COLOUR Top of body yellow or brown, with dark, rhomboidal markings; black median line from snout to nape; oblique postocular stripe reaches corner of mouth; belly yellow with small brown spots.

HABITAT AND BEHAVIOUR Associated with forest edges, especially at the water's edge, and also common in towns and cities, where it inhabits sewers. Mostly terrestrial or at the water's edge, but can climb trees. Nocturnal. Diet comprises warm-blooded animals such mammals and birds; lizards also consumed. Large individuals have been recorded attacking and even consuming human adults. Oviparous, producing clutches of 14–124 eggs, each 90–93 × 58–62mm.

DISTRIBUTION Myanmar, Thailand, Peninsular Malaysia, Singapore, Borneo, Sumatra, Pulau Bangka, Pulau Belitung, Pulau Weh, Pulau Enggano, Pulau Nias, Mentawai, Natuna

and Riau archipelagos, Java, Bali, Laos, Cambodia, Vietnam, Nicobar Islands, Ambon, Anambas Archipelago, Babi, Batjan, Banda Besar, Bankak, Boano, Buru, Butung, Flores, Halmahera, Haruku, Lang, Lombok, Obira, Saparua, Seram, Sula Archipelago, Sulawesi, Sumba, Sumbawa, Tanimbar, Ternate, Timor, Basilan, Bohol, Calamian Islands, Cebu, Leyte, Luzon, Mindanao, Mindoro, Negros, Palawan, Panay, Polillo, Samar, Tawi-Tawi and the Sulu Archipelago. Widespread on Borneo, with records from Bako National Park, Lambir Hills National Park, Baleh National Park, Sungei Rajang, Sungei Baram, Sungei Pesu and Sungei Mengiong, Maludam National Park, Lawas, Oya, Lundu, Gunung Mulu National Park, Gunung Santubong National Park, Tinbarap Palm Oil Estate, Loagan Bunut National Park, and Niah National Park in Sarawak; Bandar Seri Begawan and Ulu Temburong in Brunei Darussalam; Gunung Kinabalu Park (Poring), Pulau Gaya, Sandakan, Kalabakan, Sungei Malutut in Sabah, and Balikpapan, Banjarmasin, Sungei Bulungan, Muara Jawa in Kutai, Pontianak District, Samarinda, and Singkawan in Kalimantan.

IUCN THREAT STATUS Least Concern.

Python Typical Pythons

Large species of python, characterized by thermoreceptive pits more distinct than infralabial pits; dark suborbital patch; medial anterior process of ectopterygoid subequal to lateral anterior process; suborbital portion of maxilla with lateral flare; mandibular foramen of compound bone posterior to dentary tooth row; frontal not medially divided.

BORNEAN SHORT PYTHON *Python breitensteini* Steindachner, 1881

(Bahasa Malaysia/Indonesia: Ular Sawa Darah. Iban: Ripong)

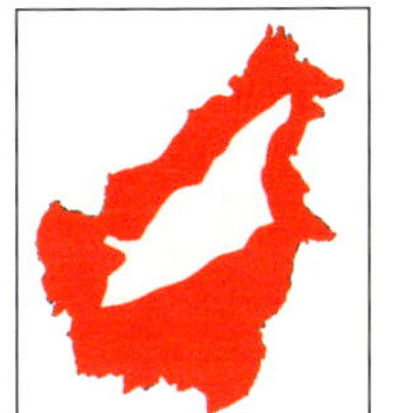

SIZE 2.1m

IDENTIFICATION FEATURES Body short and relatively stout; head elongated, flattened and distinct from neck; eyes small with vertical pupils; vertebral region ridged; no supralabials contact orbit, eye separated from them by 1–5 subocular scales (rarely absent); some infralabials have weak pits; tail short; cloacal spurs present; midbody scale rows 50–57; ventrals 154–165; subcaudals 27–33, paired; cloacal scute entire.

COLOUR Top of body pale yellow or tan, with dark, subrectangular blotches, turning darker towards tail, or with a fully dark top that turns black posteriorly; scattered pale spots on vertebral region, becoming elongated posteriorly to form vertebral stripe; black stripe between internasals and occipital, fusing with dark pattern on neck; sides of head darker than forehead, with dark flecks and dark postocular stripe; pale postocular stripe runs to angle of jaws; chin and belly plain cream, sometimes brown spotted.

HABITAT AND BEHAVIOUR Found in lowland rainforests, peat swamps, heath forests and submontane forests at < 1,000m asl, at edges of rivers, swamps and marshes. Diet includes small mammals and birds. Oviparous, with clutches comprising 12 eggs.

DISTRIBUTION Endemic to Borneo. Widespread on the island, with records from Gunung Gading and Wilmar Plantations, Lawas and Baleh National Park in Sarawak (with historical records from the Kuching area); Jalan Tutong in Brunei Darussalam; Gunung Kinabalu (Poring), Danum Valley, Ulu Dusun and Sungei Purulon in Sabah, and Telang, Batang Alai Utara in Kalimantan.

IUCN THREAT STATUS Least Concern.

Head and forebody

Full view of body

Family Xenopeltidae Sunbeam Snakes

Two living representatives of this family are known, both with highly iridescent dorsals, depressed snout, subcylindrical body, short tail, large scales on forehead and reduced ventrals. They are found in lowland forests and are subfossorial, living under loose leaf litter or fallen objects on the forest floor, and feeding on small vertebrates that they kill by constriction. They are known from South-east Asia and eastern China.

Xenopeltis Sunbeam Snakes

Medium-sized terrestrial species, characterized by flattened, weakly distinct head; cranial elements strongly united; small eyes; large occipital, in contact with frontal; preocular present; loreal present; smooth body scales; tail short.

Sunbeam Snake *Xenopeltis unicolor* Reinwardt, 1827

(Bahasa Malaysia: Ular Pelangi)

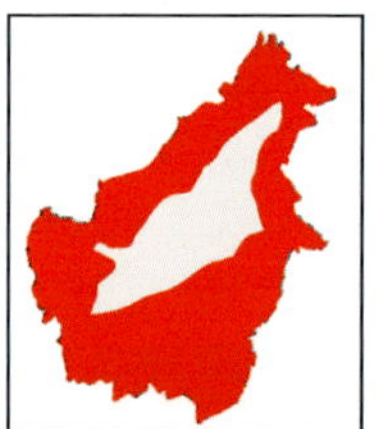

SIZE 1.14m

IDENTIFICATION FEATURES Body robust and cylindrical; head slightly distinct from neck; snout rounded and distinctly flattened; eyes relatively small with vertical pupil; tail short; dorsals very smooth and reflective; midbody scale rows 15; ventrals 164–196; subcaudals 24–31; cloacal scute divided.

COLOUR Top of body iridescent mid-brown, with each scale light edged; juveniles have a pale collar; belly white or cream.

HABITAT AND BEHAVIOUR Found in lowland dipterocarp forests, reaching submontane limits at < 1,402m asl. Terrestrial and subfossorial. Diet comprises rodents, birds, lizards and frogs. Cannibalism also known. Oviparous, with clutches including 3–17 eggs, each 18 × 58mm.

DISTRIBUTION China, Myanmar, Thailand, Laos, Cambodia, Vietnam, Peninsular Malaysia, Singapore, Borneo, Sumatra, Pulau Bangka, Pulau Belitung, Riau Archipelago, Pulau Natuna Besar, Pulau Sipura, Pulau Simeulue, Pulau Siberut, Java, Nicobar Archipelago, Palawan and the Sulu Archipelago. Widespread on Borneo, being recorded from Kubah National Park, Gunung Mulu National Park, Gunung Penrissen, Sebungan Oil Palm Estate in Bintulu, Samunsam Wildlife Sanctuary, Gunung Santubong National Park and Batu Kawa township west of Kuching in Sarawak; Gadong in Bandar Seri Begawan in Brunei Darussalam; Danum Valley, Penampang, Gunung Kinabalu Park (Poring, along Sungei Kipungit) and Sandakan in Sabah (with old records from the Kota Kinabalu area), and Banjarmasin, Samarinda, Singkawang, Sungei Bulangan and Sungei Kapuas in Kalimantan.

IUCN THREAT STATUS Least Concern.

Adult

Juvenile

Family Colubridae 'Typical' Snakes

Most of the snakes in South-east Asia belong to the family Colubridae, in a restricted sense – until recently, the family included water snakes, slug-eating snakes and several other snake groups. They can be told apart by their enlarged forehead scales, solid maxillary teeth, lateral nostrils and well-developed ventrals. These snakes tend to be oviparous, and are cosmopolitan in distribution, being widespread in temperate, subtropical and tropical parts of the world. Current knowledge of the morphology of this diverse family in South-east Asia does not permit the construction of a working dichotomous key to the genera. Further, the use of multiple characteristics that diagnose some of them, such as internal organs and teeth, in such a key would render it not useful in the field.

Ahaetulla Vine Snakes

Medium-sized, slender, arboreal species; head distinct from neck; snout long and narrow; eyes large; pupils horizontal; loreal region concave; maxillary teeth increase progressively, with 1–2 enlarged, grooved teeth at the end; body scales smooth; ventrals with lateral keels; subcaudals paired; tail long.

Key to Bornean species of *Ahaetulla*

1a. Forehead green (yellow in juveniles), lacking dark speckling pattern; cloacal divided; snout terminates in curled rostral.............. *A. fasciolata* (below)
1b. Forehead pale brown with distinct dark specking pattern; cloacal single; snout does not terminate in curled rostral................. *A. prasina* (p. 206)

SPECKLE-HEADED VINE SNAKE *Ahaetulla fasciolata* (Fischer, 1885)

(Iban: Ular Bunga Merisian)

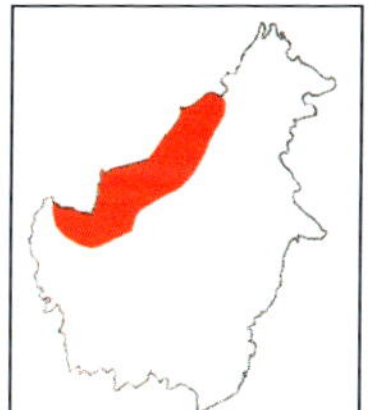

SIZE 1.69m

IDENTIFICATION FEATURES Body slender; snout elongated, ending in curled rostral; eyes enlarged, with horizontal pupils; tail long with prehensile tip; vertebrals enlarged; dorsals smooth; midbody scale rows 15; ventrals 211–240; subcaudals 178–197, paired; cloacal scute single.

COLOUR Top of body light brown, grey or pinkish-tan, with numerous narrow, oblique, dark bands on anterior; forehead has elongated or curved dark markings; belly dark grey.

HABITAT AND BEHAVIOUR Inhabits lightly forested and semi-urban habitats at < 900m asl. Arboreal; encountered in thick undergrowth and other low vegetation. Diet includes lizards and frogs. Ovoviviparous (numbers and sizes of neonates unknown).

DISTRIBUTION Thailand, Peninsular Malaysia, Singapore, Sumatra, Borneo, Natuna and Riau archipelagos. Isolated Bornean localities in Kuching, Bintulu, Niah National Park, the Kelabit Highlands and Sungei Balui, Sungei Baram, Sungei Mengion and Sungei Segaham in Sarawak, and Bandar Seri Begawan in Brunei Darussalam.

IUCN THREAT STATUS Least Concern.

Oriental Vine Snake *Ahaetulla prasina* (Boie, 1827)

(Bahasa Malaysia: Ular Pucuk. Bahasa Brunei: Ular Daun, Ular Kunyet. Bahasa Indonesia: Ular Pucuk. Iban: Ular Bungai, Puchok Pisang. Kelabit: Selangoi Bata)

SIZE 1.97m

IDENTIFICATION FEATURES Body slender; snout elongated, with parallel groove; eyes enlarged, with horizontal pupils; tail long with prehensile tip; dorsals smooth; midbody scale rows 15; ventrals 194–235; subcaudals 141–207, paired; cloacal scute divided.

COLOUR Top of body typically green in adults, and occasionally brown; juveniles yellow, dark grey or golden-yellow, speckled with black; yellow stripe along lower flanks; belly pale green or dark grey.

HABITAT AND BEHAVIOUR Inhabits forest edges as well as parks and gardens, from lowlands to about 2,100m asl. Arboreal, in low vegetation such as ferns. Diet comprises lizards and birds. Ovoviviparous, producing 4–10 neonates, each 240–490mm.

DISTRIBUTION Bhutan, eastern India, Bangladesh, Myanmar, Thailand, Laos, Cambodia, Vietnam, China, Peninsular Malaysia, Singapore, Sumatra, Borneo, Mentawai, Riau and Natuna archipelagos, Pulau Bangka, Pulau Belitung, Pulau Sibutu, Java, Bali, Sulu Archipelago and other islands of the Philippines. Widespread on Borneo, and often encountered at Kubah National Park near the Frog Pond, off the Summit Trail to Gunung Serapi. Other Bornean localities include Bako National Park, Baleh National Park, Banting, Bario, Belaga, Kapit, Gunung Gading National Park, Gunung Mulu National Park, Gunung Santubong National Park, Sadong, the Kelabit Highlands, Samunsam Wildlife Sanctuary, Sungei Baram, Sungei Pesu and Sungei Mengion in Sarawak; Bandar Seri Begawan and Ulu Temburong in Brunei Darussalam; Danum Valley, Kota Kinabalu, Kudat, Ranau, Sandakan Bay, Sepagaya Forest Reserve, Tawau Hills Park, as well as localities around Gunung Kinabalu, such as Bundu Tuhan, Kiau and Marak-Parak and Sungei Purulon in Sabah, and Buntok, Busam, Banjermasin, Landak, Pontianak, Samarinda, Sebruang Valley, Sintang, Sinkawang, Tegora, Sungei Bulangan and Sungei Mahakam in Kalimantan.

IUCN THREAT STATUS Least Concern.

Boiga Cat Snakes

Small to large snakes that are mostly arboreal, with some terrestrial activity seen in larger individuals; body slender to robust; head wider than neck; eyes large with vertically elliptic pupils; snout short; body compressed; scales smooth, with apical pits; vertebrals enlarged; subcaudals paired; 2–3 enlarged, grooved teeth at back of maxilla.

Key to Bornean species of *Boiga*

1a. Dorsum of body black with narrow yellow rings .... *B. dendrophila* (p. 209)
1b. Dorsum of body not back with narrow yellow rings **2**

2a. Dorsum of body pale brown or grey; distinct dark postocular stripe .... *B. cynodon* (p. 208)
2b. Dorsum of body not pale brown or grey; no distinct dark postocular stripe **3**

3a. Forehead scales yellow edged .... *B. jaspidea* (p. 211)
3b. Forehead scales unicoloured **4**

4a. Midbody scale rows 19; pale spots on lowest dorsal rows .... *B. drapiezii* (p. 210)
4b. Midbody scale rows 21; no pale spots on lowest dorsal rows .... *B. nigriceps* (p. 212)

Dog-toothed Cat Snake (p. 208)

Dog-toothed Cat Snake *Boiga cynodon* (Boie, 1827)

(Bahasa Malaysia: Ular Telor. Bahasa Indonesia: Ular Kucing Bergigi Panjang. Iban: Ular Blidah, Ular Kengkang Mas)

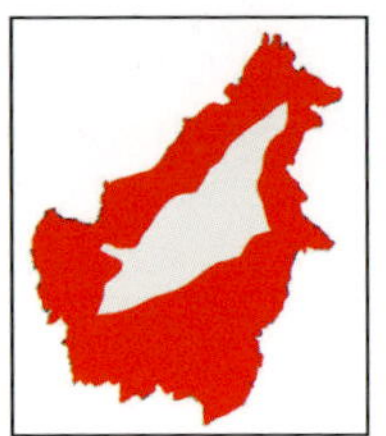

SIZE 2.8m

IDENTIFICATION FEATURES Body slender, elongated and laterally compressed; head distinct from neck; snout short and rounded; eyes enlarged, with vertical pupils; vertebrals distinctly enlarged; dorsals smooth; midbody scale rows 23 or 25; ventrals 248–290; subcaudals 114–165, paired; cloacal scute entire.

COLOUR Top of body brownish-tan or yellowish-brown, with dark brown or reddish-brown bands that darken posteriorly; juveniles paler; dark postocular stripe.

HABITAT AND BEHAVIOUR Inhabits lowland forests and forest edges. Arboreal, climbing low trees and moving in dense undergrowth. Diet includes lizards, birds and their eggs, and small mammals. Oviparous, laying 6–23 eggs (size unknown).

DISTRIBUTION Thailand, Peninsular Malaysia, Singapore, Sumatra, Borneo, Pulau Nias, Pulau Belitung, Pulau Bangka, Mentawai Archipelago, Java, Bali, Lesser Sunda Islands and the Philippines. Widespread on Borneo, with records from Kubah National Park, Gunung Penrissen, Bintulu, Busau, Niah National Park, Gunung Santubong National Park, Sungei Baram, Sungei Segaham, Sungei Seran and Sungei Tangap in Sarawak (in addition to historical records from Kuching); Brunei Darussalam (without a precise locality); Baturong, Kalabakan, Tawau Hills Park, Gunung Kinabalu Park (at Poring) and Kota Kinabalu in Sabah, and Balikpapan, Samarinda, Sebruang Valley and Sungei Kapuas in Kalimantan.

IUCN THREAT STATUS Least Concern.

Close-up of head

MANGROVE CAT SNAKE *Boiga dendrophila* (Boie, 1827)

(Bahasa Malaysia/Indonesia: Ular Bakau, Ular Taliwangsa. Iban: Bangkit, Chinchin Mas)

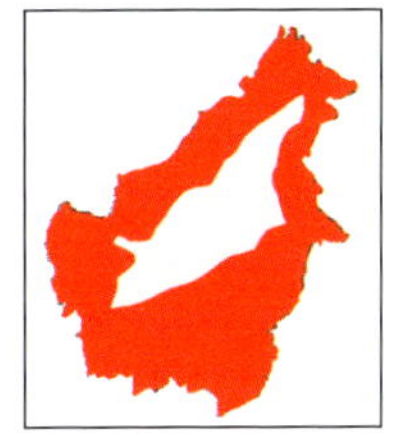

SIZE 2.5m

IDENTIFICATION FEATURES Body large, robust and compressed; head distinct from neck; snout short and rounded; eyes large with vertical pupils; dorsals smooth; midbody scale rows 21 (rarely 23); ventrals 209–253; subcaudals 89–118, paired; cloacal scute entire.

COLOUR Top of body black, with 35–45 narrow yellow transverse rings on body and 10 on tail; labials and gular region yellow; belly grey.

HABITAT AND BEHAVIOUR Inhabits lowland forests, including dipterocarp forests, mangroves and peat swamps, and occasionally seen in forest edges close to human settlements. Found on trees and in dense undergrowth. Diet comprises birds and their eggs and nestlings, frogs, lizards, other snakes, and even mammals such as mouse deer and tree shrews. Oviparous, laying 4–15 eggs, each 45.5–51 × 24.5–25mm.

DISTRIBUTION Peninsular Malaysia, Singapore, Borneo and Sumatra. Records from Borneo, where it is widespread, include Anna Rais at foothills of Gunung Penrissen, Baleh National Park, Bako National Park, Bintulu, Bukit Berayong in Lawas, Kota Samarahan, Lambir Hills National Park, Niah National Park, Gunung Santubong National Park, Pelagus National Park, Loagan Bunut National Park, Sungei Subis, Sungei Labang, Sungei Baloi, Sungei Mengion, Sungei Seran and Sungei Pesu in Sarawak; Labuan; Serasa mangroves in Brunei Darussalam; Pulau Gaya, Pulau Tiga Park, Danum Valley, Deramakot, Dewhurst Bay, Kalabakan, Kota Kinabalu, Kudat, Gunung Kinabalu (Poring and Ranau), Sandakan, Sepagaya Forest Reserve, Tawau Hills Park and Telupid in Sabah, and Balikpapan, Banjarmasin, Muara Jawa in Kutai, Samarinda, Pontianak, Sebruang Valley, Sintang, Sinkawang, and Sungei Bulangan and Sungei Mahakam in Kalimantan.

IUCN THREAT STATUS Not Evaluated.

REMARKS Bornean population may receive recognition as a distinct species within the *Boiga dendrophila* complex.

White-spotted Cat Snake *Boiga drapiezii* (Boie, 1827)

(No vernacular names recorded)

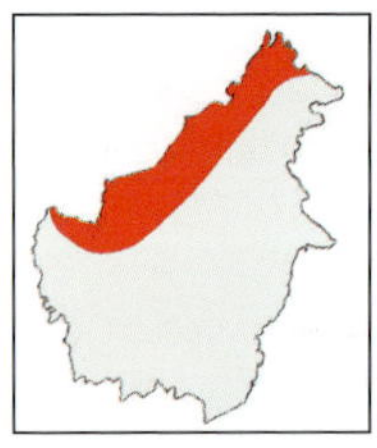

SIZE 2.10m

IDENTIFICATION FEATURES Body relatively long and slender; head distinct from neck; eyes enlarged, with vertical pupils; vertebrals enlarged; dorsals smooth; midbody scale rows 19; ventrals 250–287; subcaudals 114–173, paired; cloacal scute entire.

COLOUR Top of body olive-grey to reddish-brown; vertebral region has paired pink spots anteriorly that may be fused to form a line; pink or cream spots on flanks; forehead has dark speckling.

HABITAT AND BEHAVIOUR Inhabits lowland forests and mid-hills at < 1,000m asl. Arboreal; associated with trees and shrubs. Diet comprises birds and their eggs, frogs, lizards and large insects. Oviparous, with eggs produced in termite-infested wood.

DISTRIBUTION Thailand, Peninsular Malaysia, Singapore, Borneo, Sumatra, Pulau Bangka, Mentawai Archipelago, Vietnam, central islands of Indonesia and the Philippines. Bornean localities include Kubah National Park, Niah National Park, Loagan Bunut National Park, Pangkalan Ampat at the foothills of Gunung Penrissen, Gunung Mulu National Park, Pelagus National Park, Tanjung Datu National Park, Gunung Santubong National Park, Sungei Baram, Sungei Mengiong, Sungei Pesu, Sungei Landak and Sungei Seran in Sarawak; Ulu Temburong in Brunei Darussalam, and Danum Valley, Ranau and Sandakan in Sabah.

IUCN THREAT STATUS Least Concern.

JASPER CAT SNAKE *Boiga jaspidea* (Duméril, Bibron & Duméril, 1854)

(Iban: Ular Banjang)

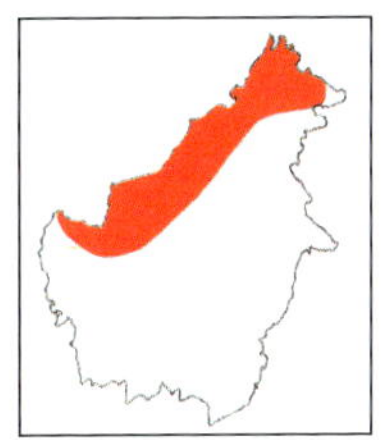

SIZE 1.5m

IDENTIFICATION FEATURES Body slender, laterally compressed; head large, distinct from neck; eyes large with vertical pupils; vertebrals enlarged; dorsals smooth; midbody scale rows 21; ventrals 243–267; subcaudals 128–166, paired; cloacal scute enlarged.

COLOUR Top of body brown, reddish-brown or grey-brown, with paired rows of dark spots or bars on flanks and greyish-red vertebral stripe.

HABITAT AND BEHAVIOUR Inhabits lowland forests and peat swamps at < 1,524m asl. Associated with trees and dense undergrowth. Diet includes lizards, small mammals, birds and their eggs, as well as other snakes. Oviparous, producing six eggs, each 38–39 × 18–19mm, in nests of tree-dwelling termites.

DISTRIBUTION Thailand, Vietnam, Peninsular Malaysia, Singapore, Borneo, Sumatra, Pulau Nias, Mentawai Archipelago, Pulau Bangka and Java. Within Borneo, known from Kubah National Park, Gunung Mulu National Park, Loagan Bunut National Park, Pangkalan Ampat at the foothills of Gunung Penrissen, Pelagus National Park, Saribas, Sungei Baram, Sungei Labang and Sungei Segaham in Sarawak; Labuan; Bukit Puan and Jalan Bukok, Puni, Temburong in Brunei Darussalam; Danum Valley, Gunung Kinabalu (Poring) and Sandakan in Sabah, and Balikpapan in Kalimantan.

IUCN THREAT STATUS Least Concern.

BLACK-HEADED CAT SNAKE *Boiga nigriceps* (Günther, 1863)

(Iban: Ular Banjang)

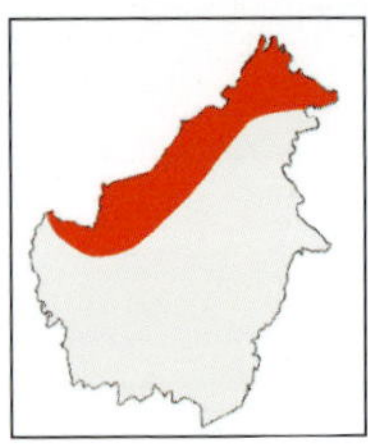

SIZE 2m

IDENTIFICATION FEATURES Body robust and laterally compressed; head large and distinct from neck; eyes large with vertical pupils; dorsals smooth; midbody scale rows 21; ventrals 240–293; subcaudals 134–164, paired; cloacal scute entire.

COLOUR Top of body straw-brown, olive-brown or reddish-brown; forehead and tail often darker; labials cream or yellow; belly cream, darkening posteriorly.

HABITAT AND BEHAVIOUR Inhabits lowland to submontane forests at about 1,100m asl. Associated with trees and dense forest undergrowth. Diet includes birds and other snakes. Oviparous, laying three eggs, each 48 × 17mm.

DISTRIBUTION Thailand, Peninsular Malaysia, Borneo, Sumatra, Pulau Nias, Pulau Simeulue, Java and the Mentawai Archipelago. Bornean localities include the Kelabit Highlands, Samunsam Wildlife Sanctuary, Sungei Pesu, Sungei Labang and Singei Seran in Sarawak; Brunei Darussalam (without a precise locality), and Danum Valley, Gunung Kinabalu (at Poring) and Bongon in Sabah.

IUCN THREAT STATUS Least Concern.

Calamaria Reed Snakes

Small, mostly slender, cylindrical snakes with small head and eyes, and short tail; head and neck scarcely distinct; pupils rounded; internasals fused with prefrontals; nostrils placed within small nasal scute; middorsal rows 13; loreal scute absent; subcaudal scutes paired; tail short.

Key to Bornean species of *Calamaria*

1a. Supralabials III, or both II and III contact orbit.......2
1b. Supralabial III and IV contact orbit.......4

2a. Preocular absent.......3
2b. Preocular present.......5

3a. Supraocular and postocular fused.......*C. gracillima* (p. 218)
3b. Supraocular and postocular distinct.......4

4a. Mentals contact anterior chin shields.......*C. lovii* (p. 222)
4b. Mentals do not contact anterior chin shields.......*C. schmidti* (p. 227)

5a. Two infralabials contact anterior chin shields.......*C. borneensis* (p. 216)
5b. Three infralabials contact anterior chin shields.......*6*

6a. Middorsal region has narrow longitudinal stripes.......*C. battersbyi* (p. 215)
6b. Middorsal region lacks narrow longitudinal stripes.......*C. melanota* (p. 224)

7a. Light longitudinal stripe on rows 2–3 bordered below by dark stripe on first row.......*C. lumholtzi* (p. 224)
7b. Stripe, if present, not as above.......8

8a. Preoculars absent.......9
8b. Preoculars present.......10

9a. Dorsal scales typically dark brown with pale network, or yellow with dark network.......*C. rebentischi* (p. 226)
9b. Dorsal scales above second row dark brown without light network.......*C. schlegeli* (p. 227)

10a. Mental not in contact with anterior chin shields.......11
10b. Mental in contact with anterior chin shields.......18

11a. Dorsum has dark transverse cross-bands.......*C. bicolor* (p. 215)
11a. Dorsum lacks cross-bands.......12

12a. Dorsal and ventral colouration similar.......*C. lateralis* (p. 220)
12b. Dorsal colours different from that of venter.......13

13a. Ventrals below neck region lack dark pigmentation.......14
13b. Ventrals below neck region with dark pigmentation.......16

14a. Orbit larger than eye–mouth distance.......*C. leucogaster* (p. 221)
14b. Orbit smaller than eye–mouth distance.......15

15a. Prefrontals fail to contact supralabial III .... *C. schegeli* (p. 227)
15b. Prefrontals contact supralabial III .... *C. bicolor* (opposite)

16a. Dorsal scale row reduction to four on tail at 14 or over subcaudals .... *C. everetti* (p. 216)
16b. Dorsal scale row reduction to four on tail at 13 or less subcaudals **17**

17a. Tail-tip blunt .... *C. virgulata* (p. 228)
17b. Tail-tip acute .... *C. modesta* (p. 225)

18a. Venter bears dark cross-bands, each over one ventral wide .... *C. lumbricoidea* (p. 223)
18b. Venter either not dark banded or each band under one ventral wide **19**

19a. Orbit 2/3 eye–mouth distance .... *C. hilleniusi* (p. 219)
19b. Orbit ≥ eye–mouth distance **20**

20a. Dorsum bears longitudinal stripes .... *C. griswoldi* (p. 218)
20b. Dorsum unstriped **21**

21a. Tail abruptly ending in blunt point .... *C. prakkei* (p. 226)
21b. Tail tapering gradually **22**

22a. In males, ventrals < 145; in females < 164 .... *C. suluensis* (p. 228)
22b. In males, ventrals > 145; in females > 164 .... *C. grabowskyi* (p. 217)

Battersby's Reed Snake *Calamaria battersbyi* Inger & Marx, 1965

(No vernacular names recorded)

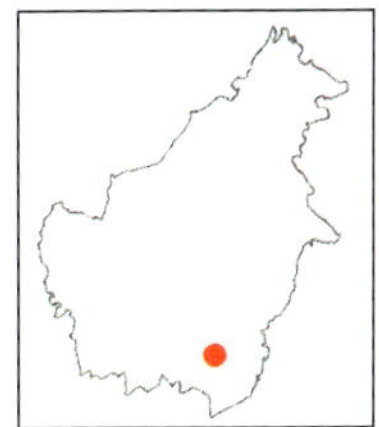

SIZE 92mm

IDENTIFICATION FEATURES Body slender, cylindrical; head short, indistinct from neck; loreal absent; single preocular; single postocular; eyes small with rounded pupils; tail short, abruptly tapered to a point; dorsals smooth; subcaudals paired; midbody scale rows 13; ventrals 171; subcaudals 16, paired; cloacal scute entire.

COLOUR Top of body brown, scales yellow with brown margins; narrow dark stripe on adjacent edges of scale rows 1–2; wider dark stripe on adjacent edges of scale rows 2–3 and 4–5; narrow dark stripe on adjacent edges of scale row six and vertebral rows; narrow yellow ring behind head and one at tail-base; lower half of supralabials yellow; belly has dark stippling anteriorly, unpatterned yellow posteriorly.

HABITAT AND BEHAVIOUR Nothing known of its biology.

DISTRIBUTION Endemic to Borneo. The sole locality for the species is Tanjung in southern Kalimantan.

IUCN THREAT STATUS Data Deficient.

Bicoloured Reed Snake *Calamaria bicolor* Duméril, Bibron & Duméril, 1854

(No vernacular names recorded)

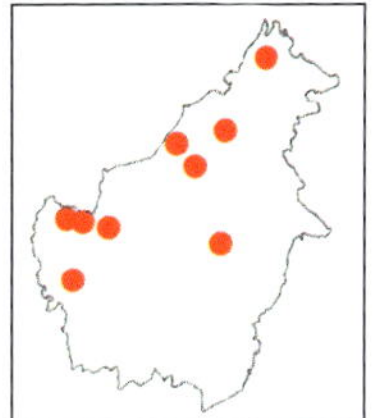

SIZE 450mm

IDENTIFICATION FEATURES Body slender and cylindrical; head short and indistinct from neck; nasals point forwards; eyes small with rounded pupils; tail thick, tapering from base; dorsals smooth; midbody scale rows 13; ventrals 139–169; subcaudals 18–28, paired; cloacal scute entire.

COLOUR Top of body blue-black or dark brown, unpatterned or with dark cross-bands; forehead dark brown, sometimes with two oblique dark bands crossing yellow labials; belly typically plain yellow or spotted with black.

HABITAT AND BEHAVIOUR Inhabits forested mid-hills up to 1,200m asl. Terrestrial. Diet and reproductive habits unstudied.

DISTRIBUTION Borneo and Java. Bornean records are from the Kelabit Highlands, Gunung Matang, Niah National Park, Paku, Kampung Tapuh and Sungei Entout in the Baram basin in Sarawak; Gunung Kinabalu in Sabah, and Singkawang, Tayan in Hilir, Kedup, Muarateweh, Nanga Padang, and Sungei Kapuas and Sungei Mahakam in Kalimantan.

IUCN THREAT STATUS Least Concern.

Bornean Reed Snake *Calamaria borneensis* Bleeker, 1860

(No vernacular names recorded)

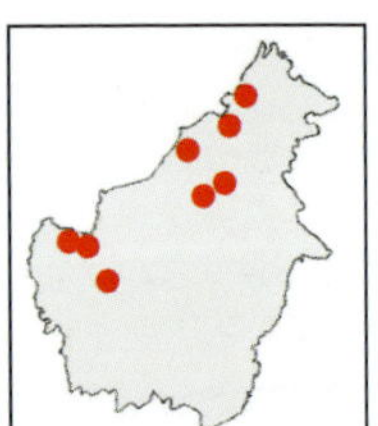

SIZE 374mm

IDENTIFICATION FEATURES Body slender and cylindrical; head short and indistinct from neck; eyes small with rounded pupils; tail short; dorsals smooth; midbody scale rows 13; ventrals 126–192; subcaudals 13–26, paired; cloacal scute entire.

COLOUR Top of body greyish-brown, each scale pale with dark reticulation; scattered dark spots or stripes on scales on middorsum; first scale row yellow, sometimes with dark cross-bars; head greyish-brown with indistinct dark spots; 1–3 yellow caudal rings; belly dark striped along ventral edges, or with checkered pattern of yellow and black.

HABITAT AND BEHAVIOUR Inhabits lowlands and submontane regions. Diet and reproductive biology unstudied.

DISTRIBUTION Endemic to Borneo. Localities include the Kelabit Highlands, Marudi, Simanggang, Gunung Gading National Park, the Matang Range, Gunung Mulu National Park, Gunung Santubong National Park, Paku and the drainage of Sungei Pesu in Sarawak; Kimanis in Sabah, and Sintang in Kalimantan.

IUCN THREAT STATUS Least Concern.

Calamaria borneensis Bleeker, 1860 (ZRC 2.2679) Long Mujan, Ulu Baram, Sarawak.

Everett's Reed Snake *Calamaria everetti* Boulenger, 1893

(No vernacular names recorded)

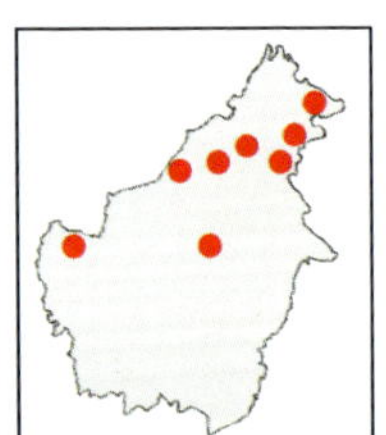

SIZE 330mm

IDENTIFICATION FEATURES Body slender; head short, indistinct from neck; loreal absent; single preocular; single postocular; three infralabials contact anterior chin shields; mental fails to contact anterior chin shields; tail short; dorsals smooth; subcaudals paired; midbody scale rows 13; ventrals 136–157; subcaudals 16–25, paired; cloacal scute entire.

COLOUR Top of body brown, scale sometimes with fine dark network; dark longitudinal stripe on scales anteriorly; pale brown or yellow vertical bar at back of head and collar of the same colour; forehead brown with darker spots; belly yellow.

HABITAT AND BEHAVIOUR Inhabits forested lowlands to mid-hills, at elevations up to 1,500m asl. Diet and reproductive biology unstudied.

DISTRIBUTION Endemic to western and northwestern Borneo. Localities include the Kelabit Highlands, Niah National Park, Pangkalan Ampat at the foothills of Gunung Penrissen, Bukit Tiban National Park and Sungei Baram and Sungei Senah in Sarawak (with a historical record from Kuching); Dewhurst Bay in Sabah, and Sungei Nunukan and Sembakung in Kalimantan.

IUCN THREAT STATUS Least Concern.

GRABOWSKY'S REED SNAKE *Calamaria grabowskyi* Fischer, 1885

(No vernacular names recorded)

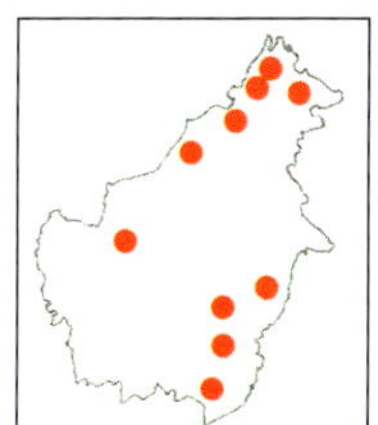

SIZE 480mm

IDENTIFICATION FEATURES Body slender, cylindrical; head short, slightly distinct from neck; loreal absent; single preocular; single postocular; three infralabials contact anterior chin shields; eyes small with rounded pupils; tail long, tapering to blunt tip; dorsals smooth; midbody scale rows 13; ventrals 150–190; subcaudals 20–29, paired; cloacal scute entire.

COLOUR Top of body dark brown, each scale with a darker network; scattered dark brown or yellow spots on back; lateral bands composed of elongated dark spots; labials yellow; belly plain yellow or with varying amounts of dark pigmentation; subcaudals yellow with a dark median band.

HABITAT AND BEHAVIOUR Inhabits submontane forests at elevations of 1,000–1,400m asl. Diet and reproductive biology unstudied.

DISTRIBUTION Endemic to Borneo. Localities include the Kelabit Highlands, Bukit Berayong in Lawas, Long Mujan, Niah National Park, Sarikei and Sungei Labang and Sungei Seran in Sarawak; Danum Valley, Deramakot, Kedamaian, Kenokok, Malutut, Quoin Hill, Tenompok, Gunung Kinabalu (at Sayap, Sungei Kadamaian, Tenompok and Kenokok), Malutut, Mendolong, Ulu Dusun, Sungei Purulon in Sabah, and Long Petah, Muarateweh, Tamiang Layang and Telang in Kalimantan.

IUCN THREAT STATUS Least Concern.

Slender Reed Snake *Calamaria gracillima* (Günther, 1872)

(No vernacular names recorded)

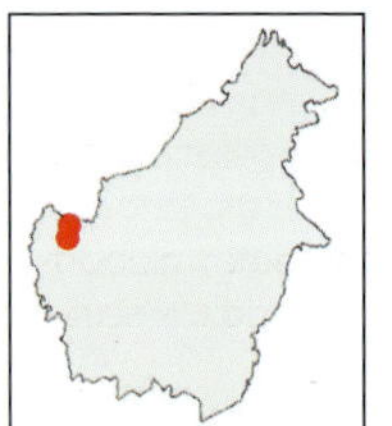

SIZE 290mm

IDENTIFICATION FEATURES Body slender, cylindrical; head short, indistinct from neck; loreal absent; preocular absent; single postocular; supraocular fused with postocular; three infralabials contact anterior chin shields; eyes small with rounded pupils; tail short, non-tapering, with blunt tip; dorsals smooth; midbody scale rows 13; ventrals 290–304; subcaudals 12–15, paired; cloacal scute entire.

COLOUR Top of body dark brown, with widely spaced, vertical yellow bars on anterior flanks and around tail-base; belly dark brown, ventrals and subcaudals paler on posterior half.

HABITAT AND BEHAVIOUR Apparently associated with lowland forests. Diet and reproductive biology unstudied.

DISTRIBUTION Endemic to western Borneo. Known only from Gunung Matang and Tegora in Sarawak State.

IUCN THREAT STATUS Data Deficient.

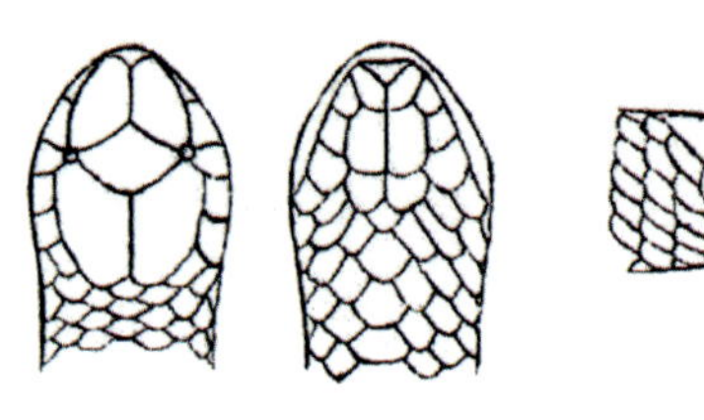

After Günther (1872).

Lined Reed Snake *Calamaria griswoldi* Loveridge, 1938

(No vernacular names recorded)

SIZE 490mm

IDENTIFICATION FEATURES Body slender and cylindrical; head short, indistinct from neck; eyes small with rounded pupils; tail short, tapering to sharp point; dorsals smooth; midbody scale rows 13; ventrals 155–192; subcaudals 13–18, paired; cloacal scute entire.

COLOUR Top of body dark brown, with blackish-brown and yellow stripes; forehead dark brown; lower portions of supralabials yellow; oblique pale bar between parietals to gular region; ventrals plain yellow; subcaudals yellow with indistinct medial zigzag mark.

HABITAT AND BEHAVIOUR Inhabits submontane forests at 1,200–1,800m asl. Subfossorial and active in leaf litter. Diet and reproductive habits unstudied.

DISTRIBUTION Endemic to Borneo. Records are from the Kinabalu Massif (including Lumu-Lumu, Bundu Tuhan, Mesilau, Tenompok, Ranau and Sungei Luidan) in Sabah.

IUCN THREAT STATUS Least Concern.

Hillenius's Reed Snake *Calamaria hilleniusi* Inger & Marx, 1965

(No vernacular names recorded)

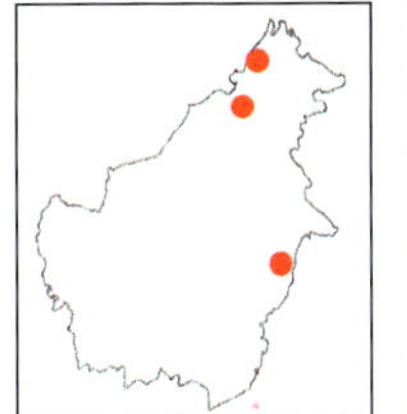

SIZE 370mm

IDENTIFICATION FEATURES Body slender, cylindrical; head short, indistinct from neck; loreal absent; single preocular; single postocular; three infralabials contact anterior chin shields; mental contacts anterior chin shields; tail short, tapering to tip gradually; dorsals smooth; midbody scale rows 13; ventrals 147–151; subcaudals 14–21, paired; cloacal scute entire.

COLOUR Top of body brown, with dark pigmentation abruptly ending on centre of third scale row; forehead brown, with dark pigments ending in oblique line from upper edge of second supralabial to lower third or fifth; supralabials yellow; throat yellow; belly unpatterned yellow.

HABITAT AND BEHAVIOUR Inhabits lowland hill forests and submontane forests. Diurnal and terrestrial, associated with leaf litter. Diet and reproductive biology unstudied.

DISTRIBUTION Endemic to Borneo. Recorded from Tuaran in Sabah; Gunung Murud in Sarawak, and Samarinda in Kalimantan.

IUCN THREAT STATUS Least Concern.

White-striped Reed Snake *Calamaria lateralis* Mocquard, 1890

(No vernacular names recorded)

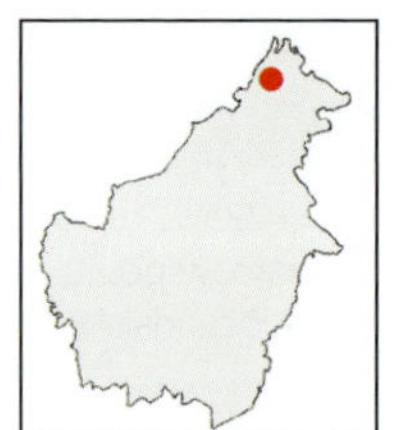

SIZE 290mm

IDENTIFICATION FEATURES Body slender, cylindrical; head short, indistinct from neck; loreal absent; single preocular; single postocular; three infralabials contact anterior chin shields; eyes small with rounded pupils; tail short, tapering gradually to a point; dorsals smooth; midbody scale rows 13; ventrals 146–151; subcaudals 26–23, paired; cloacal scute entire.

COLOUR Top of body brownish-black, with longitudinal white line from eye to along flanks of body; belly dark brown.

HABITAT AND BEHAVIOUR Natural history unstudied.

DISTRIBUTION Borneo and Java. Sole Bornean locality is Gunung Kinabalu in Sabah.

IUCN THREAT STATUS Data Deficient.

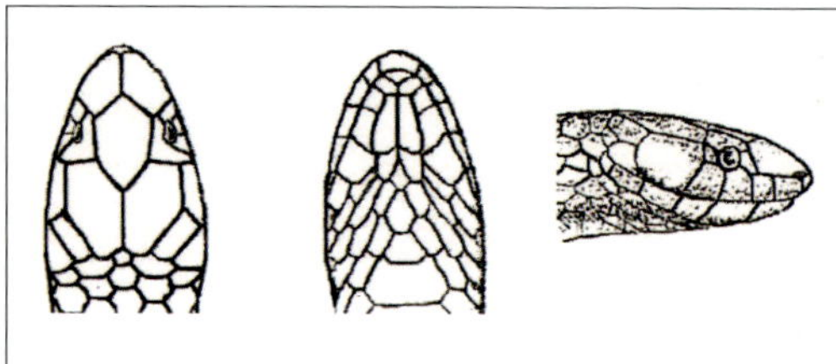

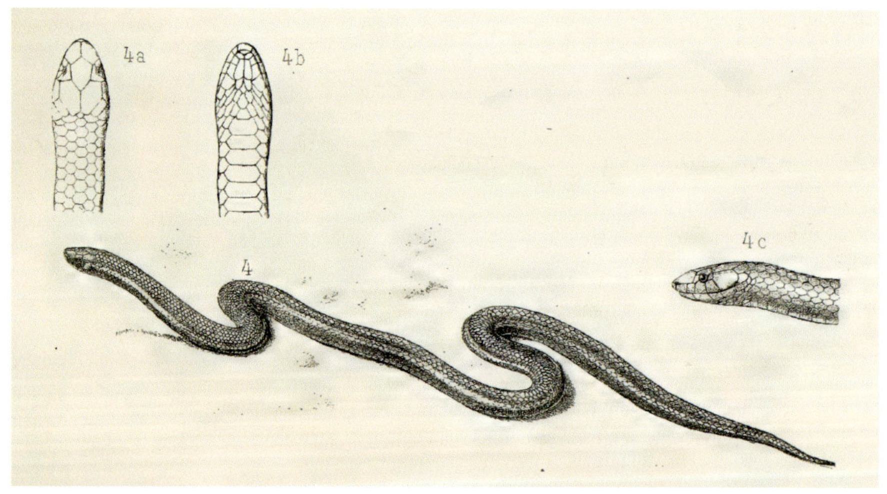

After Mocquard (1890).

COLLARED REED SNAKE *Calamaria leucogaster* Bleeker, 1860

(No vernacular names recorded)

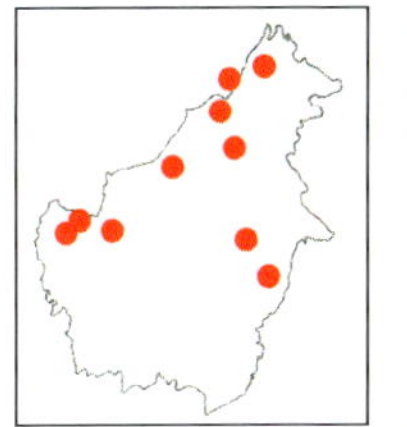

SIZE 223mm

IDENTIFICATION FEATURES Body slender, cylindrical; head short, slightly distinct from neck; loreal absent; single preocular; single postocular; infralabials contact anterior chin shields; mental does not contact anterior chin shield; eyes small with rounded pupils; tail tapers to a narrow point; dorsals smooth; midbody scale rows 13; ventrals 126–157; subcaudals 12–26, paired; cloacal scute entire.

COLOUR Top of body bright orange-red, through olive to brown, with dark longitudinal stripes; black collar, 3–5 scales wide; half ring around tail-base; belly cream.

HABITAT AND BEHAVIOUR Inhabits lowland and mid-hill forests. Terrestrial or fossorial, hiding under logs. Diet and reproductive biology unstudied.

DISTRIBUTION Borneo and Sumatra. Bornean records are from Bidi near Bau (now Deded Krian National Park), Lawas, Long Mujan, Lubok Antu, the Matang Range, Niah National Park, Gunung Santubong National Park, as well as Sungei Ja'ong, Sungei Mengiong and Sungei Tangap in Sarawak; Labuan; Danum Valley, Malutut, Gunung Kinabalu (Kiau) and Sungei Purulon in Sabah, and Long Petah and the upper reaches of Sungei Mahakam in Kalimantan.

IUCN THREAT STATUS Least Concern.

Low's Reed Snake *Calamaria lovii* Boulenger, 1887

(No vernacular names recorded)

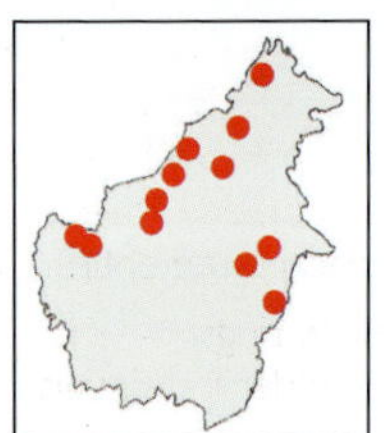

SIZE 320mm

IDENTIFICATION FEATURES Body slender and cylindrical; head short, indistinct from neck; eyes small with rounded pupils; tail short and thick, tapering abruptly; dorsals smooth; midbody scale rows 13; ventrals 205–256; subcaudals 11–26, paired; cloacal scute entire.

COLOUR Top of body dark brown with yellow spots or with narrow, light stripes; yellow ring around vent; forehead dark brown with indistinct light markings; belly cream anteriorly; posterior ventrals dark brown or yellow, with squarish black blotches.

HABITAT AND BEHAVIOUR Inhabits lowland hill forests and forested mid-hills (under 750m asl). Terrestrial, inhabiting the leaf litter. Diet and reproductive habits unstudied.

DISTRIBUTION Thailand, Vietnam, Peninsular Malaysia, Borneo, Sumatra and Java. On Borneo, known from Long Mujan, Niah National Park, as well as Sungei Pasang, Sungei Pesu, Sungei Baram, Sungei Rajang, Sungei Labang, Sungei Mengiong and Sungei Seran in Sarawak; Kota Kinabalu in Sabah, and Balikpapan and Sungei Merah in Kalimantan.

IUCN THREAT STATUS Least Concern.

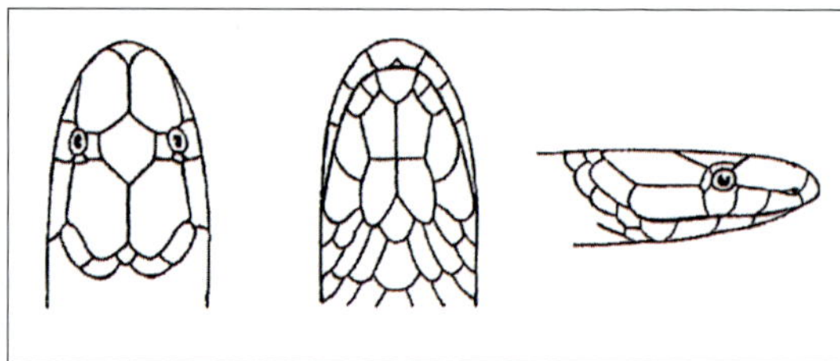

After Boulenger (1887).

Variable Reed Snake *Calamaria lumbricoidea* Boie, 1827

(No vernacular names recorded)

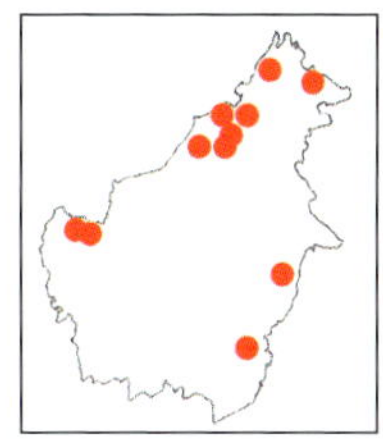

SIZE 640mm

IDENTIFICATION FEATURES Body quite robust and cylindrical; head short, indistinct from neck; eyes small with rounded pupils; tail short and thick, tapering abruptly to narrow point; dorsals smooth; midbody scale rows 13; ventrals 137–210; subcaudals 14–26, paired; cloacal scute entire.

COLOUR Top of body black, with narrow cream or yellow rings; forehead red or pink in juveniles, turning dark or black in older individuals; belly yellow with black ventral scales that form bands.

HABITAT AND BEHAVIOUR Inhabits lowland and submontane forests, as well as gardens (under 1,676m asl). Terrestrial, and encountered in leaf litter. Diet comprises earthworms and insect larvae. Reproductive habits unstudied.

DISTRIBUTION Thailand, Peninsular Malaysia, Singapore, Borneo, Sumatra, Pulau Nias, Mentawai Archipelago, Java, Mindanao, Basilan and Leyte. Bornean records are from Batu Song, Bidi near Bau (currently, Deded Krian National Park), Kuching, Lawas, the Matang Range, Niah National Park, Gunung Santubong National Park, Pa Brayong, Saribas, Sungei Pesu, Sungei Nyabau, Sungei Mengiong and the Baram Division in Sarawak; Sandakan, Tenompok Forest Reserve, the Gunung Kinabalu region (including Bundu Tuhan, Sungei Kenokok, Lumu-Lumu, Tenompok and Kiau) and Kota Kinabalu in Sabah, and Boven, Sungei Mahakam and Tanjung in Kalimantan.

IUCN THREAT STATUS Least Concern.

Lumholtz's Reed Snake *Calamaria lumholtzi* Andersson, 1923

(No vernacular names recorded)

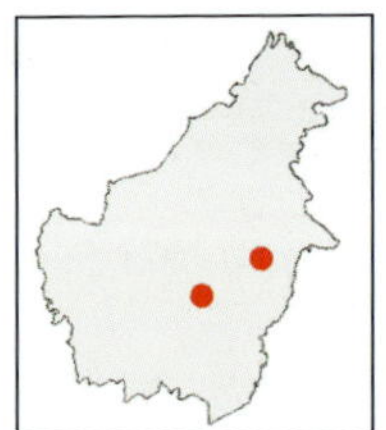

SIZE 210mm

IDENTIFICATION FEATURES Body slender and cylindrical; head short, indistinct from neck; loreal absent; single preocular and postocular; three infralabials contact anterior chin shields; eyes small with rounded pupils; tail short, tapering to blunt tip; dorsals smooth; midbody scale rows 13; ventrals 167–171; subcaudals 13–15, paired; cloacal scute entire.

COLOUR Top of body dark brown, with longitudinal stripe on scale rows 2–3, bordered at bottom by dark line; forehead brown with black spots; belly white, sometimes with dark speckling along anterior margins.

HABITAT AND BEHAVIOUR Inhabits lowland forests. Presumably terrestrial/ subfossorial. Diet and reproductive habits unstudied.

DISTRIBUTION Endemic to central Borneo. Records are from Sungei Merah and Tumbang Maruwai in Kalimantan.

IUCN THREAT STATUS Data Deficient.

Specimen of *Calamaria lumholtzi* Andersson, 1923 (ZRC 2.5322)from Pa'Umor, Kelabit Highlands, Sarawak.

Kapuas Reed Snake *Calamaria melanota* Jan, 1862

(No vernacular names recorded)

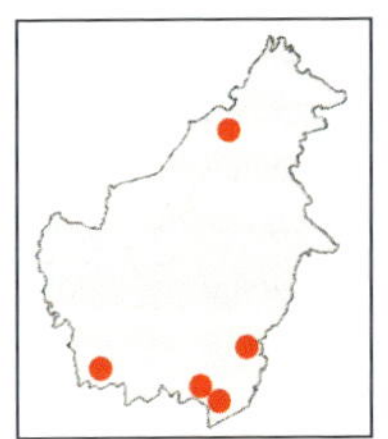

SIZE 260mm

IDENTIFICATION FEATURES Body slender and cylindrical; head short, indistinct from neck; loreal absent; single preocular and postocular; three infralabials contact anterior chin shields; eyes small with rounded pupils; tail short, tapering to a point; dorsals smooth; midbody scale rows 13; ventrals 121–154; subcaudals 16–26, paired; cloacal scute entire.

COLOUR Top of body dark brown, each dorsal with light-spotted or pale reticulated tip; forehead dark brown; supralabials dark, yellow centred; chin yellow with dark areas; belly and subcaudals have dark anterior and pale posterior.

HABITAT AND BEHAVIOUR Apparently inhabits lowland rainforests. Diet and reproductive habits unstudied.

DISTRIBUTION Endemic to southern and southeastern Borneo. Localities include Gunung Mulu in Sarawak, and Kuala Kapuas, Martapura, Sungei Pasir and Tanjung in Kalimantan.

IUCN THREAT STATUS Least Concern.

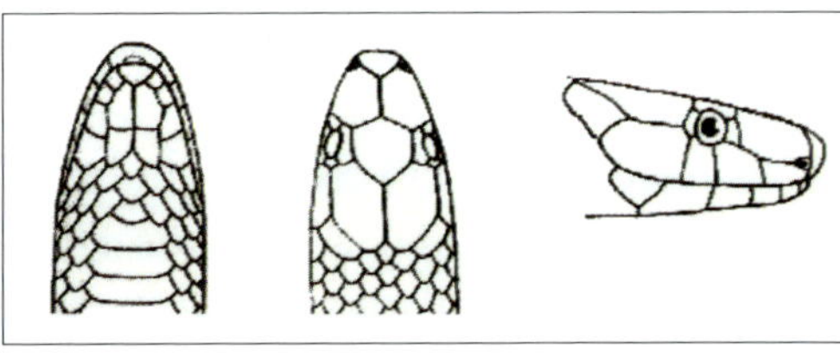

After Jan & Sordelli (1862).

Yellow-spotted Reed Snake *Calamaria modesta* Duméril, Bibron & Duméril, 1854

(No vernacular names recorded)

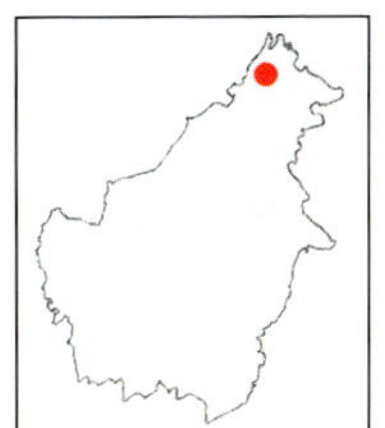

SIZE 460mm

IDENTIFICATION FEATURES Body slender and cylindrical; head short, indistinct from neck; loreal absent; single preocular and postocular; three infralabials contact anterior chin shields; eyes small with rounded pupils; tail short, tapering gradually to a point; dorsals smooth; midbody scale rows 13; ventrals 131–202; subcaudals 12–31, paired; cloacal scute entire.

COLOUR Top of body dark brown or black; dorsal scales unpatterned or with light network and dark longitudinal streak; yellow scattered spots; forehead black with yellow spots, or yellowish-brown with black spots; belly dark with pale areas medially and laterally.

HABITAT AND BEHAVIOUR Inhabits submontane forests at 1,370–1,430m asl. Diet and reproductive habits unstudied

DISTRIBUTION Sumatra, Borneo, Pulau Simeuleu and Java. Bornean localities include Gunung Kinabalu (at Bundu Tuhan and Tenompok) and Quoin Hill in Sabah.

IUCN THREAT STATUS Least Concern.

Specimen of *Calamaria modesta* Duméril et al., 1854 (ZRC 2.2674)from Gunung Kinabalu, Sabah.

PRAKKE'S REED SNAKE *Calamaria prakkei* van Lidth de Jeude, 1893

(No vernacular names recorded)

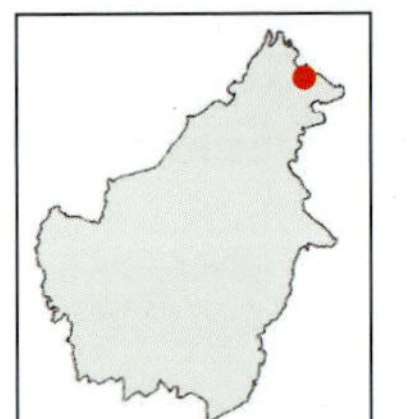

SIZE 260mm

IDENTIFICATION FEATURES Body slender and cylindrical; head short, indistinct from neck; loreal absent; single preocular and postocular; three infralabials contact anterior chin shields; eyes small with rounded pupils; tail long and thick, tapering to a sharp point; dorsals smooth; midbody scale rows 13; ventrals 126–144; subcaudals 24–32, paired; cloacal scute entire.

COLOUR Top of body brown; dorsal scales above first row with dark network; middorsal scales dark centred; yellow ventrolateral stripe along first dorsal scale row and along adjacent halves of first and second scale rows; pale brown nuchal collar; forehead brown with darker spots; belly yellow with brown lateral edges; subcaudals yellow.

HABITAT AND BEHAVIOUR Apparently inhabits lowland forests. Diet and reproductive habits unstudied.

DISTRIBUTION Borneo and Singapore. Sole Bornean record is from Sandakan Bay in Sabah.

IUCN THREAT STATUS Critically Endangered.

REBENTISCH'S REED SNAKE *Calamaria rebentischi* Bleeker, 1860

(No vernacular names recorded)

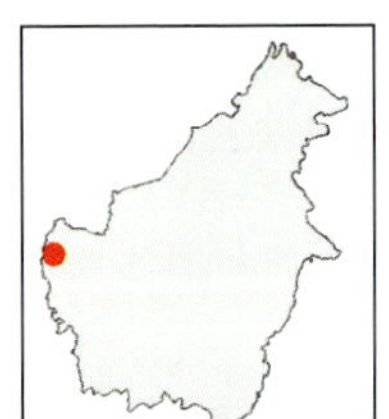

SIZE 270mm

IDENTIFICATION FEATURES Body slender and cylindrical; head short, indistinct from neck; loreal and preocular absent; single postocular; three infralabials contact anterior chin shields; eyes small with rounded pupils; tail short, tapering; dorsals smooth; midbody scale rows 13; ventrals 140; subcaudals 29, paired; cloacal scute entire.

COLOUR Top of body dark brown; dorsal scales yellowish-brown with dark network; forehead dark brown; supralabials yellow with brown spots; belly yellow with brown lateral edges; subcaudals yellow with small brown spots.

HABITAT AND BEHAVIOUR Apparently inhabits lowland forests. Diet and reproductive habits unstudied.

DISTRIBUTION Endemic to western Borneo. The sole locality known is Sinkawang in Kalimantan.

IUCN THREAT STATUS Data Deficient.

RED-HEADED REED SNAKE *Calamaria schlegeli* Duméril, Bibron & Duméril, 1854

(No vernacular names recorded)

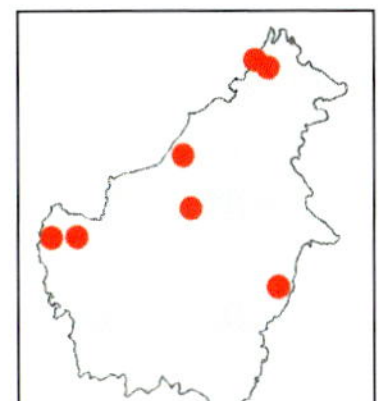

SIZE 450mm

IDENTIFICATION FEATURES Body slender and cylindrical; head short and indistinct from neck; nasal as large as eye; eyes small with rounded pupils; tail long and tapering; dorsals smooth; midbody scale rows 13; ventrals 130–180; subcaudals 19–44, paired; cloacal scute entire.

COLOUR Top of body dark brown or black; belly plain yellow, with red or orange forehead.

HABITAT AND BEHAVIOUR Inhabits lowland forests and low hills. Associated with leaf litter. Diet comprises frogs and slugs. Oviparous.

DISTRIBUTION Thailand, Peninsular Malaysia, Singapore, Borneo, Sumatra, Java and Bali. On Borneo, recorded from Anna Rais at the foothills of Gunung Penrissen, Ulu Baleh and Baleh National Park, Sungei Balui, Sungei Labang, Sungei Mengiong, Sungei Nyabau and Sungei Seran in Sarawak, Kota Kinabalu and Gunung Kinabalu in Sabah; and Singkawang and Saramanda in Kalimantan.

IUCN THREAT STATUS Least Concern.

Specimen of *Calamaria schlegeli* Duméril et al., 1854 (ZRC 2.7331) from Sri Aman, Sarawak.

SCHMIDT'S REED SNAKE *Calamaria schmidti* Marx & Inger, 1955

(No vernacular names recorded)

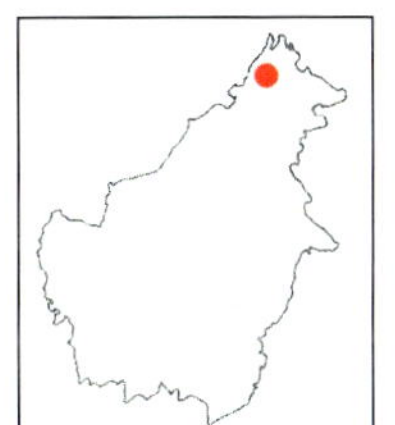

SIZE 280mm

IDENTIFICATION FEATURES Body slender and cylindrical; head short and indistinct from neck; eyes small with rounded pupils; tail short with blunt point; dorsals smooth; midbody scale rows 13; ventrals 127–150; subcaudals 14–22, paired; cloacal scute entire.

COLOUR Top of body blackish-grey, with green and blue iridescence, scales pale margined; belly light grey or yellow, darkening posteriorly to purple.

HABITAT AND BEHAVIOUR Restricted to montane forests at 1,370–1,570m asl. Presumably terrestrial and/or subfossorial, and associated with rocky hill streams. Diet includes earthworms. Reproductive habits unstudied.

DISTRIBUTION Endemic to Borneo. Known from Gunung Kinabalu (Silau-Silau Trail, Lumu-Lumu, Kamborangoh and Bundu Tuhan) in Sabah.

IUCN THREAT STATUS Least Concern.

Yellow-bellied Reed Snake *Calamaria suluensis* Taylor, 1922

(No vernacular names recorded)

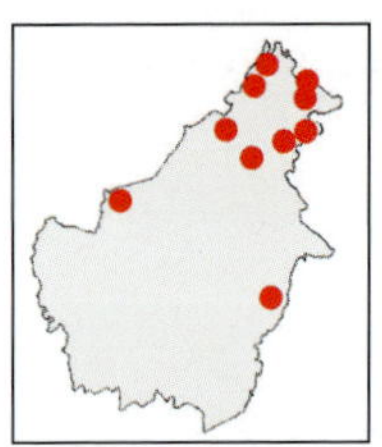

SIZE 300mm

IDENTIFICATION FEATURES Body slender and cylindrical; head short and indistinct from neck; eyes small with rounded pupils; tail short; dorsals smooth; midbody scale rows 13; ventrals 129–168; subcaudals 14–20, paired; cloacal scute entire.

COLOUR Top of body brown, each scale with dark network; some scattered scales with dark-centred spot; forehead dark brown, with scattered darker spots; belly yellow with dark spots; under surface of tail yellow, sometimes with dark median line.

HABITAT AND BEHAVIOUR Inhabits hill dipterocarp and submontane forests at 915–1,430m asl. Diet and reproductive biology unstudied.

DISTRIBUTION Borneo and the Sulu Archipelago of the Philippines. Bornean records are from Sarikei and Sungei Mengiong in Sarawak, and Gunung Kinabalu Park (including Bundu Tuhan, Kiau, Sayap and Lumu-Lumu), Bongon, Danum Valley, Deramakot, Sandakan and Tawau Hills Park in Sabah.

IUCN THREAT STATUS Least Concern.

Short-tailed Reed Snake *Calamaria virgulata* Boie, 1827

(No vernacular names recorded)

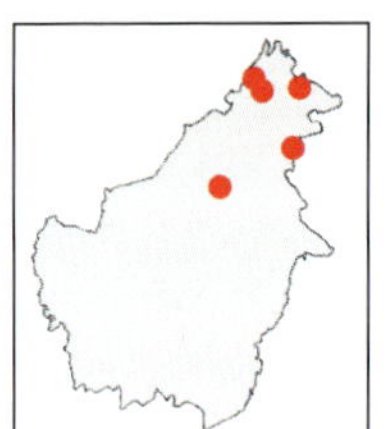

SIZE 370mm

IDENTIFICATION FEATURES Body slender, cylindrical; head short, indistinct from neck; eyes small with rounded pupils; tail thick and tapering; dorsals smooth; midbody scale rows 13; ventrals 160–260; subcaudals 8–30, paired; cloacal scute entire.

COLOUR Top of body dark brown; dorsal scales have light network, with or without dark longitudinal stripes; yellow nuchal collar sometimes present; forehead dark brown; supralabials yellow; belly cream or brownish-black, with dark pigments on lateral edges of ventrals; subcaudals dark edged, and dark median stripe.

HABITAT AND BEHAVIOUR Restricted to submontane forests. Terrestrial and found in leaf litter. Diet remains unknown. Oviparous, producing three eggs, each 26–30 × 8–8.5mm.

DISTRIBUTION Borneo, Sumatra, Java, the Sulu Archipelago, Mindanao, Palawan and Sulawesi. Bornean localities include Bukit Tiban in Sarawak; Sapagaya Protection Forest Reserve and Gunung Kinabalu (including Bundu Tuhan, Kenokok, Sungei Luidan, Sayap and Mesilau) in Sabah, and Pulau Nunukan in Kalimantan.

IUCN THREAT STATUS Least Concern.

Chrysopelea Flying Snakes

Small to medium-sized arboreal snakes; body slender; head slightly depressed and distinct from neck; eyes large with rounded pupils; midbody scale rows 17; scales smooth or with weak keels; snout rounded in front; ventrals have a lateral keel; posterior 3–4 teeth of maxilla enlarged and grooved; tail long.

Key to Bornean species of *Chrysopelea*

1a. Middorsum either with longitudinal row of red or orange spots covering four scales, or pale green or olive spots; belly yellowish-green .... *C. paradisi* (below)

1b. Middorsum red or orange, edged by narrow black bands that are separated by a cream or yellow band; belly grey .... *C. pelias* (p. 230)

GARDEN FLYING SNAKE *Chrysopelea paradisi* H. Boie, in F. Boie, 1827

(Bahasa Malaysia: Ular Petola. Bahasa Indonesia: Ular Pohon Paradise)

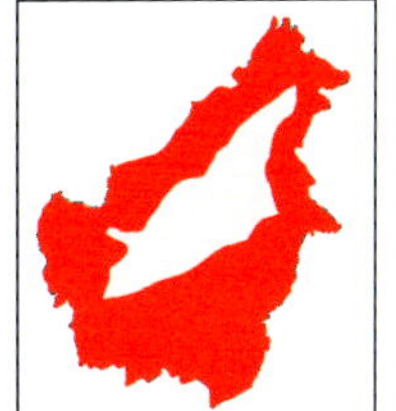

SIZE 1.5m

IDENTIFICATION FEATURES Body slender; head depressed, distinct from neck; eyes large with rounded pupils; tail long and slender; ventrals have pronounced lateral keels; dorsals smooth or weakly keeled; midbody scale rows 17; ventrals 198–239; subcaudals 106–149, paired; cloacal scute divided.

COLOUR Top of body black, scale centre has green spot; vertebral region sometimes has row of 3–4 pink or red spots; forehead yellow banded; belly green with black edges

HABITAT AND BEHAVIOUR Inhabits lowland forests including park and gardens, ascending to submontane forests at < 1,524m asl. Arboreal, and found in trees and shrubs. Famous for its extended leaps. Diet includes lizards, and perhaps bats and small birds. Oviparous, producing 5–8 eggs.

DISTRIBUTION Andaman Islands, Myanmar, Thailand, Peninsular Malaysia, Singapore, Borneo, Sumatra, Mentawai Archipelago, Java, Sulawesi and the Philippines archipelago. Bornean records are from foothills of Bako National Park, Lambir Hills National Park, Gunung Singgai, Loagan Bunut National Park, Kubah National Park, Sama Jaya Nature Reserve, Oya, Niah National Park, Kampung Braang Payang, Gunung Dulit, Kampung Sebako at foothills of Gunung Pueh, Gunung Santubong National Park, Sungei Baram, Sungei Ja'ong, Sungei Lupar, Sungei Rajang and Sungei Tubau in Sarawak; Tasek Merimbun in Brunei Darussalam; Kota Kinabalu, Penampang, Pulau Gaya, Pulau Sebatik, Pulau Balambangan, Ulu Dusun, Gunung Kinabalu (Poring), and Quoin Hill in Sabah, and Labuan.

IUCN THREAT STATUS Least Concern.

TWIN-BARRED FLYING SNAKE *Chrysopelea pelias* (Linnaeus, 1758)

(Bahasa Malaysia: Ular Pokok Belang)

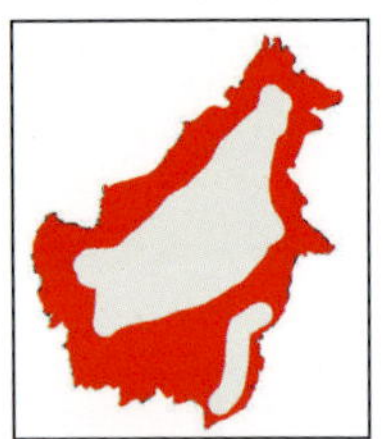

SIZE 740mm

IDENTIFICATION FEATURES Body slender; head depressed and distinct from neck; eyes large with rounded pupils; tail long and slender; ventrals have pronounced lateral keels; dorsals smooth or weakly keeled, with apical pits; midbody scale rows 17; ventrals 181–201; subcaudals 89–120, paired; cloacal scute divided.

COLOUR Top of body red or orange with yellow or cream cross-bars, edged with black bars; forehead has three red cross-bars; belly pale.

HABITAT AND BEHAVIOUR Inhabits relatively open forests, parks, gardens and plantations; sometimes enters human habitation around forest edges at < 600m asl. Arboreal; can make extended leaps. Diet comprises lizards. Breeding habits unstudied.

DISTRIBUTION Myanmar, Thailand, Peninsular Malaysia, Singapore, Borneo, Sumatra, Pulau Bangka, Mentawai, Natuna and Riau archipelagos and Java. Bornean records are from Bako National Park, Kuching, Gunung Gading National Park, the Matang Range, Gunung Mulu National Park and Nyabau Forest Reserve in Sarawak; Ulu Temburong in Brunei Darussalam, and Gunung Kinabalu (Poring) and Sepagaya Protection Forest Reserve in Sabah.

IUCN THREAT STATUS Least Concern.

Coelognathus South-east Asian Rat Snakes

Medium-sized to large terrestrial snakes (formerly allocated to the genus Elaphe), and identified through the possession of a distinct head and neck; large eyes with rounded pupils; three supralabials in contact with orbit; anterior subocular absent; paravertebral reductions of dorsal scale rows; uniform teeth, showing no enlargement; scales smooth or keeled; hemipenis with basal spines; tracheal lung absent or rudimentary; cloacal scute single; tail short.

Key to Bornean species of *Coelognathus*

1a. Dark postocular stripe encircles neck.......... *C. radiatus* (p. 233)
1b. Dark postocular stripe does not encircle neck.......... **2**

2a. Dorsum with pale vertebral stripe; juveniles lack pale bands.......... *C. flavolineatus* (p. 232)
2b. Dorsum without pale vertebral stripe; juveniles with pale bands.......... *C. erythrurus* (below)

PHILIPPINE RAT SNAKE *Coelognathus erythrurus* (Duméril, Bibron & Duméril, 1854)

(No vernacular names recorded)

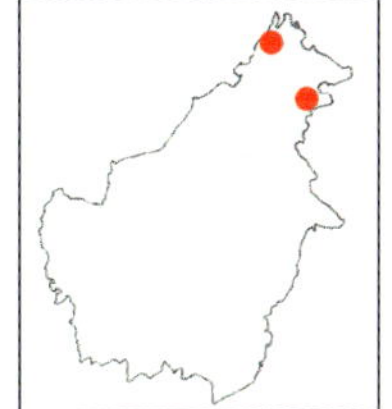

SIZE 1.67m

IDENTIFICATION FEATURES Body slender; head elongated and distinct from neck; eyes large with rounded pupils; tail long and slender; dorsals keeled; midbody scale rows 19 or 21; ventrals 219–238; subcaudals 102–114 (for subspecies *C. e. philippinus* that occurs in northeastern Borneo).

COLOUR Top of body brown to olive; narrow, dark postocular stripe to angle of jaws; posterior third of body darker; tail reddish-brown; belly plain cream or white.

HABITAT AND BEHAVIOUR Inhabits lowlands to mid-hills, including forest edges and cultivated areas at < 850m asl. Terrestrial and diurnal. Diet comprises rodents, birds and lizards. Oviparous, with clutches including 6–10 eggs.

DISTRIBUTION Borneo, Palawan, Sulu Archipelago, Sulawesi, Mindanao, Samar, Leyte, Negros, Mindoro, Luzon and other Philippines islands. Bornean records are from Kiulu, Tawau Hills Park, and a general one without a precise locality from Gunung Kinabalu in Sabah.

IUCN THREAT STATUS Least Concern.

Yellow-striped Rat Snake *Coelognathus flavolineatus* (Schlegel, 1837)

(Bahasa Malaysia: Ular Laju Ekor Hitam, Ular Sawa. Bahasa Indonesia: Sawa Angin; Ular Babi)

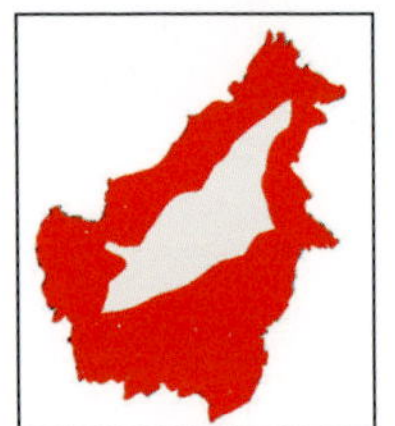

SIZE 1.8m

IDENTIFICATION FEATURES Body slender; head nearly distinct from neck; snout long; eyes large with rounded pupils; tail long and slender; dorsals keeled; midbody scales 19; ventrals 193–242; subcaudals 80–116, paired; cloacal scute entire.

COLOUR Top of body brownish-grey or brownish-olive; dark postocular stripe to above back of mouth and another along nape; several short, dark stripes or elongated blotches on dorsum and flanks; belly pale yellow anteriorly, some ventrals dark grey edged; tail darker posteriorly; subcaudals dark grey or black.

HABITAT AND BEHAVIOUR Inhabits lowland forest edges, as well as parks and gardens at < 900m asl. Terrestrial but can climb low trees. Diet comprises rodents, birds, frogs and lizards. Oviparous, producing 5–12 eggs, each 51–62 × 23–25.5mm.

Close-up of head

DISTRIBUTION Andaman Islands, Myanmar, Thailand, Cambodia, Vietnam, Peninsular Malaysia, Singapore, Borneo, Sumatra, Pulau Nias, Mentawai and Riau archipelagos, Pulau Weh, Pulau Bangka, Pulau Belitung, Java and Sulawesi. Widespread on Borneo, and likely to be found across the island. Known localities include Bako National Park, Kuching, Kota Samarahan, the Kelabit Highlands, Niah National Park, Gunung Mulu National Park, Gunung Santubong National Park, Loagan Bunut National Park, Sarikei, Anna Rais at foothills of Gunung Penrissen, Sungei Baram, Sungei Tangap and Sungei Trusan in Sarawak; Bandar Seri Begawan and Ulu Temburong in Brunei Darussalam; Kota Kinabalu, including Bukit Padang, Bongan, Kalabakan, Kasigui, Kanibongan, Sandakan, Sandakan Bay, Sepilok, Tawau Hills Park and Gunung Kinabalu (localities including Bundu Tuhan, Kiau and Ranau) in Sabah, and Balikpapan, Buntok, Kutai, Long Iram, Singkawang, Sungei Bulangan, Sungei Mahakam and Sungei Duri in Kalimantan.

IUCN THREAT STATUS Least Concern.

COPPERHEADED RACER *Coelognathus radiatus* (Boie, 1827)

(Bahasa Malaysia: Ular Rusuk Kerbau. Bahasa Indonesia: Ular Racer Berkepala Tembaga. Mandarin Chinese: Sangianjing She)

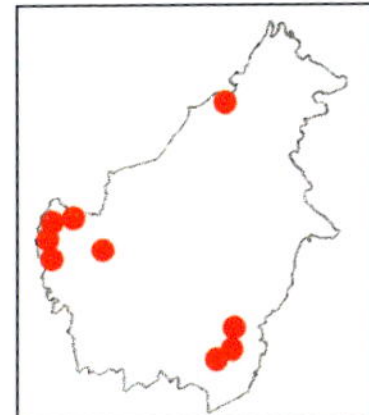

SIZE 2.3m

IDENTIFICATION FEATURES Body slender; head slightly distinct from neck; eyes large with rounded pupils; tail long and slender; dorsals smooth anteriorly and on flanks, weakly keeled posteriorly; midbody scale rows 19; ventrals 207–250; subcaudals 80–108, paired; cloacal scute entire.

COLOUR Top of body greyish-brown or yellowish-brown, with four (comprising two broad and two narrow) black stripes along anterior; cream stripe runs along upper stripes; lower stripes narrower and typically broken up; forehead coppery-brown, with three black lines radiating from eyes; belly grey to yellowish-grey; iris golden-yellow; tongue dark brown to violet.

HABITAT AND BEHAVIOUR Inhabits lowland forests as well as grassland, from lowlands to submontane limits at < 1,400m asl. Diet comprises frogs, birds and rats. Oviparous, with clutches comprising 5–23 eggs, each 40–53 × 20–26mm.

DISTRIBUTION Northern and eastern India, Nepal, Bangladesh, southern and eastern China, Myanmar, Thailand, Cambodia, Laos, Vietnam, the Malay Peninsula, Borneo, Sumatra and Java. Bornean records are from unspecified localities in Brunei Darussalam, the Bau Caves of western Sarawak, and western and southern Kalimantan.

IUCN THREAT STATUS Least Concern.

Dendrelaphis Bronzeback Tree Snakes

Medium-sized arboreal snakes, characterized by slender body; head distinct from neck; large eyes; rounded pupils; smooth body scales; enlarged vertebral series (except in *D. caudolineatus*); keeled ventral scales; long tail.

Key to Bornean species of *Dendrelaphis*

1a. Midbody scale rows 13 .... *D. caudolineatus* (opposite)
1b. Midbody scale rows 25 .... **2**

2a. Light ventrolateral stripe present **3**
2b. Light ventrolateral stripe absent **4**

3a. Postocular stripe covers > 50 per cent of temporal region .... *D. pictus* (p. 239)
3b. Postocular stripe covers < 50 per cent of temporal region .... *D. haasi* (p. 237)

4a. Postocular stripe covers lower temporal region and does not enter neck .... *D. kopsteini* (p. 238)
4b. Postocular stripe covers most of temporal region and enters neck **5**

5a. Ventrals < 170 .... *D. striatus* (p. 240)
5b. Ventrals > 170 .... *D. formosus* (p. 236)

Juvenile Stripe-tailed Bronzeback Tree Snake (opposite)

STRIPE-TAILED BRONZEBACK TREE SNAKE *Dendrelaphis caudolineatus* (Gray, 1834)

(Bahasa Malaysia/Indonesia: Ular Padang. Iban: Ular Bendira)

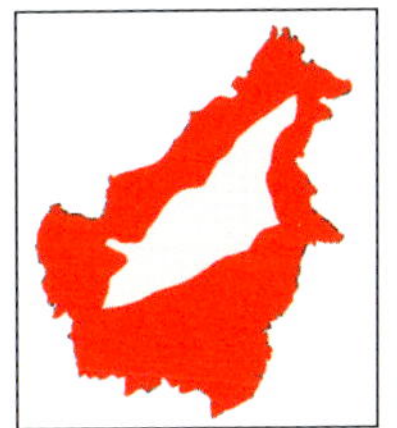

SIZE 1.52m

IDENTIFICATION FEATURES Body slender; head distinct from neck; snout bluntly rounded; tail long; eyes large with rounded pupils; dorsals smooth; ventrals and subcaudals with sharp keels on outer edges; midbody scale rows 13; ventrals 174–189; subcaudals 105–116, paired; cloacal scute divided.

COLOUR Top of body olive-brown; pale green stripe along lower flanks of body, edged dorsally by narrow black stripe and ventrally by broad black stripe; stripes most distinct on tail; belly pale green.

HABITAT AND BEHAVIOUR Inhabits lowland and submontane forests, in addition to more open habitats such as parks and gardens at < 1,524m asl. Arboreal. Diet includes frogs and lizards. Oviparous, with clutches comprising 5–8 eggs, each 12 × 48mm.

DISTRIBUTION Myanmar, Thailand, Laos, Peninsular Malaysia, Singapore, Borneo, Sumatra, Pulau Babi, Pulau Bangka, Batu, Natuna, Mentawai and Riau archipelagos, Pulau Belitung, Pulau Nias, Palawan, Luzon, Camiguin and Sulu archipelagos. On Borneo, common around Kuching, Sarawak and Kota Kinabalu in Sabah. Other records from the island include Bako National Park, Tanjung Datu National Park, Lawas, Limbang, Oya, Simanggang, Gunung Gading National Park, Kubah National Park, Niah National Park, Loagan Bunut National Park, Gunung Santubong National Park, Sungei Baram, Sungei Rejang, Sungei Labang, Sungei Mengiong, Sungei Nyabau, Sungei Pesu, Sungei Seran and Sungei Tibas in Sarawak; Jalan Muara in Bandar Seri Begawan and Ulu Temburong in Brunei Darussalam; Danum Valley, Kalabakan, Dewhurst Bay, Lahad Datu, Ranau, Sandakan, Tawau Hills Park, Gunung Kinabalu Park and Sungei Brantian in Sabah, and Balikpapan, Buntok, Muara Jawa, Pontianak and Sungei Kapuas in Kalimantan.

IUCN THREAT STATUS Least Concern.

Beautiful Bronzeback Tree Snake *Dendrelaphis formosus* (Boie, 1827)

(No vernacular names recorded)

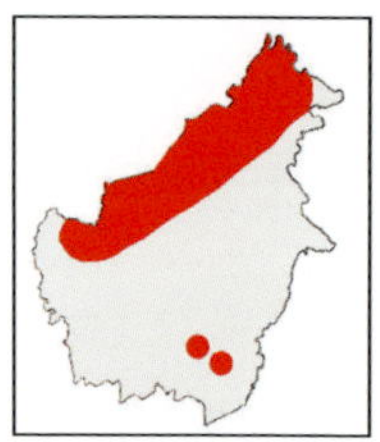

SIZE 1.4m

IDENTIFICATION FEATURES Body slender; head wide behind snout and distinct from neck; eyes large with rounded pupils; vertebrals enlarged; tail slender; midbody scale rows 15; ventrals 176–194; subcaudals 142–162, paired; cloacal scute divided.

COLOUR Top of body reddish-brown with blue interscale areas; black postocular stripe between rostral, across temporal region and on to neck; three dark lateral stripes on posterior third of body; dorsals dark edged; no light ventrolateral stripe; subcaudals have black medial point.

HABITAT AND BEHAVIOUR Inhabits primary lowland forests and open secondary forests. Arboreal. Diet comprises frogs and lizards. Oviparous, clutches comprising 6–8 eggs, each 31–42.5 × 11.5– 13mm.

DISTRIBUTION Thailand, Peninsular Malaysia, Borneo, Sumatra, Mentawai Archipelago, Pulau Bangka, Pulau Belitung and Java. Isolated Bornean records from Kuching, Bako National Park, Baleh National Park, Gunung Santubong National Park, Lambir Hills National Park, Sungei Baram, Sungei Mengiong, Sungei Pesu, Sungei Seran and Sungei Tangap in Sarawak; Sandakan Bay, Danum Valley and Dewhurst Bay in Sabah, and Tanjung and Sungei Kapuas in Kalimantan.

IUCN THREAT STATUS Least Concern.

Haas's Bronzeback Tree Snake *Dendrelaphis haasi* van Rooijen & Vogel, 2008

(No vernacular names recorded)

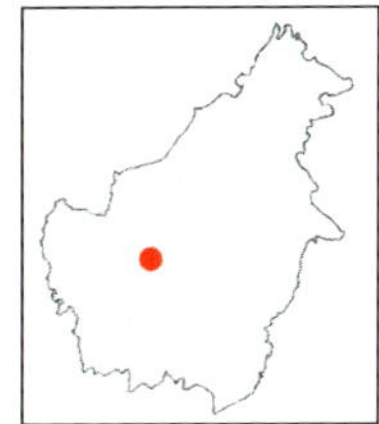

SIZE 950mm

IDENTIFICATION FEATURES Body slender; head distinct from neck; eyes small with rounded pupils; tail long; vertebrals greatly enlarged, larger than first row of middorsals; dorsals smooth; midbody scale rows 15; ventrals 161–173; subcaudals 126–153, divided; cloacal scute divided.

COLOUR Top of body olive-brown; narrow postocular stripe covers ventral part of lower temporals and ends at back of jaws or extends on to neck, where it is broken; oblique black bars on sides of neck; pale cream ventrolateral stripe not bordered by black lines; belly light yellow or green.

HABITAT AND BEHAVIOUR Inhabits lowland forests. Diet and reproductive habits unstudied.

DISTRIBUTION Peninsular Malaysia, Singapore, Borneo, Sumatra, Pulau Nias, Pulau Belitung, Mentawai Archipelago and Java. On Borneo, has been recorded in Sebruang in Kalimantan, as well as in an unspecified east-central portion of the island.

IUCN THREAT STATUS Least Concern.

Kopstein's Bronzeback Tree Snake *Dendrelaphis kopsteini*

Vogel & van Rooijen, 2007

(Bahasa Brunei: Ular Kasau)

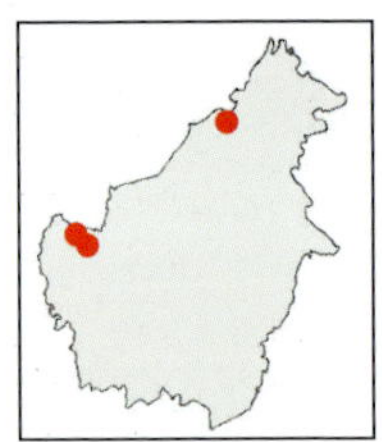

SIZE 1.42m

IDENTIFICATION FEATURES Body slender; head distinct from neck; 2 supralabials touch orbit of eye; eyes moderate with rounded pupils; vertebrals larger than lowest dorsal row; midbody scale rows 15; ventrals 167–181; subcaudals 140–154, paired; cloacal scute divided.

COLOUR Top of body bronze-brown; black postocular stripe runs across lower half of temporal region to end of jaws; vertebrals have broad black posterior margin; interstitial skin on anterior of body brick-red; belly grey.

HABITAT AND BEHAVIOUR Inhabits lowland dipterocarp forests at < 700m asl. Arboreal, living in shrubs and on low tree branches. Diet includes geckos. Oviparous, with clutches comprising eight eggs.

DISTRIBUTION Thailand, Peninsular Malaysia, Singapore, Borneo, Sumatra and Mentawai Archipelago. Isolated Bornean records in Bako National Park, Bau and Kuching in Sarawak, and Sungei Rampayoh in Brunei Darussalam.

IUCN THREAT STATUS Least Concern.

PAINTED BRONZEBACK TREE SNAKE *Dendrelaphis pictus* (Gmelin, 1789)

(Bahasa Malaysia: Ular Lidi. Bahasa Indonesia: Ular Tampar Jawa; Ular Tali. Iban: Ular Meresian)

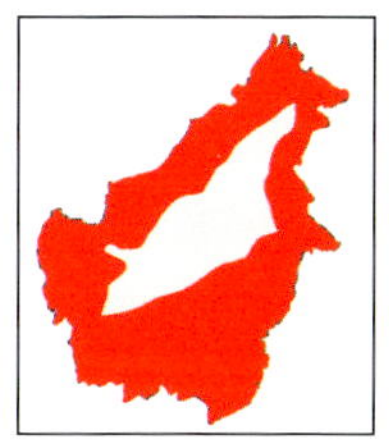

SIZE 1.25m

IDENTIFICATION FEATURES Body slender; head distinct from neck; eyes large with rounded pupils; vertebrals smaller than or equal to first row of dorsals; dorsals smooth; midbody scales 15; ventrals 167–200; subcaudals 109–169, paired; cloacal scute divided.

COLOUR Top of body bronze-brown or brownish-olive; black-edged yellow or cream ventrolateral stripe along flanks; forehead brown with black postocular stripe covering over half of temporal region and extending to neck; blue or greenish-blue neck patch used in threat displays.

HABITAT AND BEHAVIOUR Associated with lowland and submontane forests; often encountered in parks, gardens, plantations and human habitation at < 1,524m asl. Arboreal; associated with trees and shrubs. Diet comprises frogs and lizards. Oviparous, producing 3–8 eggs, each 22–38.5 × 8.5–11mm.

DISTRIBUTION Eastern China, Myanmar, Thailand, Laos, Cambodia, Vietnam, Peninsular Malaysia, Singapore, Borneo, Sumatra, Pulau Belitung, Mentawai Archipelago, Java, Bali and the Philippines. Perhaps the most common Bornean snake, with records from across the island: Kuching, Kubah National Park, Gunung Dulit, Niah National Park, Gunung Santubong National Park, Gunung Gading National Park, Lambir Hills National Park, Sungei Baram, Sungei Ja'ong, Sungei Mengiong, Sungei Seran, Sungei Subis, Sungei Tangap in Sarawak; Labuan; Bandar Seri Begawan in Brunei Darussalam; Kota Kinabalu, Petagas, Gunung Kinabalu (Lumu-Lumu and Ranau) and Sandakan in Sabah, and Balikpapan, Banjarmasin, Kutai, Long Iram, Montrado in Bengkayang, Sinkawang, Sintang, Tanjung, and Sungei Bulangan and Sungei Kapuas in Kalimantan.

IUCN THREAT STATUS Least Concern.

Striated Bronzeback Tree Snake *Dendrelaphis striatus* (Cohn, 1905)

(No vernacular names recorded)

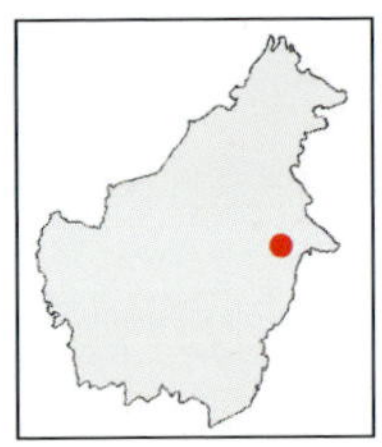

SIZE 1.02mm

IDENTIFICATION FEATURES Body slender; head distinct from neck; snout short and rounded; eyes moderate with rounded pupils; vertebrals slightly enlarged; dorsals smooth; midbody scale rows 15; ventrals 152–163; subcaudals 103–142, paired; cloacal scute divided.

COLOUR Top of body bronze-brown; labials and gular region yellow; narrow, dark stripe between nostril to orbit; dark postocular stripe across temporal region and on to neck; neck yellow when inflated; black oblique bars on body flanks.

HABITAT AND BEHAVIOUR Inhabits lowland forests. Arboreal, associated with trees and shrubs. Diet and reproductive habits unstudied.

DISTRIBUTION Thailand, Peninsular Malaysia, Borneo and Sumatra. The sole Bornean record is from Kutai in Kalimantan.

IUCN THREAT STATUS Least Concern.

Dryophiops Whip Snakes

Slender, medium-sized arboreal snakes; body elongated; head distinct from neck; snout longer than eye diameter; distinct groove along snout-tip in front of eyes; eyes moderate with horizontal pupils; midbody scale rows 15; scales smooth; tail long, slender and about a third of total body length.

KEEL-BELLIED WHIP SNAKE *Dryophiops rubescens* (Gray, 1835)

(No vernacular names recorded)

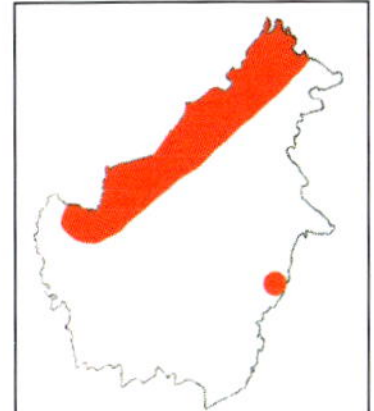

SIZE 750mm

IDENTIFICATION FEATURES Body slender and compressed; head distinct from neck; eyes large with horizontal pupils; tail long and slender; dorsals smooth; midbody scale rows 15; ventrals 186–199; subcaudals 111–136, paired; cloacal scute divided.

COLOUR Top of body reddish-brown with small black spots; forehead dark, streaked with dark postocular streak; labials dark spotted; belly yellow or olive.

HABITAT AND BEHAVIOUR Inhabits lowland forests and forest edges (under 300m asl). Arboreal, on shrubs and low branches of trees. Diet comprises small vertebrates such as lizards and probably small birds. Oviparous, with clutches comprising 2–3 eggs.

DISTRIBUTION Thailand, Cambodia, Peninsular Malaysia, Singapore, Borneo, Sumatra, Mentawai and Natuna archipelagos and Java. Isolated Bornean records from Bako National Park, Gunung Mulu National Park, Gunung Santubong National Park and the base of Gunung Pueh in Sarawak; Bandar Seri Begawan in Brunei Darussalam; Kota Kinabalu, Ulu Dusun, Sandakan and Sepilok in Sabah, and Samarinda in Kalimantan.

IUCN THREAT STATUS Least Concern.

Gongylosoma Ground Snakes

Small, stout terrestrial/subfossorial snakes; eyes large with rounded pupils; midbody scale rows 13; scales smooth; nasal divided; tail short; colouration of anterior portion of body red-brown, fading posteriorly to brown-grey.

Key to Bornean species of *Gongylosoma*

1a. Anterior of dorsum yellow spotted or striped; subcaudals 58–79 .... *G. baliodeirum* (below)
1b. Anterior of dorsum has five orange stripes; subcaudals 77–103 .... *G. longicauda* (opposite)

Orange-bellied Snake *Gongylosoma baliodeirum* Boie, 1827

(Bahasa Indonesia: Ular Tanah Bertotol-totol. Kelabit: Depong)

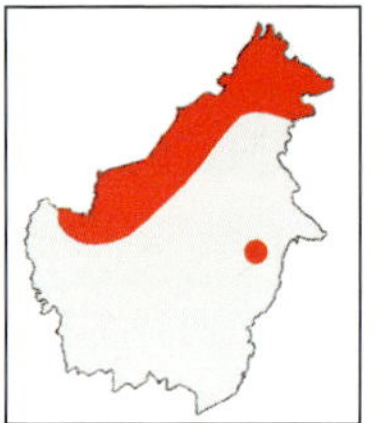

SIZE 450mm

IDENTIFICATION FEATURES Body slender and cylindrical; head slightly wider than neck; nasals divided; eyes small with rounded pupils; dorsals smooth; midbody scale rows 13; ventrals 115–141; subcaudals 42–75; cloacal scute divided.

COLOUR Top of body dark brown to reddish-brown, with paired rows of cream spots; upper labials edged with dark grey; belly yellowish-cream, sometimes with fine dark spots.

HABITAT AND BEHAVIOUR Associated with lowland and submontane forests at up to 1,525m asl. Terrestrial and lives in leaf litter, hiding under fallen tree trunks and other plant debris. Diet comprises spiders, other arthropods and lizards. Oviparous, producing 2–3 eggs, each 22–24 × 7.5mm.

DISTRIBUTION Thailand, Peninsular Malaysia, Singapore, Borneo, Sumatra, Pulau Nias, Natuna Archipelago and Java. Rarely encountered species on Borneo, with records from the Kelabit Highlands, the Bau region, Lundu, Niah National Park, Gunung Penrissen, Sungei Baram, Sungei Mengiong, Sungei Pesu, Sungei Segaham, Sungei Seran and Sungei Trusan in Sarawak; Ulu Temburong in Brunei Darussalam; Danum Valley, Maliau Basin, Mendolonh, Sandakan, Gunung Kinabalu (Kiau, Poring and Marak-Parak), and Sungei Malutut in Sabah, and Kutai in Kalimantan.

IUCN THREAT STATUS Least Concern.

STRIPED GROUND SNAKE *Gongylosoma longicauda* (Peters, 1871)

(No vernacular names recorded)

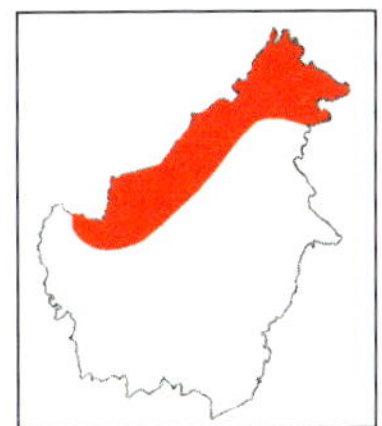

SIZE 500mm

IDENTIFICATION FEATURES Body slender; head wider than neck; loreal present; preoculars 1–2; postoculars two; supralabials eight; supralabials III–V contact orbit; four infralabials contact anterior chin shields; eyes large; tail long and slender; dorsals smooth; midbody scale rows 13; ventrals 110–138; subcaudals 71–105; cloacal scute divided.

COLOUR Top of body brownish-red or black, with yellow or cream chevron at back of head and five orange or yellow stripes on dorsum, most distinct anteriorly; belly unpatterned red or cream, or each ventral with brown spot on outer edge.

HABITAT AND BEHAVIOUR Inhabits lowland rainforests, and occasionally encountered near human habitation. Diet includes spiders and lizards. Reproductive habits unstudied.

DISTRIBUTION Peninsular Malaysia, Borneo, Sumatra and Java. Bornean records are from Bidi (currently, Deded Krian National Park), Busau, the Matang Range, Batu Song, Sungei Mengiong and Sungei Tangap in Sarawak, and Danum Valley, Ulu Dusun, Kalabakan, Sungei Malutut and Sungei Purulon in Sabah.

IUCN THREAT STATUS Least Concern.

Gonyosoma Racers

Large diurnal snakes; body slender, laterally compressed; head elongated, distinct from neck; eyes large with rounded pupils; midbody scale rows 19, smooth; ventrals have lateral keels; tail long.

Key to Bornean species of *Gonyosoma*

1a. Ventrals and subcaudals rounded on sides; midbody scale rows 23, 25 or 27 .... *G. oxycephalum* (opposite)
1b. Ventrals and subcaudals squarish on sides; midbody scale rows 19 .... *G. margaritatum* (below)

ROYAL TREE SNAKE *Gonyosoma margaritatum* Peters, 1871

(No vernacular names recorded)

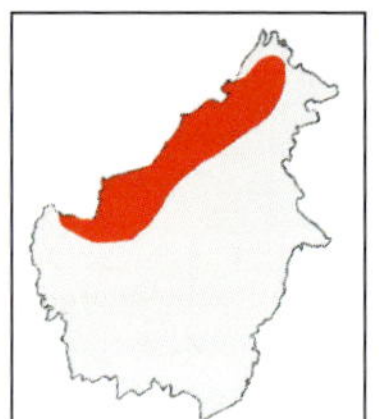

SIZE 2m

IDENTIFICATION FEATURES Body robust, elongated, compressed; head long, distinct from neck; snout elongated, squarish; loreal present; single preocular; postoculars two; supralabials 8–9; supralabials IV–VI or III–V contact orbit; infralabials 12; eye large with rounded pupils; tail long, tapering; dorsals smooth; midbody scale rows 19; ventrals 230–249; subcaudals 108–130, paired; cloacal scute divided.

COLOUR Top of body bright green, each scale edged with black; 5–11 orangish-red bands on posterior of body and tail; black postocular stripe; forehead has black streaks; venter yellowish-pink.

HABITAT AND BEHAVIOUR Inhabits lowland forests at up to 700m asl. Diurnal and arboreal; known from forest canopies. Diet and reproductive habits unstudied.

DISTRIBUTION Peninsular Malaysia, Singapore and Borneo. Bornean records are from Baleh National Park, Sungei Mengiong, Gunung Dulit, Kubah National Park, Gunung Mulu National Park, Gunung Merinjak, Gunung Singgai near Bau in Sarawak (with a historical record from Kuching), and Gunung Kimabalu (Poring), Sungei Purulon, Ranau and Long Pasia in Sabah.

IUCN THREAT STATUS Least Concern.

RED-TAILED RACER *Gonyosoma oxycephalum* (Boie, 1827)

(Bahasa Malaysia: Ular Laju Ekor Merah, Ular Pucuk, Ular Selenseng. Bahasa Indonesia: Ular Bangka Laut, Ular Hidjau, Ular Racer Berekor Merah. Iban: Ular Mati Iko, Ular Mati Elok)

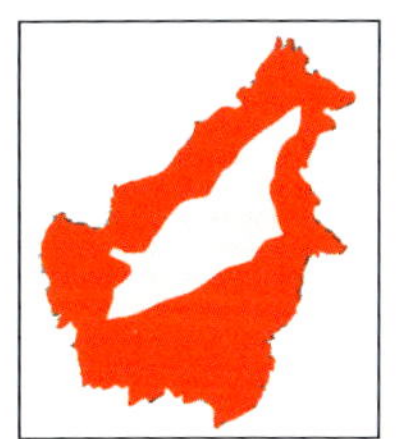

SIZE 2.4m

IDENTIFICATION FEATURES Body slender, elongated and compressed; head distinct from neck; snout elongated and squarish; eyes large with rounded pupils; tail long and tapering; dorsals smooth or weakly keeled; midbody scale rows 23, 25 or 27; ventrals 229–263; subcaudals 120–157, paired; cloacal scute divided.

COLOUR Top of body of adults emerald-green or pale green, with pale green throat and black stripe from nostril to above level of upper jaw; tail-tip yellowish-brown or reddish-orange; belly yellow; tongue bluish-black. Juveniles olive-brown with narrow white bars towards posterior.

HABITAT AND BEHAVIOUR Common in lowland forests, including dipterocarp, coastal and mangrove forests, as well as plantations and gardens at < 750m asl. Arboreal, on trees but with some terrestrial activity. Diet comprises rodents and birds. Oviparous, laying 5–12 eggs, each 65mm diameter.

DISTRIBUTION Thailand, Laos, Cambodia, Vietnam, Peninsular Malaysia, Singapore, Borneo, Sumatra, Pulau Nias, Pulau Bangka, Pulau Belitung, Riau, Mentawai and Natuna archipelagos, Java, Andaman Islands, Lombok, Sulawesi, Balabac, Bohol, Bongao, Dinagat, Lubang, Luzon, Mindoro, Negros and Palawan. Widespread on Borneo; the best place to see it is Gunung Gading National Park in Sarawak. Other localities include Bako National Park, Kampung Braang Payang, Tanjung Datu National Park, Niah National Park, Gunung Santubong National Park, Samunsam Wildlife Sanctuary, Pangkalan Lobang, Sungei Baram, Sungei Labang, Sungei Mengiong, Sungei Rajang, Sungei Segaham, Sungei Seran in Sungei Trusan in Sarawak (with a historical record from Kuching); Labuan; Brunei Darussalam (without a precise locality); Danum Valley, Deramakot, Kudat, Pulau Gaya, Ranau, Sandakan, Sepagaya Protection Forest Reserve, Tamparuli, Tawau Hills Park, Gunung Kinabalu (Poring) and Sungei Malutut in Sabah, and Balikpapan, Banjarmasin, Pondianak, Samarinda, Singkawan, Sungei Kapuas and Sungei Kendawangan in Kalimantan.

IUCN THREAT STATUS Least Concern.

Close-up of head

Full view

Close-up of head and forebody

Liopeltis Ringnecks

Small, somewhat slender arboreal (on saplings) snakes, diagnosed by rounded body; head indistinct from neck; no nuchal band or triangular postocular patch; nasal single; anterior and posterior chin shields not subequal; midbody scale rows 15 or 17; hemipenes unforked; colouration of anterior portion of body not different from that of posterior.

Tricoloured Ringneck *Liopeltis tricolor* (Schlegel, 1827)

(No vernacular names recorded)

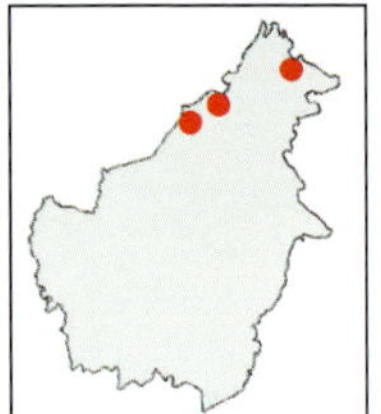

SIZE 560mm

IDENTIFICATION FEATURES Body slender; head indistinct from neck; snout long; eyes large with rounded pupils; tail long; dorsals smooth; ventrals rounded; midbody scale rows 15 or 17; ventrals 140–187; subcaudals 103–137; cloacal scute divided.

COLOUR Top of body yellowish-olive; dark postocular streak beyond neck; belly yellowish-cream, with olive streak on sides of each scale.

HABITAT AND BEHAVIOUR Associated with lowland forests. Arboreal, inhabiting small trees and other low vegetation. Diet comprises insects and spiders. Reproductive biology unstudied.

DISTRIBUTION Thailand, Vietnam, Peninsular Malaysia, Singapore, Borneo, Sumatra, Mentawai Archipelago, Java and Palawan. Bornean records are from Sungei Tangap and Gunung Santubong National Park in Sarawak; Ulu Temburong in Brunei Darussalam, and Ulu Dusun in Sabah.

IUCN THREAT STATUS Least Concern.

On sapling

Close-up of head

Lycodon Wolf Snakes

Long, slender arboreal snakes, characterized by head distinct from neck; rounded body; arched maxillary bone angled inwards anteriorly; 3–6 anterior fang-like teeth increasing in size and separated by diastema; largest teeth, the posterior-most, 2–3; large eyes with vertically elliptic pupils; scales smooth or keeled; midbody scale rows 15–19; tail long and slender.

Key to Bornean species of *Lycodon*

1a. Midbody scale rows 15 .... **2**
1b. Midbody scale rows 17 .... **3**

2a. Dorsum has dark brown cross-bars .... *L. subannulatus* (p. 251)
2b. Dorsum has yellow stripes .... *L. tristrigatus* (p. 252)

3a. Dorsals keeled .... *L. albofuscus* (below)
3b. Dorsals smooth **4**

4a. Two scales between posterior nasal and orbit .... *L. capucinus* (p. 248)
4b. Single scale between posterior nasal and orbit **5**

5a. Preocular absent; loreal present .... *L. sealei* (p. 250)
5b. Preocular present; loreal absent .... *L. effraenis* (p. 249)

Dusky Wolf Snake *Lycodon albofuscus* (Duméril, Bibron & Duméril, 1854)

(No vernacular names recorded)

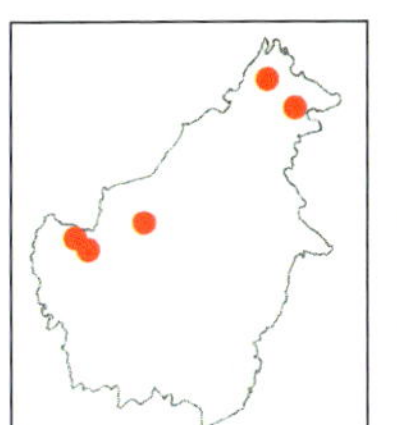

SIZE 2.07m

IDENTIFICATION FEATURES Body slender, subcylindrical; head wider than neck; snout short, blunt, depressed; eyes small with vertical pupils; tail long; dorsals keeled; midbody scale rows 17; ventrals 225–259; subcaudals 148–208, paired; cloacal scute divided.

COLOUR Top of body plain dark brown or brownish-black in adults; supralabials yellow; belly plain cream or yellow; in juveniles, 30–40 narrow white or yellow bands on back.

HABITAT AND BEHAVIOUR Inhabits open lowland forests and forest edges, usually with streams, at < 500m asl. Terrestrial, and may also climb low vegetation. Diet comprises lizards and frogs. Reproductive habits unstudied.

DISTRIBUTION Thailand, Peninsular Malaysia, Sumatra, Borneo and Pulau Nias. Bornean records are from Kuching, Pelagus National Park, Baleh National Park, Gunung Penrissen, Sungei Labang, Sungei Mengiong, Sungei Pesu and Sungei Seran in Sarawak, and Danum Valley, Sepagaya Protection Forest Reserve and Gunung Kinabalu (including Marak-Parak) in Sabah.

IUCN THREAT STATUS Least Concern

Island Wolf Snake *Lycodon capucinus* (Boie, 1827)

(Bahasa Malaysia: Ular Rumah Biasa. Bahasa Indonesia: Ular Cecak, Ular Tanah)

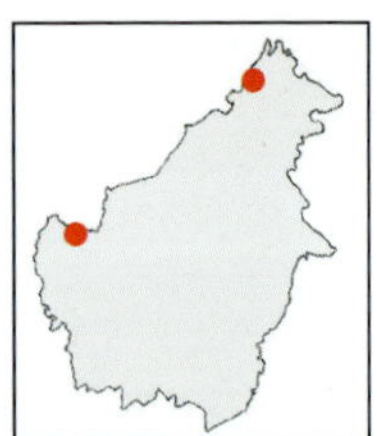

SIZE 760mm

IDENTIFICATION FEATURES Body relatively slender and subcylindrical; snout rounded; head flattened; eyes small with vertical pupils; tail long; dorsals smooth; midbody scale rows 17; ventrals 176–224; subcaudals 53–80, paired; cloacal scute entire.

COLOUR Top of body brown, grey-brown or purple, with narrow yellow or cream band at back of head that may be brown spotted; interstitial skin yellow or grey; scales of body light edged and form indistinct cross-bars or reticulated pattern; belly cream or light yellow.

HABITAT AND BEHAVIOUR Associated with lowland forests and mid-hills at < 600m asl, including human habitation. Terrestrial and arboreal. Diet includes lizards. Oviparous, producing 3–11 eggs, each 20–30 × 10mm.

DISTRIBUTION Myanmar, Thailand, Vietnam, Peninsular Malaysia, Singapore, Sumatra, Java, Bali, Sulawesi, Lesser Sundas, China and the Philippines. Isolated recent records from Borneo, including Kota Kinabalu in Sabah, and Demak Laut, Taman Sepakat Jaya, near Kuching in Sarawak, are suspected to be based on human-mediated dispersals.

IUCN THREAT STATUS Least Concern.

BROWN WOLF SNAKE *Lycodon effraenis* Cantor, 1847

(No vernacular names recorded)

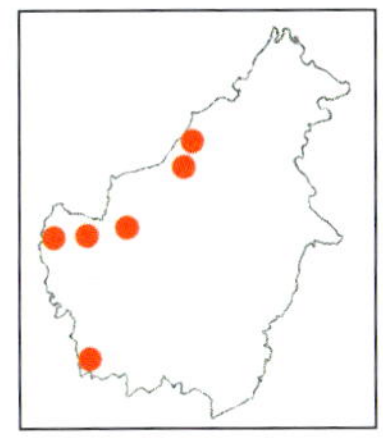

SIZE 1m

IDENTIFICATION FEATURES Body slender, subcylindrical; head flattened; rounded snout; single preocular; loreal absent; single preocular; eyes small with vertical pupils; tail long; dorsals smooth or weakly keeled; midbody scale rows 17; ventrals 215–233; subcaudals 72–100, paired; cloacal scales entire.

COLOUR Top of body reddish-brown or dark brown, with 18–20 pale yellow or cream-coloured bands; belly unpatterned brown.

HABITAT AND BEHAVIOUR Inhabits lowland forests and more open habitats, from sea level to c. 700m asl. Nocturnal as well as diurnal, and mostly terrestrial. Diet unknown and presumably lizards and small snakes. Reproductive habits unstudied.

DISTRIBUTION Peninsular Malaysia, Borneo and Sumatra. Records from Borneo are from Ranchan Pool in Serian, Sungei Baram, Sungei Labang, Sungei Seran and Sungei Tangap, and Serian in Sarawak, and Kendawangan and Sinkawang in Kalimantan.

IUCN THREAT STATUS Least Concern.

White-banded Wolf Snake *Lycodon sealei* Leviton, 1955

(Bahasa Indonesia: Ular Cecak Belang)

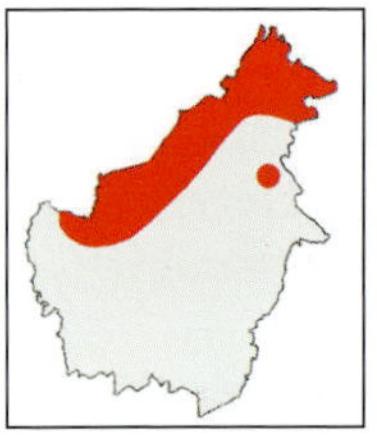

SIZE 1.02m

IDENTIFICATION FEATURES Body slender; head flattened; snout rounded; eyes small with vertical pupils; tail long; dorsals have weak keels; midbody scale rows 17; ventrals 198–211; subcaudals 59–69, paired; cloacal scute divided.

COLOUR Top of body black or dark brown, with 8–9 cream bands, 3–5 scales wide; pattern more distinctive in juveniles that in adults, fading with growth; in adults, remains of bands on grey or cream belly.

HABITAT AND BEHAVIOUR Inhabits lowland forests and mid-hills at < 1,000m asl. Arboreal and terrestrial. Diet includes geckos and skinks. Oviparous; other details unknown.

DISTRIBUTION Borneo, as well as Balabac and Palawan, and possibly Mindanao, in the Philippines. Bornean records are from Bintulu, Kapit, Niah National Park, Gunung Mulu National Park, Sungei Mengiong and Sungei Pesu in Sarawak (with historical records from Kuching); Ulu Temburong in Brunei Darussalam; Gunung Kinabalu (Bundu Tuhan), Danum Valley, Sandakan, Sepagaya Forest Reserve, Lahad Datu and Tawau Hills Park in Sabah, and Sungei Bulangan in Kalimantan. Allocation of Bornean specimens to *sealei* is tentative, and the species here were formerly allocated to *L. subcinctus*, which, in its expanded sense, had a distribution from China, Nicobar islands, Myanmar, Thailand, Laos, Cambodia, Vietnam, Peninsular Malaysia, Singapore, Borneo, Sumatra, Pulau Nias, Mentawai Archipelago, Java, Lombok, Sumbawa and the Philippines.

IUCN THREAT STATUS Least Concern.

HALF-BANDED WOLF SNAKE *Lycodon subannulatus* (Duméril, Bibron & Duméril, 1854)

(No vernacular names recorded)

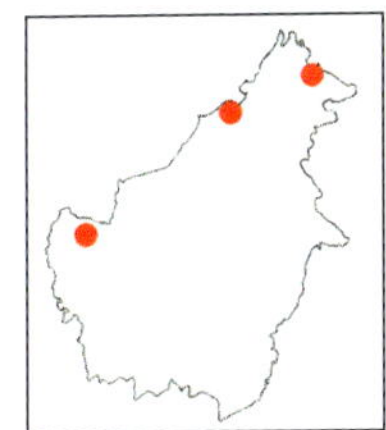

SIZE 600mm

IDENTIFICATION FEATURES Body slender and compressed; head depressed, distinct from neck; eyes large with vertical pupils; dorsals smooth; midbody scale rows 15; ventrals 225–244; subcaudals 88–107, paired; cloacal scute entire.

COLOUR Top of body tan or light brown, with large brown transverse spots; smaller spots on flanks; two dark postocular streaks; belly plain yellow.

HABITAT AND BEHAVIOUR Inhabits lowland forests and disturbed areas. Terrestrial as well as arboreal. Diet comprises small vertebrates. Reproductive biology unstudied.

DISTRIBUTION Myanmar, Thailand, Peninsular Malaysia, Singapore, Borneo, Sumatra, Mentawai and Riau archipelagos, and Palawan. The only Bornean records are from Gunung Penrissen in Sarawak; Jalan Merdeka in Bandar Seri Begawan, in Brunei Darussalam, and Sandakan in Sabah.

IUCN THREAT STATUS Least Concern.

REMARKS A recent paper synonymized this species with *Lycodon tristrigatus*.

THREE-BANDED WOLF SNAKE *Lycodon tristrigatus* (Günther, 1858)

(No vernacular names recorded)

SIZE 650mm

IDENTIFICATION FEATURES Body slender and compressed; head depressed and distinct from neck; eyes large with vertical pupils; dorsals smooth; midbody scale rows 15; ventrals 218–231; subcaudals 86–96, paired; cloacal scute entire.

COLOUR Top of body dark brown with three white stripes; forehead shields white edged; supralabials white; belly cream.

HABITAT AND BEHAVIOUR Inhabits lowland forests, particularly on rocky habitats and trees. Diurnal; diet includes lizards. Breeding habits unstudied.

DISTRIBUTION Natuna Archipelago, Balabac, Palawan and Borneo. Bornean records are from Bako National Park, Marudi, Labang Camp, Gunung Mulu National Park, Niah National Park, Sungei Mengiong, Sungei Nyabau, Sungei Segaham, Sungei Seran, the Paya Maga Highlands, Ravenscourt at the foothills of Gunung Murud in Sarawak (and two historical records, from Kuching and Jalan Penrissen); Labuan; Sungei Liang in Brunei Darussalam, and Danum Valley, Deramakot, Ulu Dusun, Tawau Hills Park and Sungei Rompon in Trust Madi in Sabah.

IUCN THREAT STATUS Least Concern.

Oligodon Kukri Snakes

Medium-sized snakes with the following morphological characteristics: 2–3 posterior maxillaries enlarged and compressed; head short, weakly distinct from neck; eyes moderately large with rounded pupils; scales smooth; rostral large; ventrals rounded; subcaudals paired; tail short.

Key to Bornean species of *Oligodon*

1a. Dorsum striped 2
1b. Dorsum unstriped 3

2a. Dark areas of dorsum without pale spots 6
2b. Dark area of dorsum with red, yellow or white spots 5

3a. Ventrals dark barred .... *O. purpurascens* (p. 258)
3b. Ventrals not barred 4

4a. Dorsum has narrow red cross bars .... *O. signatus* (p. 259)
4b. Dorsum has c. 25 black rings .... *O. annulifer* (p. 254)

5a. Cloacal scute divided .... *O. vertebralis* (p. 260)
5b. Cloacal scute single .... *O. everettii* (p. 255)

6a. Dorsum has seven distinct dark stripes .... *O. octolineatus* (p. 257)
6b. Dorsum has single distinct vertebral stripe .... *O. meyerinkii* (p. 256)

Spotted Kukri Snake (page 254)

Spotted Kukri Snake *Oligodon annulifer* (Boulenger, 1893)

(No vernacular names recorded)

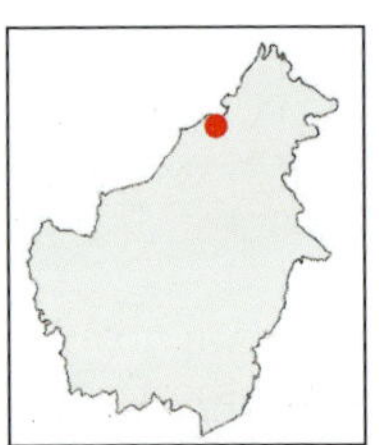

SIZE 450mm

IDENTIFICATION FEATURES Body robust, subcylindrical; head short, barely distinct from neck; snout rounded; eyes moderate; dorsals smooth; midbody scale rows 15; ventrals 151–162; subcaudals 44–64, paired; cloacal scute divided.

COLOUR Top of body dark brown, lighter on flanks, with 20–26 orangish-brown blotches edged with greyish-brown; chevron at nape; belly cream with black spots on each side; under surface of tail brick-red.

HABITAT AND BEHAVIOUR Inhabits hill dipterocarp forests at about 120m asl. Nocturnal, and arboreal and terrestrial. Diet includes lizards. Reproductive biology unstudied.

DISTRIBUTION Endemic to Borneo. Records exist for Sabah (without a precise locality), and Ulu Temburong in Brunei Darussalam.

IUCN THREAT STATUS Least Concern.

Head and forebody

Grey Kukri Snake *Oligodon cinereus* (Günther, 1864)

(No vernacular names recorded)

SIZE 730mm

IDENTIFICATION FEATURES Body robust, subcylindrical; head short, indistinct from neck; loreal present; single preocular; eyes moderate and pupils rounded; tail short and blunt; dorsals smooth; midbody scale rows 17; ventrals 155–186; subcaudals 28–42, paired; subcaudals paired; cloacal scale entire.

COLOUR Top of body reddish-brown or red, unpatterned, or with white or grey, black-edged cross-bars; forehead unpatterned brown; belly cream.

HABITAT AND BEHAVIOUR Inhabits forested lowlands and mid-hills at 457–700m asl. Nocturnal and terrestrial. Diet includes spiders and insects. Oviparous; clutches comprise 4–5 eggs.

DISTRIBUTION Myanmar, Thailand, Laos, Cambodia, Vietnam and eastern China. The record from Borneo (Bongon, Sabah State) requires confirmation.

IUCN THREAT STATUS Least Concern.

Jewelled Kukri Snake *Oligodon everetti* Boulenger, 1893

(No vernacular names recorded)

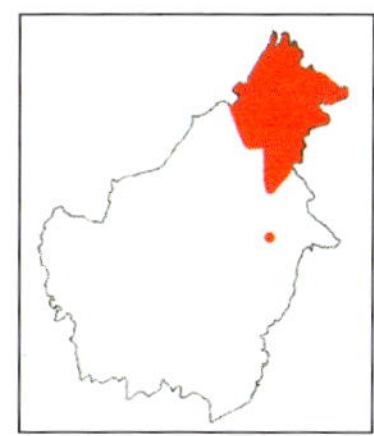

SIZE 420mm

IDENTIFICATION FEATURES Body moderately robust and subcylindrical; head short, indistinct from neck; eyes moderate with rounded pupils; dorsals smooth; midbody scale rows 15; ventrals 132–154; subcaudals 46–72, paired; cloacal scute entire.

COLOUR Top of body greyish-brown with three blackish-brown stripes, broadest (three scales wide) along vertebral region, enclosing short, white and red or orange bars; dark stripe on each lower flank that encloses white spots; dark, backwards directed, 'V'-shaped mark on forehead, another over snout crosses orbit; belly plain coral-red.

HABITAT AND BEHAVIOUR Inhabits hill dipterocarp forests at < c. 1,000m asl. Subfossorial and associated with leaf litter. Diet comprises skinks, and possibly other small vertebrates. Reproductive biology unknown.

DISTRIBUTION Endemic to Borneo. Records are from the foothills of Gunung Kinabalu, Tawau Hills Park, Danum Valley, Ulu Dusun and Malutut in Sabah, and Banjaran and Tanjung in Kalimantan.

IUCN THREAT STATUS Least Concern.

Meyerink's Kukri Snake *Oligodon meyerinkii* (Steindachner, 1891)

(No vernacular names recorded)

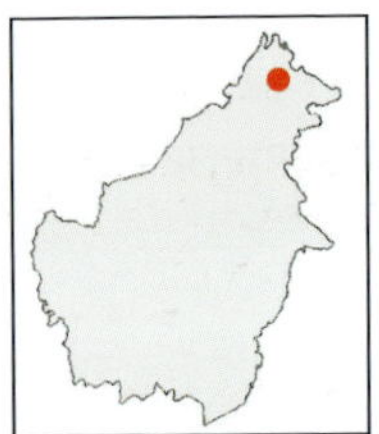

SIZE 379mm

IDENTIFICATION FEATURES Body robust, subcylindrical; head short, indistinct from neck; loreal present; single preocular; eyes moderate and pupils rounded; dorsals smooth; midbody scale rows 17; ventrals 154–169; subcaudals 38–57, paired; cloacal scale entire.

COLOUR Top of body reddish-brown, with 5–7 salmon-pink longitudinal stripes, bordered by grey, or dark brown stripes on scale rows 4–5; forehead pale brown with two black stripes, including interocular bar to sides of head to cover parts of supralabials; oblique stripe from angle of jaws to parietals; dark stripe along median plane from parietals to frontal; belly rosy-pink with fine dark brown dots; subcaudals pink with brown spots.

HABITAT AND BEHAVIOUR Appears to be a lowland forest species. Diet and reproductive habits unstudied.

DISTRIBUTION The Philippines (Bongao, Jolo, Papahang, Sibutu and Tawi-Tawi), and eastern Borneo (Sabah State).

IUCN THREAT STATUS Vulnerable.

EIGHT-LINED KUKRI SNAKE *Oligodon octolineatus* (Schneider, 1801)

(Bahasa Brunei: Ular Kalibantang. Bahasa Indonesia: Ular Birang. Iban: Ular Emparo, Ular Matahari)

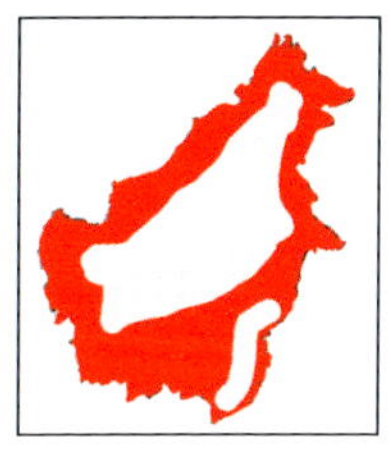

SIZE 700mm

IDENTIFICATION FEATURES Body slender and subcylindrical; head short and rounded, indistinct from neck; eyes moderate in size with rounded pupils; dorsals smooth; midbody scale rows 17; ventrals 155–197, paired; subcaudals 43–61, paired; cloacal scute entire.

COLOUR Top of body brown or reddish-brown with 5–7 cream or white longitudinal stripes, bordered by 6–8 dark brown or black stripes; forehead has two black stripes; belly yellow.

HABITAT AND BEHAVIOUR Found in lowland and hill dipterocarp forests at < 500m asl. Mostly terrestrial, with occasional arboreal activity. Diet comprises birds' eggs, frogs and their eggs, lizards and other snakes. Oviparous, producing clutches of 4–5 eggs, each 16 × 30mm.

DISTRIBUTION Thailand, Peninsular Malaysia, Singapore, Borneo, Sumatra, Pulau Nias, Mentawai and Riau archipelagos, Pulau Bangka, Pulau Belitung, Java, Sulu Archipelago and Sulawesi. Widespread on Borneo, and found in localities in Bako National Park, Baleh National Park, Gunung Dulit, Gunung Gading National Park, Gunung Santubong National Park, Matang Wildlife Centre, Kampung Sebako at foothills of Gunung Pueh, Sungei Baram in Sarawak (as well as historical records from Kuching); Bukit Patoi in Brunei Darussalam; Apin-Apin, Kiansom, Kota Kinabalu, Batu Putih, Gunung Kinabalu Park and Sandakan in Sabah, and Banjarmasin, Montrado in Bengkayang, Samarinda, Singkawang, Sintang, and Sungei Duri and Sungei Mahakam in Kalimantan.

IUCN THREAT STATUS Least Concern.

Purple Kukri Snake *Oligodon purpurascens* (Schlegel, 1837)

(Bahasa Malaysia: Ular Kebun Perang. Iban: Ular Paut)

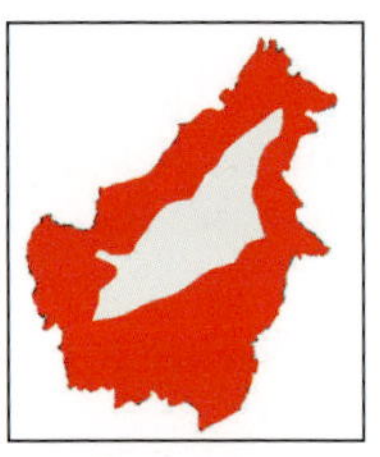

SIZE 950mm

IDENTIFICATION FEATURES Body thick and subcylindrical; head short and indistinct from neck; eyes small with rounded pupils; dorsals smooth; midbody scale rows 19 or 21; ventrals 160–210; subcaudals 40–60, paired; cloacal scute entire.

COLOUR Top of body brownish-purple with dark, wavy bands, or transverse, dark-edged yellow bands or blotches, separated by narrow, dark cross-bars; dark chevron on forehead; belly pink or red, with dark, squarish spots.

HABITAT AND BEHAVIOUR Inhabits hill dipterocarp forests and peat-swamp forests at <1,200m asl. Diet includes frog and lizard eggs, and frogs and tadpoles. Oviparous, with clutches comprising 8–13 eggs, each 18–22 × 27–33mm.

DISTRIBUTION Thailand, Peninsular Malaysia, Singapore, Borneo, Sumatra, Mentawai Archipelago and Java. Bornean records are from the base of Gunung Pueh, Paku, Pangkalan Ampat at the foot of Gunung Penrissen, Gunung Santubong National Park, Kudi in Simunjan, Gunung Matang, Saribas, Kota Samarahan and Jalan Penrissen in Kuching in Sarawak; Labuan; Bandar Seri Begawan, Brunei Darussalam; Gunung Kinabalu Park (Sayap, Bundu Tuhan and Poring), Tambunan in Crocker Range and Ulu Dusun in Sabah, and Buntok and Samarinda in Kalimantan.

IUCN THREAT STATUS Least Concern.

HALF-KEELED KUKRI SNAKE *Oligodon signatus* (Günther, 1864)

(No vernacular names recorded)

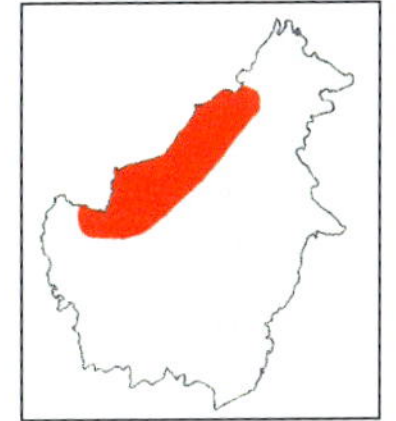

SIZE 390mm

IDENTIFICATION FEATURES Body slender and subcylindrical; head short, indistinct from neck; eyes small with rounded pupils; dorsals weakly keeled; midbody scale rows 17; ventrals 154–166; subcaudals 50–57, paired; cloacal scute entire.

COLOUR Top of body reddish-brown with 20–30 dark-edged, pale cross-bars; chevron pattern on forehead; belly orange-red.

HABITAT AND BEHAVIOUR Inhabits lowland forests at < 200m asl, and may be found in buttresses of large trees and edges of peat-swamp forests. Diet and reproductive habits unstudied.

DISTRIBUTION Peninsular Malaysia, Singapore and Borneo. Occurs in isolated Bornean localities in Bako National Park, Gunung Santubong National Park, Gunung Matang, Kota Samarahan and Sungei Mengiong and Nyabau in Sarawak, and Mendolong in Sabah.

IUCN THREAT STATUS Least Concern.

Black-spined Kukri Snake *Oligodon vertebralis* Günther, 1865

(No vernacular names recorded)

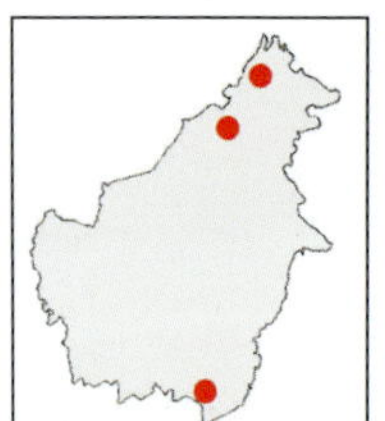

SIZE 350mm

IDENTIFICATION FEATURES Body moderate, subcylindrical; head short, indistinct from neck; loreal present; single preocular; eyes moderate with rounded pupils; dorsals smooth; midbody scale rows 15; ventrals 136–154; subcaudals 35–54, paired; cloacal scute divided.

COLOUR Top of body brown, with scattered small yellow spots with black edges; spots largest on vertebral region; head yellow with two black-edged brown bands, anterior one passing across orbit, posterior one across frontal; gular region black spotted; venter yellow.

HABITAT AND BEHAVIOUR Inhabits submontane forests across the island. Diet and reproductive habits unstudied.

DISTRIBUTION Endemic to Borneo. Localities include the Kelabit Highlands in Sarawak; Gunung Kinabalu in Sabah, and Banjarmasin in Kalimantan.

IUCN THREAT STATUS Data Deficient.

Dorsal view

Dorso-lateral view

Oreocalamus Mountain Reed Snakes

Relatively short, thick, terrestrial and subfossorial snakes, restricted to montane forests, and characterized by head indistinct from neck; midbody scale rows 17; scales smooth; loreal between nasal and preocular; short, pointed tail.

MOUNTAIN REED SNAKE *Oreocalamus hanitschi* Boulenger, 1899

(No vernacular names recorded)

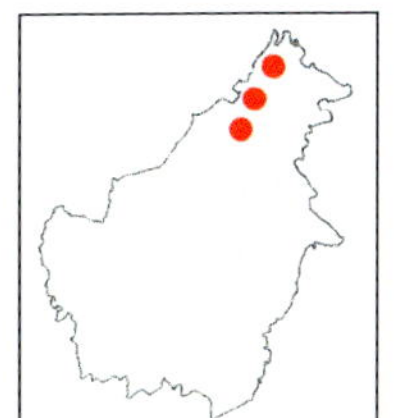

SIZE 570mm

IDENTIFICATION FEATURES Body robust, subcylindrical; head short, indistinct from neck; loreal present; single preocular; single postocular; four infralabials contact anterior chin shields; tail short, with sharp tip; dorsals smooth; midbody scale rows 17; ventrals 125–132; subcaudals 20–32, paired; cloacal scute entire.

COLOUR Top of body brownish-tan, with dark brown scales forming zigzag pattern; on lower flanks, dark scales join to form continuous line; dark stripe along eyes; dark inverted chevron on neck; belly yellowish-brown, each ventral with black spot centrally and on edges, forming dark line.

HABITAT AND BEHAVIOUR Inhabits oak forests at 1,120–1,700m asl. Terrestrial and/or subfossorial, hiding under logs and leaf litter. Diet comprises earthworms. Reproductive biology unstudied.

DISTRIBUTION Peninsular Malaysia and Borneo. Records from Borneo are from Gunung Murud in Sarawak, and Gunung Kinabalu (including Mesilau and Lumu-Lumu), Gunung Lumaku and Mendolong in Sabah.

IUCN THREAT STATUS Least Concern.

Orthriophis Cave Racers

Large, relatively slender terrestrial as well as scansorial snakes, sometimes allocated to the genus *Elaphe*, diagnosed by large eyes; long snout; midbody scale rows 25; scales smooth except the uppermost rows; cloacal scute divided; tail relatively short.

Cave Racer *Orthriophis taeniurus* (Cope, 1961)

(Bahasa Malaysia: Ular Bulan. Mandarin: Heimeijing She. Iban: Ular Bulan. Kelabit: Rari Wang)

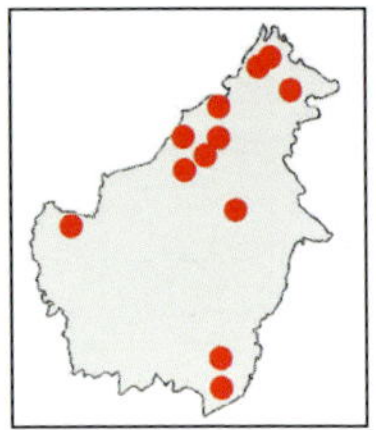

SIZE 2m

IDENTIFICATION FEATURES Body slender and elongated; head long and distinct from neck; eyes large with rounded pupils; dorsals weakly keeled; scales on flanks smooth; midbody scales 25; ventrals 271–305; subcaudals 86–112; cloacal scute divided.

COLOUR Top of body greyish-brown or greyish-black with cream or tan stripe along middle of back, especially towards posterior; forehead olive; sides of head dark striped; supralabials and chin cream; belly yellow or cream.

HABITAT AND BEHAVIOUR Inhabits lowland and submontane forests at < 2,000m asl. Arboreal and terrestrial; often encountered in caves. Diet comprises bats and swiftlets. Oviparous, producing clutches of 5–14 eggs, each 45.7–72.0 × 20.8– 31.4mm.

DISTRIBUTION Myanmar, Thailand, Laos, Cambodia, Vietnam, Peninsular Malaysia, Singapore, Borneo, Sumatra, eastern India, Bhutan, southern China, Ryukyu Archipelago of Japan and southeastern Russia. Cave and karst areas and adjacent lowland forests across Borneo, such as Gunung Mulu National Park, Niah National Park and the karst areas of Bau, as well as the sandstone region of the Kelabit Highlands in Sarawak; Brunei Darussalam (without a precise locality), and Baturong Caves, and Gunung Kinabalu (Bundu Tuhan and Lumu-Lumu) in Sabah.

IUCN THREAT STATUS Vulnerable.

Pseudorabdion Reed Snakes

Small fossorial snakes, identifiable by head being indistinct from neck; internasals present, and smaller than prefrontals; anterior temporals absent; midbody scale rows 15; scales smooth; subcaudals paired; cloacal scute entire; eyes small with rounded pupils; tail short.

Key to Bornean species of *Pseudorabdion*

1a. Loreal absent; prefrontals contact supralabials **2**
1b. Loreal present; prefrontals not in contact with supralabials **3**

2a. Preocular present .... *P. longiceps* (p. 265)
2b. Preocular absent .... *P. collaris* (p. 264)

3a. Ventrals > 125; subcaudals > 40 .... *P. albonuchalis* (below)
3b. Ventrals < 120; subcaudals < 30 .... *P. saravacense* (p. 266)

White-collared Reed Snake *Pseudorabdion albonuchalis* (Günther, 1896)

(No vernacular names recorded)

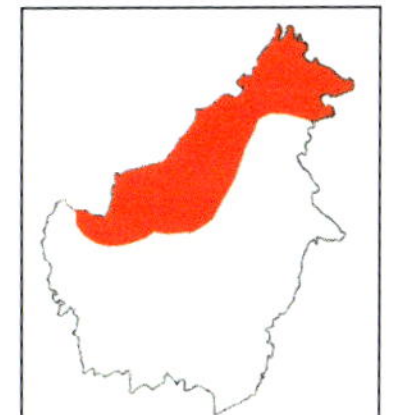

SIZE 270mm

IDENTIFICATION FEATURES Body slender; head indistinct from neck; snout pointed; single nostril between two nasals; frontal borders eye; eyes small with vertical pupils; tail short; dorsals smooth; midbody scale rows 15; ventrals 127–144; subcaudals 43–64; cloacal scale entire.

COLOUR Top of body iridescent black; broad yellow or red collar covers over half parietals; band of four scales behind parietals; belly dark brown.

HABITAT AND BEHAVIOUR Inhabits lowland dipterocarp forests at < 500m asl. Subfossorial, and associated with leaf litter. Diet unstudied and presumably small arthropods. Reproductive habits unstudied.

DISTRIBUTION Endemic to Borneo. Localities include Baleh National Park, Sungei Pesu, Sungei Baram and Kubah National Park in Sarawak; Ulu Temburong and Jalan Muara in Bandar Seri Begawan in Brunei Darussalam, and Marak Parak in Sabah.

IUCN THREAT STATUS Least Concern.

Mocquard's Reed Snake *Pseudorabdion collaris* (Mocquard, 1892)

(Iban: Ular Untup)

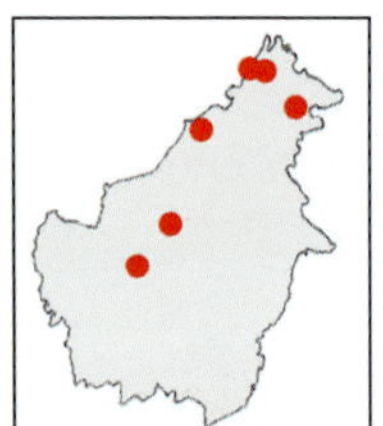

SIZE 250mm

IDENTIFICATION FEATURES Body slender; head indistinct from neck; snout pointed; nostril between two nasals; loreal and preocular absent; single nasal; preoculars not fused; eyes small; tail short; dorsals smooth; midbody scale rows 15; ventrals 110–134; subcaudals 20–41, paired; cloacal scute entire.

COLOUR Top of body shiny black; pale collar sometimes present; belly slightly lighter than back.

HABITAT AND BEHAVIOUR Inhabits lowland forests at up to 650m asl. Subfossorial, inhabiting leaf litter and areas under rocks and logs. Diet comprises earthworms. Reproductive biology unstudied.

DISTRIBUTION Endemic to Borneo. Localities include Niah National Park, Sungei Segaham and Sungei Mengiong in Sarawak; Gunung Kinabalu Park (Jalan Kamborangoh), Danum Valley and Penampang in Sabah, and Sebruang Valley in Kalimantan.

IUCN THREAT STATUS Least Concern.

DWARF REED SNAKE *Pseudorabdion longiceps* (Cantor, 1847)

(Iban: Ular Untup)

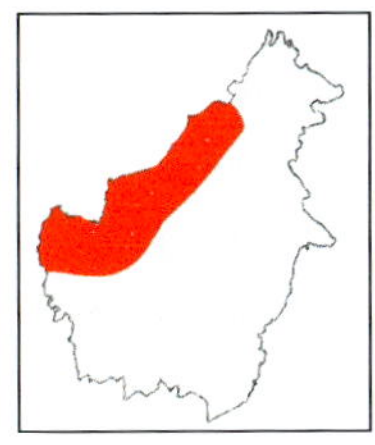

SIZE 230mm

IDENTIFICATION FEATURES Body slender; head indistinct from neck; snout pointed; nostril in single nasal; loreal absent; single preocular; single postocular; internasal separate from supralabials; eyes small; tail short; dorsals smooth; midbody scale rows 15; ventrals 129–148; subcaudals 10–31, paired; cloacal scute entire.

COLOUR Top of body iridescent black or brown; yellow collar and yellow spot above angle of mouth; belly dark brown.

HABITAT AND BEHAVIOUR Inhabits lowland rainforests. Subfossorial. Diet comprises leaf-litter invertebrates, earthworms, and small insects and larvae. Clutches comprise 2–3 narrow, elongated eggs.

DISTRIBUTION Southern Thailand, Peninsular Malaysia, Singapore, Borneo, Sumatra, Pulau Nias, the Mentawai and Riau Archipelagos and Sulawesi (Indonesia). Bornean localities include Simanggang, Gunung Santubong National Park, Jalan Penrissen in Kuching, Sungei Mengiong, Sungei Pesu and Sungei Seran in Sarawak; Ulu Temburong in Brunei Darussalam, and Pontianak and Sebruang Valley in Kalimantan.

IUCN THREAT STATUS Least Concern.

Sarawak Reed Snake *Pseudorabdion saravacense* (Shelford, 1901)

(No vernacular names recorded)

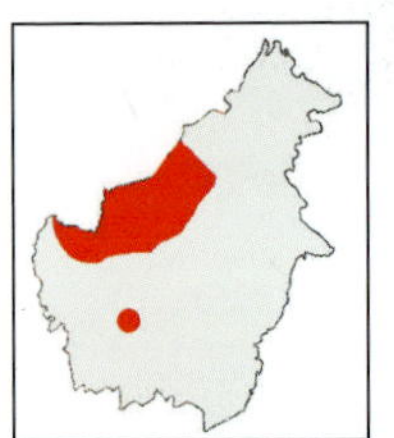

SIZE 142mm

IDENTIFICATION FEATURES Body slender; head indistinct from neck; snout pointed; nostril between two nasals; loreal present; preocular absent; single postocular; prefrontal not in contact with supralabials; nasals enlarged, not divided; anterior genials not in contact with mentals; eyes small; tail short, pointed; dorsals smooth; midbody scale rows 15; ventrals 113; subcaudals 26; cloacal scute entire.

COLOUR Top of body dark brown, iridescent; red blotches on sides of head above angle of jaws; irregular red band on neck; belly unpatterned brown.

HABITAT AND BEHAVIOUR Inhabits lowland forests, from about 30m asl. Lives in the leaf litter. Diet and reproductive habits unstudied.

DISTRIBUTION Endemic to Borneo. Localities include Kubah National Park, Gunung Santubong National Park, Sungei Mengiong and Niah National Park in Sarawak; and Bukit Baka Bukit Raya National Park, Kalimantan.

IUCN THREAT STATUS Least Concern.

Ptyas Asian Rat Snakes

Large, slender terrestrial snakes that are diurnal and common in lowland rainforests. They are characterized by a slightly laterally compressed body; head long and distinct from neck; eyes large with rounded pupils; dorsal scales either smooth or keeled; ventrals rounded; tail long.

Key to Bornean species of *Ptyas*

1a. Middorsal scale rows 15 .... *P. korros* (p. 269)
1b. Middorsal scale rows 16 or 18 .... **2**

2a. Middorsals keeled; subcaudals < 120 .... *P. carinata* (below)
2b. Middorsals smooth; subcaudals > 160 .... *P. fusca* (p. 268)

KEELED RAT SNAKE *Ptyas carinata* (Günther, 1858)

(No vernacular names recorded)

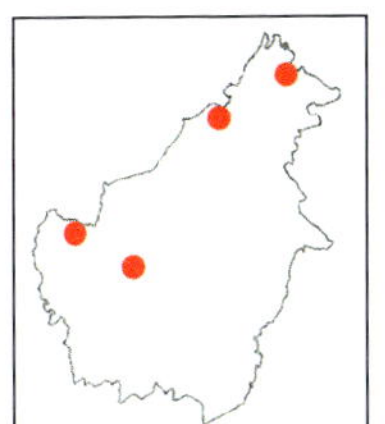

SIZE TL 4m

IDENTIFICATION FEATURES Body slender; head distinct from neck; eyes large with rounded pupils; tail long; dorsals smooth, except for 2–4 median rows that are keeled; midbody scale rows 16 or 18; ventrals 208–215; subcaudals 110–118, paired; cloacal scute divided.

COLOUR Top of body olive-brown to almost black anteriorly, sometimes with indistinct yellow cross-bars; posterior dorsum yellow with distinct checkered black pattern, ending in black tail with yellow spots; belly cream, turning grey or black posteriorly.

HABITAT AND BEHAVIOUR Inhabits lowland forests and agricultural fields. Diet comprises amphibians, lizards and rodents. Oviparous, with clutches comprising 10 eggs.

DISTRIBUTION Myanmar, Thailand, Vietnam, Peninsular Malaysia, Singapore, Borneo, Sumatra, Java, China and Palawan. Bornean records are from Gunung Mulu National Park and there exists an old report from Kuching in Sarawak; base of Bukit Patoi near Bangar in Brunei Darussalam; Ulu Dusun in Sabah, and Sebruang Valley in Kalimantan.

IUCN THREAT STATUS Least Concern.

White-bellied Rat Snake *Ptyas fusca* (Günther, 1858)

(No vernacular names recorded)

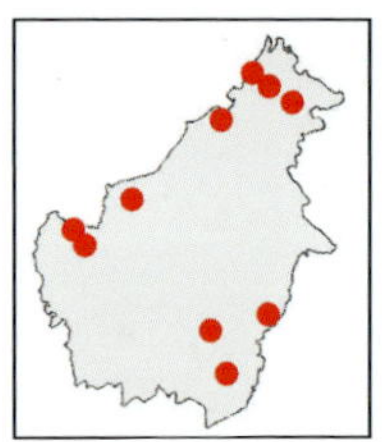

SIZE 3m

IDENTIFICATION FEATURES Body slender; head distinct from neck; eyes large with rounded pupils; tail long; dorsals smooth; midbody scale rows 16; ventrals 183–198; subcaudals 160–179, paired; cloacal scute divided.

COLOUR Top of body mid-brown to brownish-grey, typically nearly black anteriorly and lighter posteriorly; red vertebral stripe sometimes present; belly cream.

HABITAT AND BEHAVIOUR Inhabits lowland forests to mid-hills at < 1,330m asl. Diet includes birds and, probably, mammals. Oviparous.

DISTRIBUTION Peninsular Malaysia, Singapore, Borneo, Sumatra, Pulau Nias, Pulau Bangka, Pulau Belitung and Natuna Archipelago. Bornean records are from Gunung Santubong National Park, Kota Samarahan, Kuching, Sibu and Gunung Mulu National Park in Sarawak; Brunei Darussalam (without a precise locality); Danum Valley, Deramakot and Gunung Kinabalu (Ranau and Poring), besides Sungei Purulon, Menggatal and Penampang in Sabah, and Balikpapan, Buntok and Hantakan in Central Hulu Regency in Kalimantan.

IUCN THREAT STATUS Least Concern.

Close-up of head and forebody

Javanese Rat Snake *Ptyas korros* (Schlegel, 1837)

(Bahasa Malaysia: Ular Tikus Biasa. Bahasa Indonesia: Ular Koros)

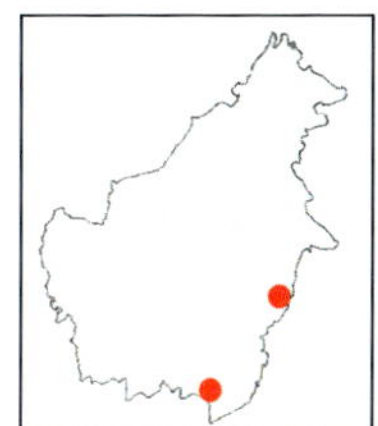

SIZE 2.68m

IDENTIFICATION FEATURES Body robust; head elongated, distinct from neck; eyes large with rounded pupils; tail long; dorsals smooth anteriorly, weakly keeled posteriorly; midbody scale rows 15; ventrals 160–187; subcaudals 120–151, paired; cloacal scute divided.

COLOUR Top of body anterior and forehead grey to olive-brown, darkening to nearly black posteriorly; scales edged with white, are distinct posteriorly, appearing as white bands on black background; belly, chin and labials brownish-cream.

HABITAT AND BEHAVIOUR Inhabits lowland and montane forests at < 3,000m asl. Terrestrial and semi-arboreal. Diet includes rodents, birds, lizards and frogs. Oviparous, laying 4–14 eggs, each 33.5–48 × 17.1–20.5mm.

DISTRIBUTION Eastern India, Bhutan, Bangladesh, eastern China, Myanmar, Thailand, Laos, Cambodia, Vietnam, Peninsular Malaysia, Singapore, Sumatra, Borneo and Java. Bornean localities include Banjarmasin and Samarinda in Kalimantan.

IUCN THREAT STATUS Near Threatened.

Sibynophis Black-headed Snakes

Small, slender terrestrial snakes of lowland rainforests and mid-hills, identifiable by head distinct from neck; body cylindrical; midbody scale rows 17, smooth; eyes with rounded pupils; compressed teeth; rounded ventrals; cloacal and subcaudals divided; tail long.

Key to Bornean species of *Sibynophis*

1a. Dark vertebral stripe, edged by red or reddish-brown stripe ....*S. geminatus* (below)
1b. Short black cross-bars over lighter band ....*S. melanocephalus* (opposite)

Striped Black-headed Snake *Sibynophis geminatus* (Boie, 1826)

(No vernacular names recorded)

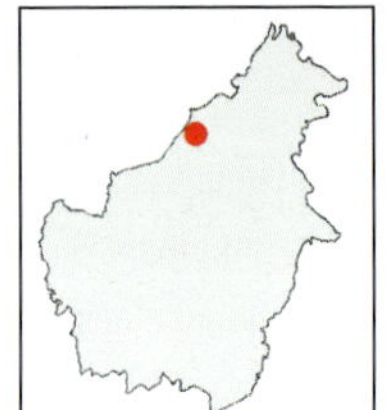

SIZE 1m

IDENTIFICATION FEATURES Body slender, cylindrical; head relatively short, slightly distinct from neck and flattened; single preocular; eye large with rounded pupils; dorsals smooth; subcaudals paired; midbody scale rows 17; ventrals 140–183; subcaudals 73–145, paired; cloacal scute divided.

COLOUR Top of body reddish-brown, darker caudally; pale nuchal collar may be present; black, bluish-grey or dark brown vertebral stripe, edged on either side by red or reddish-brown stripe, 2.5 scales wide; forehead rusty-brown, with small yellow, dark-edged spots; supralabials yellow, sutures black; venter greenish-yellow anteriorly, turning pale green caudally; outer edges of ventrals have black spot, forming narrow stripe along lower flanks.

HABITAT AND BEHAVIOUR Inhabits lowland forests, and offshore islands, under 400m asl. Diurnal and terrestrial. Diet comprises skinks. Single egg, 26.5 x 9mm, produced.

DISTRIBUTION Southern Thailand, Peninsular Malaysia, Singapore, Borneo, Sumatra, Pulau Weh, Pulau Siberut, Pulau Nias, Pulau Bangka, Pulau Belitung, Java, Bali, Tawi-Tawi and the Sulu Archipelago (the Philippines). On Borneo, recorded from Sarawak, without a precise locality.

IUCN THREAT Least Concern.

Head, dorsal view

Head, lateral view

WHITE-LIPPED BLACK-HEADED SNAKE *Sibynophis melanocephalus*

(Gray, 1834)

(No vernacular names recorded)

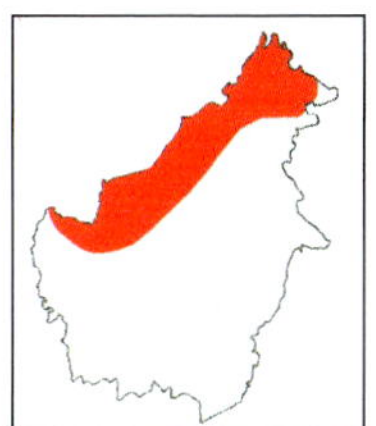

SIZE 600mm

IDENTIFICATION FEATURES Body slender and cylindrical; head relatively short, slightly distinct from neck and flattened; eyes large with rounded pupils; dorsals smooth; midbody scale rows 17; ventrals 145–177; subcaudals 132–136, paired; cloacal scute divided.

COLOUR Top of body reddish-brown or brown, with short black cross-bars over lighter band; lower flanks have yellow spots; forehead dark olive with olive-yellow spots; supralabials white, sutures with black stripes; belly yellow with orange tinge on sides posteriorly; ventrals have rounded black spot laterally.

HABITAT AND BEHAVIOUR Inhabits lowland forests at < 500m asl. Terrestrial, on leaf litter on forest floor and by streams. Diet comprises skinks. Reproductive habits unstudied.

DISTRIBUTION Thailand, Vietnam, Peninsular Malaysia, Singapore, Borneo and Sumatra. Bornean records are from Bako National Park, Sungei Baram, Sungei Mengion and Sungei Tangap in Sarawak, and Danum Valley, Sungei Malutut and Sungei Purulon, and Gunung Kinabalu (Kiau) in Sabah.

IUCN THREAT STATUS Least Concern.

Stegonotus Black Snakes

Medium-sized terrestrial and arboreal snakes; body robust; head distinct from neck; eyes large with elliptic pupils; scales smooth (Bornean species have a low keel on vertebral row); ventrals angled laterally; tail long.

Key to Bornean species of *Stegonotus*

1a. Head and body uniform dark brown to black; ventral scales 193–196; subcaudals 56–65 .... *S. borneensis* (below)

1b. Head darker than body; ventral scales 211–218; subcaudals 78 .... *S. caligocephalus* (opposite)

Bornean Lowland Black Snake *Stegonotus borneensis* Inger, 1967

(No vernacular names recorded)

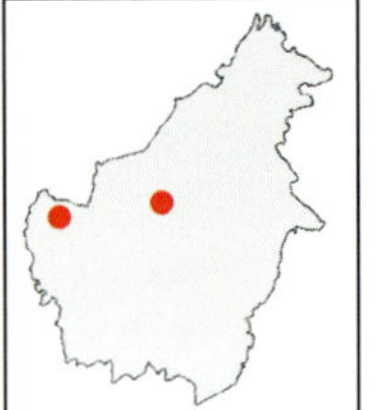

SIZE 1.37m

IDENTIFICATION FEATURES Body robust and cylindrical; head distinct from neck; vertebral ridge distinct; eyes small with vertical pupils; dorsals smooth; vertebrals enlarged; midbody scale rows 17; ventrals 193–196; subcaudals 78, entire; cloacal scute entire; tail short.

COLOUR Top of body plain grey-black, dark brown or black, with head colouration similar to that of dorsum of midbody; supralabials grey with pink tinge; each scale on belly alternately banded with dark and light grey.

HABITAT AND BEHAVIOUR Inhabits low hills of dipterocarp forests. Terrestrial. Diet unstudied. Oviparous.

DISTRIBUTION Endemic to Borneo, with a record from Nanga Takalit, along Sungei Mengiong, Kapit Division in Sarawak. A specimen assigned to this species was collected in Thailand.

IUCN THREAT STATUS Least Concern.

Close-up of head and neck

DARK-HEADED BLACK SNAKE *Stegonotus caligocephalus* Kaiser, Lapin, O'Shea & Kaiser, 2020

(No vernacular names recorded)

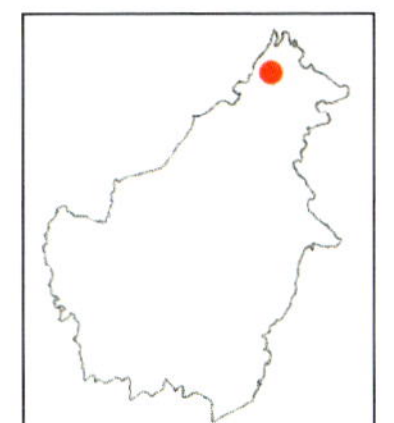

SIZE 1.25m

IDENTIFICATION FEATURES Body robust and cylindrical; head distinct from neck; vertebral ridge distinct; eyes small with vertical pupils; dorsals smooth; vertebrals enlarged; midbody scale rows 17; ventrals 211–218; subcaudals 56–65, entire; cloacal scute entire; tail short.

COLOUR Top of body dark greyish-brown, with scattered cream spots; flanks paler; supralabials pale brown; belly unpatterned pale brown.

HABITAT AND BEHAVIOUR Inhabits submontane forests at < 1,800m asl. Diet largely unknown except for an attempted predation on a gecko. Reproductive habits unstudied.

DISTRIBUTION Endemic to Borneo. Known from Gunung Kinabalu and Crocker Range Park in Sabah.

IUCN THREAT STATUS Not Evaluated.

Close-up of head and forebody

Xenelaphis Marsh Snakes

Large, fairly robust aquatic snakes, characterized by rounded body; head distinct from neck; eyes large with rounded pupils; middorsal scale rows 17, smooth; vertebrals slightly enlarged and hexagonal; ventrals rounded; tail long.

Key to Bornean species of *Xenelaphis*

1a. Dorsum has dark transverse bands anteriorly; subcaudals 140–179 .... *X. hexagonotus* (opposite)

1b. Dorsum has black-edged brown elliptic to square-like areas; subcaudals 143 .... *X. ellipsifer* (below)

Ornate Brown Snake *Xenelaphis ellipsifer* Boulenger, 1900

(No vernacular names recorded)

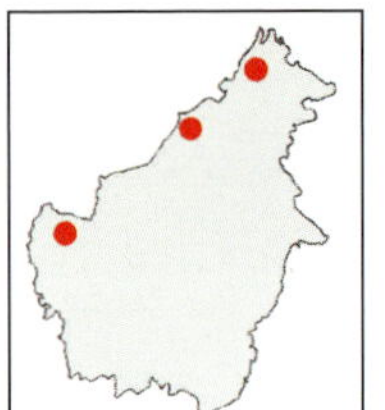

SIZE 2.51m

IDENTIFICATION FEATURES Body robust; head distinct from neck; snout rounded; eyes large with rounded pupils; tail long; dorsals smooth; midbody scale rows 17; ventrals 186–203; subcaudals 124–134, paired; cloacal scute divided.

COLOUR Top of body mid-brown, with 18–20 large, elliptic or squarish, black-edged brown blotches separated by narrow cream interspaces; forehead reddish-brown; supralabials yellow with black markings; infralabials cream; neck has dark streaks; flanks have inverted 'Y'- or 'V'- shaped markings; belly pink or white, with dark spots on outer edges of ventrals.

HABITAT AND BEHAVIOUR Inhabits mid-hills of primary forests at 800–1,000m asl. Aquatic, occupying streams and their edges. Diet presumably comprises fish. Reproductive habits unstudied.

DISTRIBUTION Peninsular Malaysia, Borneo and Sumatra. Bornean records are few, and include Pangkalan Ampat at the foot of Gunung Penrissen and Long Lellang in Sarawak, and Kundasan in the Gunung Kinabalu region in Sabah.

IUCN THREAT STATUS Least Concern.

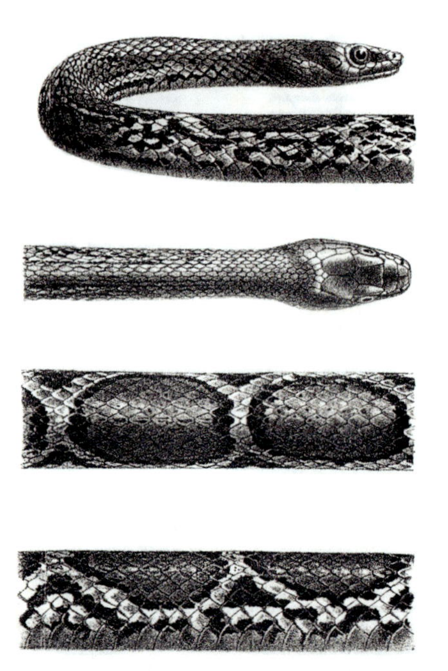

After Boulenger, 1900.

Malayan Brown Snake *Xenelaphis hexagonotus* (Cantor, 1847)

(Iban: Ular Beluai)

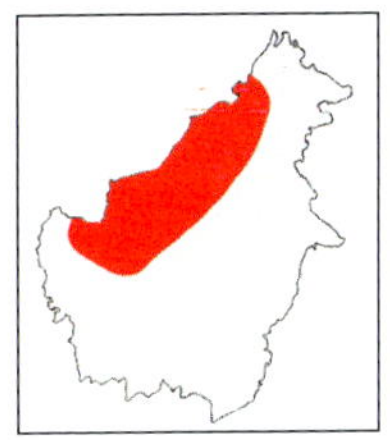

SIZE 2m

IDENTIFICATION FEATURES Body robust; head distinct from neck; snout rounded; eyes large with rounded pupils; tail long; vertebrals enlarged, hexagonal; dorsals smooth; midbody scale rows 17; ventrals 185–198; subcaudals 140–179, paired; cloacal scute divided.

COLOUR Top of body in adults brown, dark green or greenish-olive, with narrow, vertical black bars extending from neck and along body, apices reaching lower flanks; belly pale or deep yellow, ventrals have dark marginal spot; in juveniles, top of body uniformly light brown.

HABITAT AND BEHAVIOUR Inhabits lowland forests, especially coastal peat swamps and mangrove swamps at < 500m asl. Terrestrial and aquatic, in waterlogged forests, inhabiting shallow ditches. Diet comprises rodents, frog and fish. Reproductive habits unstudied.

DISTRIBUTION Myanmar, Thailand, Vietnam, Peninsular Malaysia, Singapore, Borneo, Sumatra, Pulau Bangka, Pulau Belitung and Java. Bornean records are across the island and include the Kuching region, Kota Samarahan, Kampung Bako, Bako National Park, Lawas, Lundu, Subis and Tangap near Niah, Sungei Baram, Sungei Rajang in Sarawak; Lumat in Beaufort in Sabah, and Sebruang Valley in Kalimantan.

IUCN THREAT STATUS Least Concern.

Family Natricidae Water Snakes

This family includes the large water snakes of the Old and New Worlds, which lack enlarged grooved rear fangs on the maxillary bone. Their habitats range from terrestrial to highly aquatic, and from mountain-tops to bodies of fresh water, where they feed on fish and other aquatic animals. Cosmopolitan in distribution, they may be found in both tropical and temperate regions of Asia, Europe, Africa, Australia and South America.

Elapoidis Ground Snakes

Small, slender terrestrial snakes living in submontane forests, with head scarcely distinct from neck; small eyes; distinctly keeled scales; entire cloacal; paired subcaudals.

Dark-grey Ground Snake *Elapoidis fusca* Boie, 1826

(No vernacular names recorded)

SIZE 500mm

IDENTIFICATION FEATURES Body slender; head indistinct from neck; snout short; loreal and prefrontal contact orbit; preocular absent; single postocular; eye reduced; pupil rounded; tail long; dorsals keeled; midbody scale rows 15; ventrals 146–158; subcaudals 74–91; cloacal scute entire.

COLOUR Top of body dark brown or reddish-brown, unpatterned or with yellow or red spots or flecks; sometimes, anterior of body yellow with dark brown vertebral stripe and dark spots on flanks; posterior part black and iridescent; venter yellow or cream.

HABITAT AND BEHAVIOUR Inhabits forests at 900–1,000m asl. Nocturnal and subfossorial. Diet unstudied. Clutches comprise 2–4 eggs, each 31.5–33.5 x 9mm.

DISTRIBUTION Sumatra, Borneo and Java. The Bornean record is from 'North Borneo' that may refer to either northern Sabah or northeastern Kalimantan.

IUCN THREAT STATUS Least Concern.

Hebius Keelbacks

Small to medium-sized aquatic, aquatic-margin or wetland-associated terrestrial snakes, with a distinct head; large eyes with rounded pupils; dorsal scales with a keel; short tail.

Key to Bornean species of *Hebius*

1a. Midbody scale rows 19 .... **2**
1b. Midbody scale rows 17 .... **3**

2a. White-tipped snout .... *H. flavifrons* (p. 278)
2b. Snout not white-tipped .... *H. petersii* (p. 280)

3a. Ventrals < 157 .... *H. sarawacensis* (p. 281)
3b. Ventrals > 163 .... **4**

4a. Loreal absent; ventrals 187 .... *H. arquus* (below)
4b. Loreal present; ventrals 164–166 .... *H. frenatum* (p. 279)

BORNEAN KEELBACK *Hebius arquus* (David & Vogel, 2010)

(No vernacular names recorded)

SIZE 609mm

IDENTIFICATION FEATURES Body slender; head distinct from neck; eyes large with rounded pupils; dorsals keeled; midbody scale rows 17; ventrals 187; subcaudals 110, paired; cloacal scute divided.

COLOUR Colour notes are from a century-and-half preserved specimen. Top of body dark greyish-brown; most scales edged with black; anterior third of body bears large, dark brown blotches, with white, irregular and indistinct spots in between; middle portion of dorsum unpatterned greyish-brown, with some white spots; posterior third also greyish-brown, with elongated dark brown streaks that form stripes; white ventrolateral stripe, edged with blackish-brown line below, extends from neck to around midbody; belly pale yellowish-ochre, speckled with dark brown.

HABITAT AND BEHAVIOUR Nothing known of biology of this species.

DISTRIBUTION Assumed to be endemic to Borneo. Locality within Borneo unknown. The holotype had no field data.

IUCN THREAT STATUS Data Deficient.

Holotype of *Hebius arquus* (David & Vogel, 2010) (NMW 37943) from 'Borneo', showing body in dorsal view (left) and head in lateral view (right).

WHITE-FRONTED KEELBACK *Hebius flavifrons* (Boulenger, 1887)

(Bahasa Malaysia/Indonesia: Ular Hidung Putih. Iban: Ular Ginti)

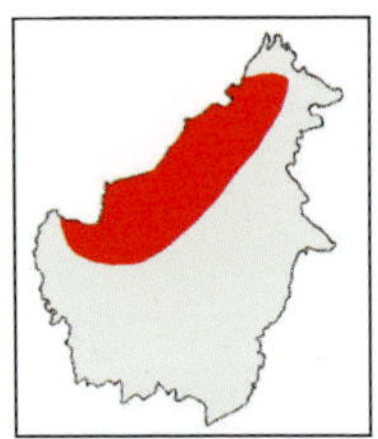

SIZE 750mm

IDENTIFICATION FEATURES Body slender; head distinct from neck; eyes large with rounded pupils; dorsals keeled; midbody scale rows 19; ventrals 146–157; subcaudals 92–101, paired; cloacal scute divided.

COLOUR Top of body of adults olive-grey with darker markings; distinctive white to yellowish-cream spot on snout; belly cream with alternating series of black spots. Juveniles have paired white spots along mid-back.

HABITAT AND BEHAVIOUR Inhabits lowlands and associated with rivers. Diet includes frogs, their eggs and tadpoles. Reproductive biology unstudied.

DISTRIBUTION Endemic to Borneo. Widespread across the island, with records from Bako National Park, Gunung Gading National Park, Gunung Penrissen, Bintulu, Nanga Segerak, Sungei Mengiong and Sungei Pesu in Sarawak; Ulu Temburong in Brunei Darussalam, and Mendolong, Gunung Kinabalu (including Kiau, Sungei Kadamaian, Sungei Luidan and Poring, along Sungei Kipungit), Rompon on Trus Madi, Sungei Purulon and Sungei Rangkam in Sabah.

IUCN THREAT STATUS Least Concern.

BRIDLED KEELBACK *Hebius frenatum* (Dunn, 1923)

(No vernacular names recorded)

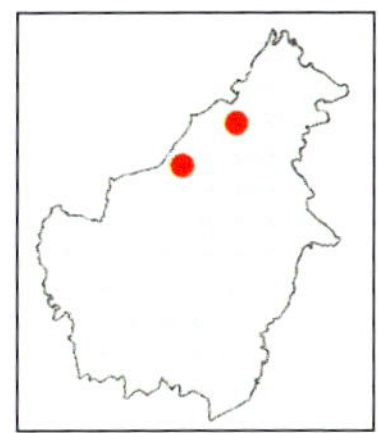

SIZE 610mm

IDENTIFICATION FEATURES Body slender; head distinct from neck; eyes large with rounded pupils; tail long; dorsals keeled; midbody scale rows 17; ventrals 164–166; subcaudals 112–116, paired; cloacal scute divided.

COLOUR Top of body brown with paired black cross-bars, enclosing cream bar on anterior of body, turning darker posteriorly; narrow white chevron on neck contacts supralabials; short black stripe covers lower parts of last three supralabials; forehead dark with small black spots; belly grey with large brown and black spots.

HABITAT AND BEHAVIOUR Inhabits submontane forests. Diurnal and terrestrial; found under fallen logs and leaf litter. Diet and reproductive habits unstudied.

DISTRIBUTION Endemic to northwestern Borneo. Localities where it occurs include Gunung Murud and Sungei Labang in Sarawak.

IUCN THREAT STATUS Data Deficient.

Peters's Keelback *Hebius petersii* (Boulenger, 1893)

(No vernacular names recorded)

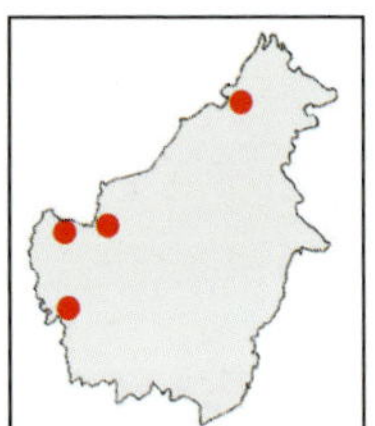

SIZE 600mm

IDENTIFICATION FEATURES Body slender; head distinct from neck; five infralabials contact anterior chin shields; eyes large with rounded pupils; dorsals keeled; midbody scale rows 19; ventrals 134–150; subcaudals 65–93, paired; cloacal scales divided.

COLOUR Top of body black or dark brown, with rounded black spots; forehead dark olive with black vermiculation; labials yellow with black sutures; belly yellow, scutes edged with black.

HABITAT AND BEHAVIOUR Inhabits lowland forests, and associated with streams. Diurnal and terrestrial/aquatic. Diet includes agamid lizards. Reproductive habits unstudied.

DISTRIBUTION Peninsular Malaysia, Singapore, Borneo and Sumatra. Bornean records include Sama Jaya Nature Reserve in Kuching, Gunung Gading National Park and Saribas in Sarawak, and Tenom District in Sabah.

IUCN THREAT STATUS Least Concern.

SARAWAK KEELBACK *Hebius sarawacensis* (Günther, 1872)

(No vernacular names recorded)

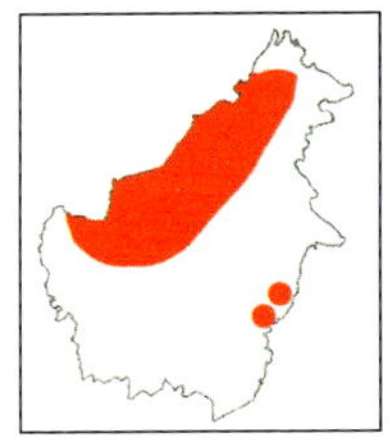

SIZE 780mm

IDENTIFICATION FEATURES Body slender; head distinct from neck; eyes large with rounded pupils; tail long; dorsals keeled; midbody scale rows 17; ventrals 134–154; subcaudals 52–112, paired; cloacal scute divided.

COLOUR Top of body olive to reddish-brown, with squarish black markings; row of light spots on flanks; supralabials yellow or cream; belly yellow and black.

HABITAT AND BEHAVIOUR Inhabits submontane forests at 640–1,700m asl. Terrestrial, near streams. Diet comprises frogs and their eggs. Oviparous, producing clutches of 4–5 eggs.

DISTRIBUTION Peninsular Malaysia and Borneo. Bornean records include Gunung Murud, the Matang Range, Gunung Dulit, Gunung Gading National Park, Gunung Penrissen, Gunung Pueh, Gunung Mulu National Park, Gunung Santubong National Park, Sungei Labang, Sungei Mengiong and Sungei Seran in Sarawak (with historical records from Kuching, including Jalan Penrissen); Penampang, Sinruron, Gunung Alap and Gunung Kinabalu (including Bundu Tuhan, Kenokok, Lumu-Lumu and Mesilau) in Sabah, and Balikpapan, Samarinda, Sebruang Valley and Sungei Kapuas in Kalimantan.

IUCN THREAT STATUS Least Concern.

Hydrablabes Bornean Water Snakes

Small, slender aquatic snakes, characterized by slender, rounded bodies, and head weakly distinct from neck; small eyes with rounded pupils; narrow snout; oblique suture between nasals; series of suboculars separating supralabials from eye; three pairs of chin shields; midbody scale rows 15 or 17; smooth scales; long tail.

Key to Bornean species of *Hydrablabes*

1a. Prefrontal single .... *H. praefrontalis* (opposite)
1b. Prefrontals 2 .... *H. periops* (below)

Yellow-spotted Water Snake *Hydrablabes periops* (Günther, 1872)

(No vernacular names recorded)

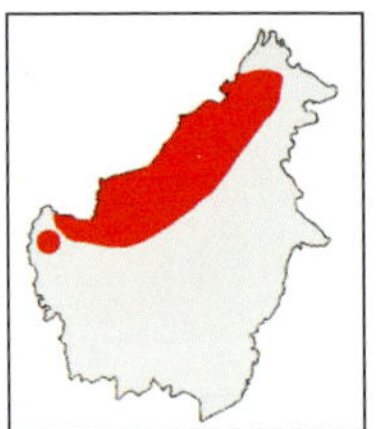

SIZE 530mm

IDENTIFICATION FEATURES Body slender; head small, slightly distinct from neck; paired prefrontals; loreal present; eyes small with rounded pupils; orbit separated from supralabials by series of suboculars; tail short; dorsals smooth; midbody scale rows 15 or 17; ventrals 179–209; subcaudals 56–76; cloacal scute divided.

COLOUR Top of body olive-brown, sometimes with pale brown stripe on flanks; some individuals unpatterned; belly yellow or grey.

HABITAT AND BEHAVIOUR Inhabits lowland and mid-hill forests. Diurnal and aquatic, associated with streams, at 150–600m asl. Diet and reproductive habits unstudied.

DISTRIBUTION Endemic to Borneo. Localities include the Matang Range, Nanga Serembuang along Sungei Skrang, Sungei Mengiong and Sungei Pesu in Sarawak; Ulu Temburong in Brunei Darussalam; Sungei Kallang, Sungei Malutut and Sungei Purulon in Sabah, and Landak in Kalimantan.

IUCN THREAT STATUS Least Concern.

Kinabalu Water Snake *Hydrablabes praefrontalis* (Mocquard, 1890)

(No vernacular names recorded)

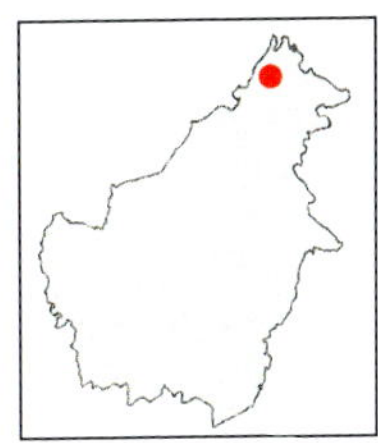

SIZE 346mm

IDENTIFICATION FEATURES Body slender; head small, slightly distinct from neck; prefrontals fused; eyes small with rounded pupils; orbit separated from supralabials by series of suboculars; tail short; dorsals smooth; midbody scale rows 15; ventrals 178–202; subcaudals 72; cloacal scute divided.

COLOUR Top of body olive-brown, with dark vertebral stripe; paired dark dorsolateral stripe, broken into spots posteriorly; belly yellow; edges of ventrals have dark brown spots; subcaudals edged with brown.

HABITAT AND BEHAVIOUR Inhabits submontane or montane forests. Presumably aquatic, associated with hill streams. Diet and reproductive habits unstudied.

DISTRIBUTION Endemic to northwestern Borneo. Sole record is from Gunung Kinabalu in Sabah.

IUCN THREAT STATUS Data Deficient.

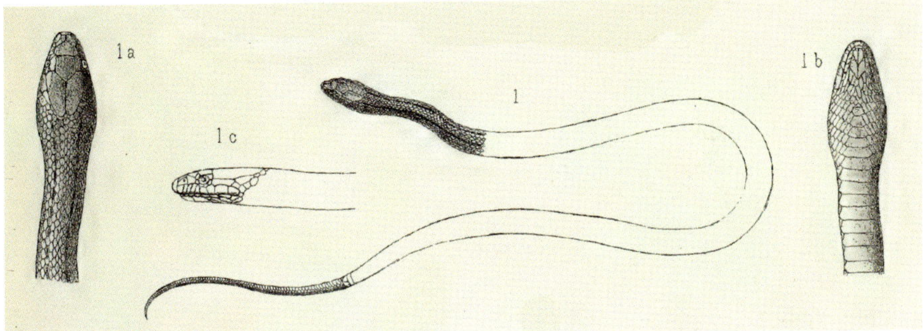

After Mocquard (1890).

Opisthotropis Corrugated Snakes

Small, slender snakes; head depressed and indistinct from neck; eyes small; maxillary teeth small; single, broad prefrontal; nostrils dorsally oriented; subcaudals paired and tail moderate.

Corrugated Water Snake *Opisthotropis typica* (Mocquard, 1890)

(No vernacular names recorded)

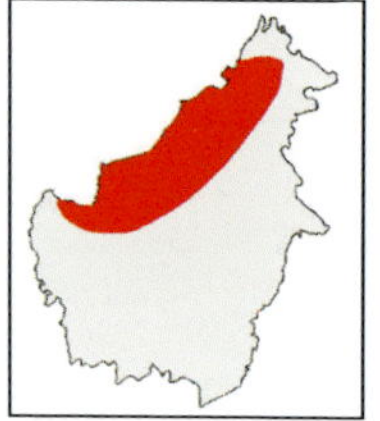

SIZE 502mm

IDENTIFICATION FEATURES Body slender; head small, depressed, indistinct from neck; prefrontal divided; forehead scales finely striated; loreal large; eyes small, and separated from labials by small scales; dorsals strongly keeled; midbody scale rows 19; ventrals 160–176; subcaudals 82–96; cloacal scales divided.

COLOUR Top of body unpatterned blackish-grey; venter unpatterned cream.

HABITAT AND BEHAVIOUR Inhabits forested plains and mid-hills, at 75–500m asl. Nocturnal and aquatic; associated with shallow, rocky streams and swamps. Diet comprises tadpoles. Reproductive biology unstudied.

DISTRIBUTION Borneo and Palawan in the Philippines. Bornean records include Kubah National Park, Baleh National Park and Pelagus National Park in Sarawak; Ulu Temburong in Brunei Darussalam, and Poring (along Sungei Kipungit II), Gunung Kinabalu (Kiau and Poring) in Sabah.

IUCN THREAT STATUS Least Concern.

Psammodynastes Mock Vipers

Small, slender, cylindrical snakes; head distinct from neck; maxillary with combination of small and large teeth; lores concave; eyes large with vertical pupils; middorsal scales smooth; subcaudals paired; tail moderate.

Key to Bornean species of *Psammodynastes*

1a. Chin shields, two pairs; forehead not streaked.......................................*P. pictus* (below)
1b. Chin shields, three pairs; forehead with dark and light streaks........*P. pulverulentus* (p. 286)

PAINTED MOCK VIPER *Psammodynastes pictus* Günther, 1858

(Bahasa Malaysia: Ular Sampah. Bahasa Indonesia: Ular Berang, Ular Percha)

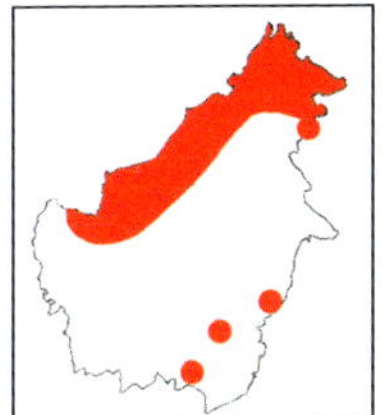

SIZE 550mm

IDENTIFICATION FEATURES Body slender; head elongated, flattened and distinct from neck; eyes large with vertical or rounded pupils; dorsals smooth; midbody scale rows 17; ventrals 152–171; subcaudals 60–80, paired; cloacal scute entire.

COLOUR Top of body brown, tan or black, with dark-edged light transverse bands; dark streak across eyes; belly cream with brown speckles.

HABITAT AND BEHAVIOUR Inhabits lowland forests. Diet comprises fish, frog, lizards and prawns. Ovoviviparous.

DISTRIBUTION Peninsular Malaysia, Singapore, Borneo, Sumatra, Riau and Mentawai archipelagos, and Pulau Belitung. Bornean records are from Semenggoh Wildlife Sanctuary, Gunung Mulu National Park, Kudi in Simunjan, Gunung Dulit, Niah and Sungei Labang, Sungei Nyabau, Sungei Pesu, Sungei Seran, Sungei Tangap and Sungei Rajang in Sarawak; Ulu Temburong in Brunei Darussalam; Danum Valley, Deramakot, Dewhurst Bay, Sepagaya Forest Reserve, Kalabakan and Kota Kinabalu in Sabah, and Balikpapan, Telang, Pulau Nunukan and Sungei Kapuas in Kalimantan.

IUCN THREAT STATUS Least Concern.

MOCK VIPER *Psammodynastes pulverulentus* (Boie, 1827)

(Bahasa Indonesia: Ular Viper Tiruan)

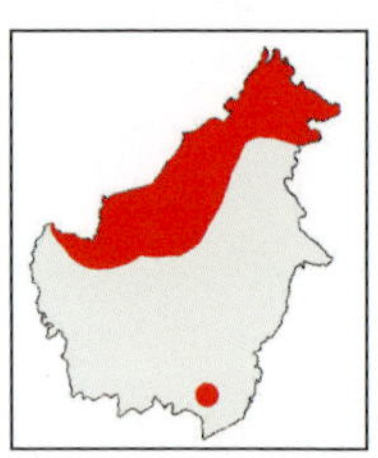

SIZE 770mm

IDENTIFICATION FEATURES Body slender; head flattened and distinct from neck; snout short; eyes large with vertical pupils; dorsals smooth, lacking apical pits; midbody scale rows 17 or 19; ventrals 146–175; subcaudals 44–70, paired; cloacal scute divided.

COLOUR Top of body reddish-brown to yellowish-grey and nearly black, with small dark spots or streaks; typically a longitudinal stripe along middorsal region and three longitudinal stripes along flanks; belly spotted with brown or grey, and with dark spots or longitudinal lines.

HABITAT AND BEHAVIOUR Inhabits lowland forests and forested hills at < 2,000m asl. Terrestrial and semi-arboreal. Diet comprises skinks, snakes, frogs and geckos. Ovoviviparous, producing 3–10 neonates, each 148–178mm.

DISTRIBUTION Eastern and central India, Bhutan, Nepal, southern China, Taiwan, Myanmar, Thailand, Laos, Cambodia, Vietnam, Peninsular Malaysia, Borneo, Sumatra, Pulau Bangka, Mentawai, Natuna and Riau archipelagos, Java, Sulawesi, Lesser Sundas and the Philippines. Bornean localities are from Pangkalan Ampat at the foothills of Gunung Penrissen, Gunung Gading National Park, Gunung Mulu National Park, Niah National Park, Sungei Seran and Sungei Mengiong in Sarawak; Brunei Darussalam (without a precise locality); Danum Valley, Deramakot, Mendolong, Kota Kinabalu, Quoin Hill, Gunung Kinabalu Park (Kiau and Sungei Kipungit I), Sinsuron, Tawau Hills Park and Sungei Purulon in Sabah, and Sungei Kapuas in Kalimantan.

IUCN THREAT STATUS Least Concern.

Rhabdophis Keelbacks

Medium-sized terrestrial species with slender bodies; large eyes; internasal scales broad anteriorly, nostrils lateral; hemipenes with divided sulcus spermaticus; two posterior-most maxillary teeth enlarged, recurved and typically preceded by diastema; vertebral glands sometimes present.

Key to Bornean species of *Rhabdophis*

1a. Nape spreads to form hood.......................................2
1b. Nape cannot be spread.......................................3

2a. Supralabials 8; red areas in neck.......................................*R. flaviceps* (p. 289)
2b. Supralabials 7; blue areas in neck.......................................*R. rhodomelas* (p. 291)

3a. Anterior temporals 2.......................................*R. chrysargos* (below)
3b. Anterior temporal single.......................................4

4a. Supralabials bright red.......................................*R. murudensis* (p. 290)
4b. Supralabials yellow.......................................*R. conspicillatus* (p. 288)

Speckle-bellied Keelback *Rhabdophis chrysargos* (Schlegel, 1837)

(Bahasa Malaysia: Ular Rabong Perut Bintik. Bahasa Indonesia: Ular Kadut)

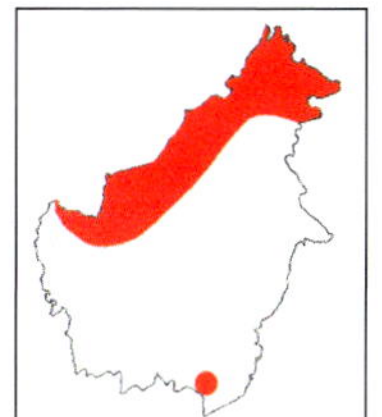

SIZE 980mm

IDENTIFICATION FEATURES Body moderate and cylindrical; head distinct from neck; eyes large with rounded pupils; tail short; dorsals keeled; subcaudals paired; midbody scale rows 19; ventrals 139–176; subcaudals 56–94, paired; cloacal scute divided.

COLOUR Top of body olive-grey or olive-brown; supralabials yellow or cream with darker smudges; cream nuchal chevron, edged with black; reddish-brown or orange band behind neck; rest of back has yellow and brown oblong marks within darker bands; belly yellow with brown mottling

HABITAT AND BEHAVIOUR Inhabits lowland to submontane forests at 100–1,676m asl. Terrestrial, and active near streams. Diet includes frogs. Oviparous, laying 3–10 eggs, each 12–21 × 19.5–34mm.

VENOM Unknown, with insufficient clinical reports; no antivenoms available; those raised for related species may be used for significant envenoming.

DISTRIBUTION Myanmar, Thailand, Laos, Cambodia, Vietnam, Peninsular Malaysia, Borneo, Sumatra, Pulau Nias, Pulau Simeulue, Mentawai and Anamba archipelagos, Java, Bali, Sulawesi, Flores, Ternate, Palawan and Balabac. Bornean records are from Kubah National Park, Pulau Satang, Bintulu, Gunung Dulit, Gunung Gading National Park, Nanga Serembuang along Sungei Skrang, Sungei Baram, Sungei Labang, Sungei Mengiong, Sungei Pesu, Sungei Rajang and Sungei Seran in Sarawak; Lawas; Jalan Tutong, Brunei Darussalam; Danum Valley, Sandakan, Mendolong, Gunung Kinabalu (including Ranau, Sungei Langanan, Lobang, Lumu-Lumu and Sungei Kipungit II), Sungei Purulon and Sungei Sugut in Sabah, and Banjarmasin in Kalimantan.

IUCN THREAT STATUS Least Concern.

RED-BELLIED KEELBACK *Rhabdophis conspicillatus* (Günther, 1872)

(No vernacular names recorded)

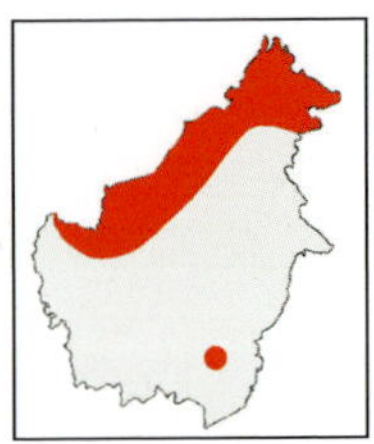

SIZE 550mm

IDENTIFICATION FEATURES Body moderately robust, cylindrical; head distinct from neck; eyes large with rounded pupils; tail short; dorsals keeled, except smooth outer rows; midbody scale rows 19; ventrals 141–152; subcaudals 51–60, paired; cloacal scute divided.

COLOUR Top of body brown to reddish-brown; sides of head have downwards curving cream postocular stripe; supralabials cream; nape and neck have paired, narrow cream collars; belly yellow, ventral dark edged.

HABITAT AND BEHAVIOUR Inhabits lowland and mid-hill forests at 100–1,000m asl. Terrestrial, near waterbodies, in leaf litter, tree buttresses or rotting logs, and under stones. Diet and reproductive habits unstudied.

VENOM Unknown, with insufficient clinical reports; no antivenoms available; those raised for related species may be used for significant envenoming.

DISTRIBUTION Peninsular Malaysia, Borneo, Sumatra, Pulau Singkep and Natuna Archipelago. Bornean records are from Sama Jaya Nature Reserve, Gunung Santubong National Park, Simanggang, the Matang Range, also, Kubah National Park, Gunung Dulit, Gunung Gading National Park, Niah National Park, Kapit, Lambir Hills National Park, Nanga Bekiat along Sungei Serembuang, Sungei Baram, Sungei Labang, Sungei Mengiong, Sungei Metallum, Sungei Nyabau, Sungei Seran, Sungei Segaham in Sarawak; Mendolong, Gunung Kinabalu Park (Poring), Sandakan Bay, Sungei Mamut, Sungei Malutut, Sungei Purulon and Sungei Tempasuk in Sabah, and Lihong Bahaya (unlocated) and Tanjung in Kalimantan.

IUCN THREAT STATUS Least Concern.

ORANGE-LIPPED KEELBACK *Rhabdophis flaviceps* (Duméril, Bibron & Duméril, 1854)

(Iban: Ular Kentu)

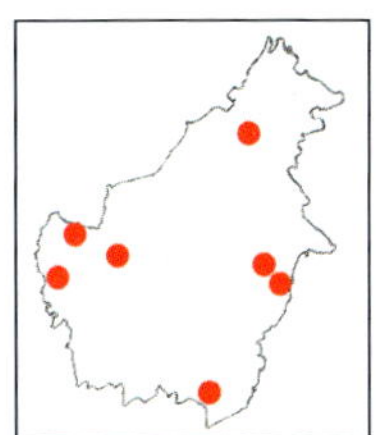

SIZE 850mm

IDENTIFICATION FEATURES Body robust; head distinct from neck; eyes large with rounded pupils; tail short; dorsals keeled; nuchal gland on nape; midbody scale rows 19; ventrals 120–138; subcaudals 49–60, paired; cloacal scute divided.

COLOUR Top of body greyish-black with faint, light cross-bars, narrow at vertebral region, wide on flanks; forehead pale brown, yellowish-brown or olive; supralabials rusty-orange; black-edged orange nuchal loop, especially in juveniles; belly black or dark green, with black bands.

HABITAT AND BEHAVIOUR Inhabits lowland and submontane forests at < 1,300m asl. Semi-aquatic. Diet includes frogs, toads, tadpoles and lizards. Reproductive biology unstudied.

VENOM Unknown, with insufficient clinical reports; no antivenoms available; those raised for related species may be used for significant envenoming.

DISTRIBUTION Thailand, Peninsular Malaysia, Borneo, Sumatra, Pulau Bangka and Pulau Nias. Bornean records are from the Kelabit Highlands and Kuching in Sarawak, and Banjarmasin, Kutai, Pontianak, Samarinda and Sebruang Valley in Kalimantan.

IUCN THREAT STATUS Least Concern.

Gunung Murud Keelback *Rhabdophis murudensis* (Smith, 1925)

(No vernacular names recorded)

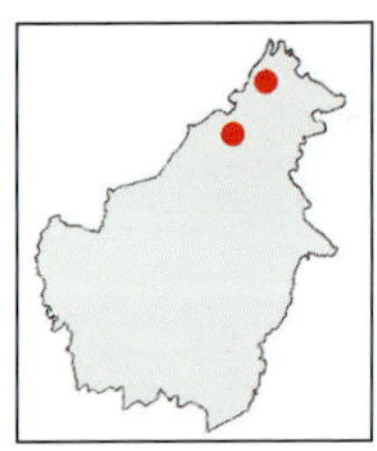

SIZE 873mm

IDENTIFICATION FEATURES Body moderately built and cylindrical; head distinct from neck; eyes large with rounded pupils; dorsals keeled, except smooth outer rows; midbody scale rows 19; ventrals 176–185; subcaudals 63–97, paired; cloacal scute divided.

COLOUR Top of body brownish-grey with indistinct, dark cross-bars, with row of light spots on their edges; supralabials bright red or brown; belly greyish-yellow with small black spots.

HABITAT AND BEHAVIOUR Inhabits submontane and montane forests at 915–2,500m asl. Diet includes frogs. Reproductive biology unstudied.

VENOM Unknown, with insufficient clinical reports; no antivenoms available; those raised for related species may be used for significant envenoming.

DISTRIBUTION Endemic to Borneo. Known only from Gunung Murud and Gunung Mulu in Sarawak, and Gunung Kinabalu (Kiau and Lumu-Lumu) and Trus Madi in Sabah.

IUCN THREAT STATUS Least Concern.

BLUE-NECKED KEELBACK *Rhabdophis rhodomelas* (Boie, 1827)

(No vernacular names recorded)

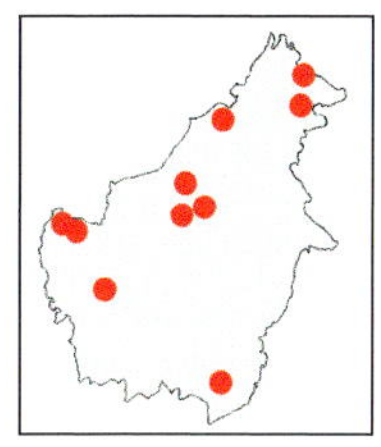

SIZE 750mm

IDENTIFICATION FEATURES Body robust; head distinct from neck; eyes large with rounded pupils; tail short; dorsals keeled; nuchal gland present on nape; midbody scale rows 19; ventrals 124–143; subcaudals 42–58, paired; cloacal scute divided.

COLOUR Top of body reddish-brown; black vertebral stripe enters nape as inverted chevron; nape pale blue posteriorly; belly pink, each ventral with small dark spot.

HABITAT AND BEHAVIOUR Associated with streams and other wetlands in lowlands and mid-hills (typically < 1,000m asl, although the Gunung Kinabalu locality is at 1,500–1,600m asl). Semi-aquatic. Diet comprises frogs, toads and tadpoles. Oviparous, laying 25 eggs.

VENOM Unknown, with insufficient clinical reports; no antivenoms available; those raised for related species may be used for significant envenoming.

DISTRIBUTION Thailand, Peninsular Malaysia, Singapore, Borneo, Sumatra and Java. Bornean localities include Kuching, Limbang, the Matang Range, Gunung Dulit, Gunung Gading National Park, Sungei Baleh, Sungei Mengiong and Sungei Pesu in Sarawak; Gunung Kinabalu (Lumu-Lumu), Danum Valley, Tawau Hills Park and Sandakan District in Sabah, and Sembawang, Bacong and Klumpang Bay in Kalimantan.

IUCN THREAT STATUS Least Concern.

Xenochrophis Keelback Water Snakes

Large aquatic snakes; body fairly robust; cylindrical; head distinct from neck; eyes large with rounded pupils; nostril within a single nasal; midbody scales with strong keels; ventrals rounded; subcaudals paired; tail moderate.

Key to Bornean species of *Xenochrophis*

1a. Upper flanks have dark triangles interspersed with red .... *X. trianguligerus* (opposite)
1b. Upper flanks lack dark triangles .... *X. maculatus* (below)

Malayan Spotted Keelback Water Snake

Xenochrophis maculatus (Edeling, 1864)

(No vernacular names recorded)

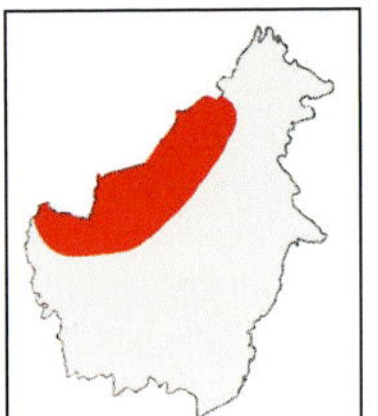

SIZE 1m

IDENTIFICATION FEATURES Body slender and cylindrical; head distinct from neck; eyes large with rounded pupils; dorsals keeled; midbody scale rows 19; ventrals 140–156; subcaudals 95–117; cloacal scute divided.

COLOUR Top of body brownish-olive, with four longitudinal series of dark, squarish marks, and paired row of yellow spots; forehead dark brown or black; supralabials yellow with darker smudges; belly yellow; ventrals edged with black

HABITAT AND BEHAVIOUR Inhabits open forests in lowlands. Terrestrial and semi-aquatic, associated with streams and ditches. Diet comprises frogs and toads. Reproductive habits unstudied.

DISTRIBUTION Peninsular Malaysia, Singapore, Borneo, Sumatra, Pulau Bangka, Pulau Belitung and Natuna and Riau archipelagos. Bornean records are from Kuching, Bako National Park, Gunung Dulit, Gunung Mulu National Park, Gunung Santubong National Park, Lundu, Lambir Hills National Park, Sungei Baram, Sungei Labang, Sungei Nyabau, Sungei Pesu, Sungei Seran, Sungei Tubau and Sungei Tangap in Sarawak; Labuan; Brunei Darussalam (without a precise locality), and Sipitang District and the Kinabalu region in Sabah.

IUCN THREAT STATUS Least Concern.

RED-SIDED KEELBACK WATER SNAKE *Xenochrophis trianguligerus* (Boie, 1827)

(Bahasa Malaysia: Ular Air. Bahasa Indonesia: Ular Cikopo Merah)

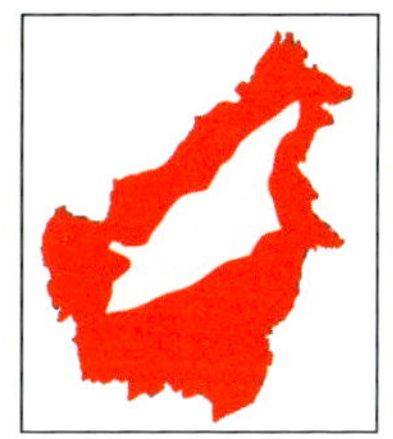

SIZE 1.35m

IDENTIFICATION FEATURES Body slender and cylindrical; head large and distinct from neck; eyes large with rounded pupils; tail short; dorsals keeled, except for outer 1–2 rows; midbody scale rows 19; ventrals 132–150; subcaudals 62–105; cloacal scute divided.

COLOUR Top of body blackish-brown, with orangish-red triangles on sides of neck and front of body; bright areas turn olive-brown or grey with age; dark, triangle-shaped marks on top of body; some scales on lips black edged; belly cream.

HABITAT AND BEHAVIOUR Inhabits lowland forests and areas with rice paddies. Semi-aquatic, near streams, standing waterbodies, small ditches and puddles at < 1,400m asl. Diet comprises frogs. Oviparous, laying 5–15 eggs, each 15–17 × 29–34mm.

DISTRIBUTION Myanmar, Thailand, Cambodia, Vietnam, Peninsular Malaysia, Singapore, Borneo, Sumatra, Pulau Nias, Pulau Bangka, Pulau Belitung, Mentawai and Riau archipelagos, Java, Sulawesi, Nicobar Archipelago. Bornean records are from Bako National Park, Semenggoh Nature Reserve, Samunsam Wildlife Sanctuary, Gunung Santubong National Park, Kuching, Kubah National Park, Nanga Segerak along Sungei Engkari, Long Lelang near Sungei Akah, Sungei Baram, Sungei Nyabau, Sungei Pesu, Sungei Rajang, Sungei Segaham, Sungei Seran, Sungei Subis and Sungei Tangap in Sarawak; Sungei Liang in Brunei Darussalam; Bongon, Danum Valley, Deramakot, Dewhurst Bay, Kasigui near Kota Kinabalu, Gunung Kinabalu (including Ranau and Bundu Tuhan), Mendolong, Sandakan, Sepagaya Protection Forest Reserve, Ulu Dusun, Tawau Hills Park, Sungei Malutut and Sungei Tibas in Sabah, and Kendawan, Pontianak, Samarinda, Singkawang and Telang in Kalimantan.

IUCN THREAT STATUS Least Concern.

Family Pseudoxenodontidae False Cobras

Members of this small family are united by morphological characteristics such as large eyes, keeled middorsal scales; two enlarged posteriormost maxillary teeth, obliquely oriented dorsals on anterior of body; and hemipenes extending to subcaudal 20, and forked at subcaudal 8. Threat display includes raising the head and spreading a small hood, presumably as a mimicry of cobras.

Pseudoxenodon False Cobras

Large, rounded terrestrial species from montane regions, diagnosable in having the following: head distinct from neck; eyes large with rounded pupils; scales on anterior of dorsum obliquely arranged; ventrals large and tail moderate.

Baram False Cobra *Pseudoxenodon baramensis* (Smith, 1921)

(No vernacular names recorded)

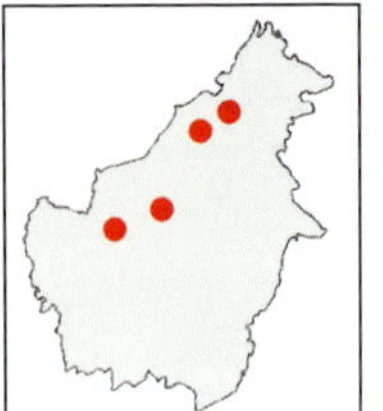

SIZE 825mm

IDENTIFICATION FEATURES Body robust; head short and distinct from neck; single preocular; nostrils and eyes large, and pupils rounded; tail moderate; dorsal scales keeled; midbody scale rows 19; ventrals 134; subcaudals 47, paired.

COLOUR Top of body greyish-olive, with indistinct dark network; belly yellow, especially towards end; outer margins of subcaudals pale lined.

HABITAT AND BEHAVIOUR Inhabits submontane forests at about 1,000m asl. Diet and reproductive habits unknown.

DISTRIBUTION Endemic to Borneo. Records exist from the Kelabit Highlands, including Gunung Murud, Gunung Dulit, Nanga Segerak in Sungei Engkari and Sungei Mengiong in Sarawak.

IUCN THREAT STATUS Data Deficient.

Family Elapidae Cobras, Kraits, Coral Snakes & Sea Snakes

Members of this family famously produce neurotoxic venom and include many of the most dangerously venomous snakes found on land and in water in South-east Asia. There are three subfamilies (recognized as families by some authors):

- **Elapinae**, comprising cobras, kraits and coral snakes, which are mostly terrestrial, have permanently erect fangs, are oviparous and have a distribution in Asia, New Guinea, Australia and Africa.
- **Hydrophiinae**, found mostly in marine environments (a few species occur in freshwater habitats), have a laterally compressed body, oar-like tail, lingual salt glands and greatly reduced ventrals. They are ovoviviparous, and are widespread in warmer seas.
- **Laticaudinae**, which are restricted to marine habitats but come ashore to rest, eat and bask, have an oar-like tail, relatively wide ventrals and laterally located nostrils. They are oviparous, and have a distribution centred around the Indo-Pacific.

Key to Bornean subfamilies of Elapidae

1a. Tail terminating into a point..Elapinae
1b. Tail flattened..**2**

2a. Ventrals > x 3 body scales; body completely encircled by ring-like dark bands..Laticaudinae
2b. Ventrals < x 3 body scales; body not completely encircled by ring-like dark bands..Hydrophiinae

Elapinae: Cobras, Kraits & Coral Snakes

This subfamily includes the terrestrial venomous elapids, including the cobras, kraits and coral snakes.

Key to Bornean genera of Elapinae

1a. Vertebrals enlarged ....*Bungarus*
1b. Vertebrals not enlarged..**2**

2a. Midbody scale rows 13..*Calliophis*
2b. Midbody scale rows > 13..**3**

3a. Enlarged occipitals..*Ophiophagus*
3b. No enlarged occipitals..*Naja*

Bungarus Kraits

Moderate to large species of elapid with the following characteristics: body triangular in cross-section; head scarcely distinct from body; tail short; eyes small with rounded pupils; loreal absent; dorsals smooth; vertebrals enlarged; tail short.

Key to Bornean species and subspecies of _Bungarus_

1a. Midbody scale rows 15 or 17; body with encircling black rings.....*B. fasciatus* (below)
1b. Midbody scale rows 13; body without encircling black rings............**2**

2a. Body black, with red and white bands at posterior; tail banded.......................*B. flaviceps baluensis* (opposite)
2b. Body uniformly black; tail lacks bands ....*B. flaviceps flaviceps* (opposite)

Banded Krait *Bungarus fasciatus* (Schneider, 1801)

(Bahasa Malaysia/Indonesia: Ular Buku Tebu, Ular Katam. Dusun: Tekamas. Iban: Ular Kendawang, Ular Kengkang Tebu, Ular Tetak Tebu)

SIZE 2.25m

IDENTIFICATION FEATURES Body robust, triangular in cross-section, with raised vertebral region; eyes small; vertebrals enlarged; tail blunt tipped; dorsals smooth; midbody scale rows 15; ventrals 200–236; subcaudals 23–39, single; cloacal scute entire.

COLOUR Top of body yellow with black bands subequal to pale interspaces; forehead has pale 'V'-shaped marking; belly pale yellow or brown, banded.

HABITAT AND BEHAVIOUR Inhabits secondary forests and swamps, and sometimes encountered at forest edges and near villages at < 2,500m asl. Terrestrial. Diet comprises snake, lizards, frogs, fish and reptile eggs. Oviparous; clutches include 3–14 eggs, each 22–38mm long.

VENOM Pre- and post-synaptic neurotoxins; may result in severe envenoming and potentially lethal. Polyvalent Anti Snake Venom Serum (Central Research Institute), Banded Krait Antivenin (Red Cross Society), Antivenin Polyvalent (Equine) (P.T. Bio Farma Persero, Bandung).

DISTRIBUTION Eastern India, Nepal, Bangladesh, China, Myanmar, Thailand, Cambodia, Laos, Vietnam, Peninsular Malaysia, Singapore, Borneo, Sumatra and Java. Bornean records are from Sadong Jaya Nature Reserve, Baleh National Park, Bako National Park, Lambir Hills National Park, Kuching, Sungei Baram and Sungei Tangap in Sarawak; Bukit Patoi and Tasek Merimbun in Brunei Darussalam; Lumat in Membakut, Menggatal, Tuaran and Gunung Kinabalu (Poring and Ranau) in Sabah, and Pontianak and Singkawang in Kalimantan.

IUCN THREAT STATUS Least Concern.

RED-HEADED KRAIT *Bungarus flaviceps* Reinhardt, 1843

(Bahasa Malaysia: Ular Katang Kepala Merah. Iban: Ular Kendawang)

SIZE 2.07m

IDENTIFICATION Features Body robust and triangular in cross-section; head large, distinct from neck; snout blunt; eyes small with rounded pupils; tail relatively short; dorsals smooth; midbody scale rows 13; ventrals 193–236; subcaudals 42–54, anterior ones entire, posterior ones divided; cloacal scute entire.

COLOUR Top of body blue-black with yellow or tan vertebral stripe; forehead and tail red, orange or yellow; belly pink or yellow; in subspecies *flaviceps* head red, orange-yellow or yellow; body lacks black rings; in subspecies *baluensis* posterior half of body and tail has eight pairs of broad black rings, separated with a narrow white ring.

HABITAT AND BEHAVIOUR Inhabits lowland and mid-hill forests. Nocturnal and terrestrial. Diet comprises other snakes, and lizards such as skinks. Reproductive biology unstudied.

VENOM Pre- and post-synaptic neurotoxins; may result in severe envenoming that is potentially lethal. Banded Krait Antivenin (Thai Red Cross Society).

DISTRIBUTION Southern Myanmar, Thailand, Peninsular Malaysia, Borneo, Sumatra, Pulau Belitung and Java. On Borneo, subspecies *flaviceps* is widespread in lowland and hill dipterocarp forests, with records from Bintulu, Baleh National Park, Kubah National Park, Sama Jaya Nature Reserve, the Kelabit Highlands, Gunung Mulu National Park, Gunung Santubong National Park, Sibu, Pangkalan Ampat at the foothills of Gunung Penrissen, Simanggang, Sungei Baram, Sungei Mengiong, Sungei Pesu and Sungei Tangap in Sarawak; Mendolong and Poring in Sabah, and Balikpapan and Samarinda in Kalimantan. Subspecies *baluensis* is restricted to submontane forests of Gunung Kinabalu (Kundasang, Kiau and Poring) and Trus Madi at 550–914m.

IUCN THREAT STATUS Least Concern.

Calliophis Coral Snakes

Small to large species of elapid, with the following characteristics: body relatively slender, cylindrical; head small and indistinct from neck; single, elongated temporal; eyes small with rounded pupils; body rounded in cross-section; midbody scale rows 13 or 15; ventrals rounded; subcaudals divided; tail short.

Key to Bornean species of *Calliophis*

1a. Belly and under surface of tail lacks cross-bars;
> 34 subcaudals .... *C. bivirgatus* (below)

1b. Belly and under surface of tail with cross-bars;
< 34 subcaudals **2**

2a. Venter white .... *C. intestinalis* (opposite)

2b. Venter red .... *C. nigrotaeniatus* (p. 300)

Blue Coral Snake *Calliophis bivirgatus* (Boie, 1827)

(Bahasa Malaysia: Ular Pantai Biru Biru, Ular Sina Mata Hari. Bahasa Indonesia: Ular Tjabeh. Iban: Ular Siran, Ular Kendawan)

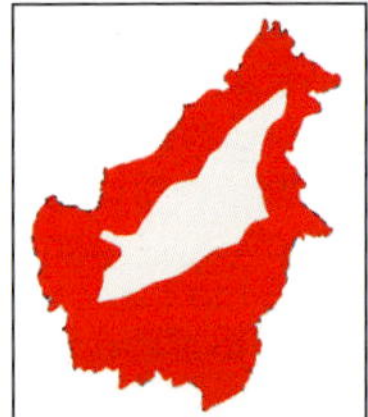

SIZE 1.85m

IDENTIFICATION FEATURES Body slender; head short, indistinct from neck; eyes small with rounded pupils; tail short, terminating in sharp point; dorsals smooth; midbody scale rows 13; ventrals 243–304; subcaudals 34–53, paired; cloacal scute divided.

COLOUR Top of body dark blue or blue-black, with distinct stripes along flanks; head, tail and belly coral-red. Subspecies: *bivirgatus* – narrow white paravertebral stripe; *flaviceps* – pale blue paravertebral stripe; *tetrataenia* – four lateral stripes, the outermost broadest.

HABITAT AND BEHAVIOUR Inhabits lowland forests, ascending up to submontane forests at < 1,200m asl. Sometimes seen in forest edges including in agricultural areas. Diet includes other snakes. Oviparous, producing 1–3 eggs, each 35–35 × 9mm.

VENOM Unknown, and potentially lethal envenoming. No antivenom raised.

DISTRIBUTION Myanmar, Thailand, Laos, Cambodia, Peninsular Malaysia, Singapore, Borneo, Sumatra, Pulau Nias, Pulau Bangka, Mentawai and Riau archipelagos and Java. Bornean records are from the Kelabit Highlands at Bario, Bako National Park, Bintulu, Banting, Pangkalan Ampat at the foothills of Gunung Penrissen, Niah National Park, Baleh National Park, Busau, Kuching, Limbang, Kubah National Park, Sibu, Tegora, Pangkalan Lobang, Sungei Labang, Sungei Mengiong and Sungei Pesu in Sarawak; Ulu Temburong in Brunei Darussalam; Bongon, Gunung Kinabalu (Lumu-Lumu), 8th mile Crocker Range Park, Dewhurst Bay, Ulu Dusun, Kampung Lawa Mandau In Menggatal and Sungei Purulon in Sabah, and Balikpapan, Banjarmasin, Buntok, Landak, Muara Jawa, Muarateweh, Pontianak, Sangkulirang, Sintang, Sungei Kapuas and Sungei Mahakam in Kalimantan.

IUCN THREAT STATUS Least Concern.

MALAYAN STRIPED CORAL SNAKE *Calliophis intestinalis* (Laurenti, 1768)

(Bahasa Malaysia/Indonesia: Ular Cabe, Ular Tali Kasut. Iban: Ular Siram)

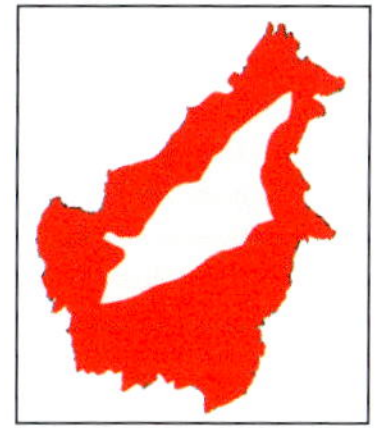

SIZE 710mm

IDENTIFICATION FEATURES Body slender; head small, indistinct from neck; loreal absent; single preocular; 3–4 infralabials contact anterior chin shields; eyes small; dorsals smooth; midbody scale rows 13 or 15; ventrals 197–273; subcaudals 15–33, divided; cloacal scales entire.

COLOUR Top of body grey-black; bifurcated yellow vertebral stripe extends to tail; dorsolateral stripes two scales wide; belly has broad black and white transverse bands, the black cross-bands not in contact with black on sides; forehead rufous-brown.

HABITAT AND BEHAVIOUR Inhabits primary lowland forests and open areas, such as parks and gardens, up to 1,100m asl. Nocturnal, and terrestrial as well as semi-fossorial in forest litter. Defence display consists of raising tail, exposing distinctly dark, pale banded underside, or turning over backwards, or dorsolateral flattening of body, turned into 'S'-shape, with head concealed under body. Diet comprises other snakes. Oviparous, laying 1–3 eggs, each 27–35 x 8–9mm. Incubation period 84 days.

VENOM Unknown, and potentially lethal envenoming. No antivenom raised. Venom gland in the species is elongated, extending into body cavity and terminating in a club-shaped end.

DISTRIBUTION Southern Thailand, Cambodia, Vietnam, Peninsular Malaysia, Singapore, Borneo, Sumatra, Pulau Nias, Pulau Bangka, Pulau Belitung, the Mentawai and Riau Archipelagos (subspecies *lineata*), eastern Sumatra and Java (subspecies *intestinalis*), Palawan, Balabac and Busuanga (subspecies *bilineata*), Luzon, Mindanao and Samar (subspecies *philippina*) and the Sulu Archipelago (subspecies *suluensis*). Subspecies *thepassi* is endemic to Borneo. Localities on the island include Kubah National Park, Baleh National Park, Bintulu, Busau, Kuching, Niah National Park, Gunung Dulit, Lio Matoh, Gunung Mulu National Park, Paku, Pangkalan Ampat at the foothills of Gunung Penrissen, Simanggang, Gunung Santubong National Park, Tanjung Datu National Park, Tegora, Trusan, Sungei Baleh, Sungei Baram, Sungei Mengiong, Sungei Labang and Sungei Rajang in Sarawak; Labuan; Ulu Temburong in Brunei Darussalam; Gunung Kinabalu (Kiau, Bundu Tuhan, Kundasang and Jalan Kamborangoh), Pulau Gaya, Kota Kinabalu, Mendolong, Sandakan Bay, Ulu Dusun and Sungei Purulon in Sabah, and Landak, Tanjung and Sungei Mahakam in Kalimantan.

Ventral view

Dorso-lateral view

IUCN THREAT STATUS Least Concern.

RED-BELLIED CORAL SNAKE *Calliophis nigrotaeniatus* (Peters, 1863)

(No vernacular names recorded)

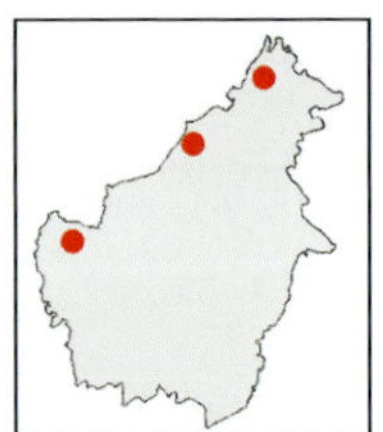

SIZE 587mm

IDENTIFICATION FEATURES Body slender; head small, indistinct from neck; loreal absent; single preocular; two infralabials contact anterior chin shields; eyes small; dorsals smooth; midbody scale rows 13; ventrals 223–236; subcaudals 21–33, paired; cloacal scute single.

COLOUR Top of body grey-black, with paired grey or dark blue paravertebral stripes; forehead yellowish-brown; belly red with black bands; under surface of tail has 2–5 dark bands.

HABITAT AND BEHAVIOUR Inhabits lower montane forests at c. 1,000–1,750m asl. One was found under a dead log, another was active at night at the edge of a partially logged forest. Diet and reproductive habits unstudied.

DISTRIBUTION Borneo, Sumatra and Peninsular Malaysia. On Borneo, known from Gunung Penrissen in Sarawak, and Gunung Kinabalu in Sabah.

IUCN THREAT STATUS Not Evaluated.

Naja Cobras

Medium-sized to large species; head distinct from neck; skin on neck dilatable; anterior ribs elongated; eyes moderately large; pupils rounded; fixed enlarged fangs followed by 2–3 smaller teeth on maxilla; loreal absent; middorsal scales smooth; subcaudals paired; tail relatively short.

Key to Bornean species of *Naja*

1a. Back of hood unicoloured .... *N. sumatrana* (p. 302)
1b. Back of hood with pale 'O'-shaped pattern .... *N. kaouthia* (below)

Monocled Cobra *Naja kaouthia* Lesson, 1831

(No vernacular names recorded)

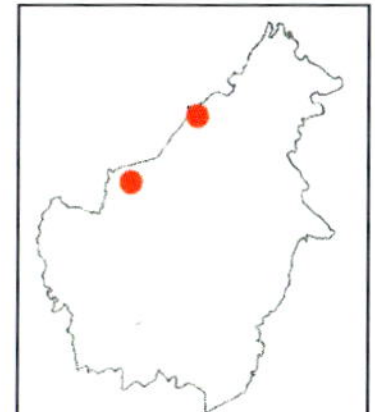

SIZE 2.3m

IDENTIFICATION FEATURES Body robust; head large, distinct from neck; frontal short, squarish; a cuneate usually present; single preocular; four infralabials contact anterior chin shields; hood rounded; eyes moderate and pupils rounded; tail short; dorsals smooth, glossy; midbody scale rows 19–21 (rarely 23); ventrals 164–197; subcaudals 43–61, paired.

COLOUR Top of body brown, greyish-brown, blackish-brown or pale yellow; some individuals have darker bands; hood marking typically a light circle, or mask shaped with dark centre; sometimes 1–2 dark spots in pale oval portion; light throat colour with paired lateral spots; rest of venter similar to dorsum or with dark pigmentation towards tail; subcaudals dark edged.

HABITAT AND BEHAVIOUR Inhabits forests, agricultural fields and plantations, from plains to 820m asl within its natural range (see distribution). Crepuscular and terrestrial; also known to swim in lakes and rivers. Diet comprises rodents, frogs, fish and snakes. Clutches comprise 15–30 eggs. Incubation period 50 days. Hatchlings measure 200–350mm.

VENOM Postsynaptic neurotoxins and necrotoxins present. Bites with severe envenoming possible and potentially lethal, and include pain, severe swelling, bruising and blistering, necrosis. Antivenin manufactured by Thai Red Cross Society.

DISTRIBUTION Natural range includes northern and eastern India, Nepal, Bangladesh and southern China (Sichuan and Yunnan Provinces), Myanmar, central, southwestern and peninsular Thailand, southern Laos, Cambodia, southern Vietnam and northern Peninsular Malaysia. Introduced through human agencies on Borneo, localities being Sibu and Miri in Sarawak.

IUCN THREAT STATUS Least Concern.

Sumatran Cobra *Naja sumatrana* Müller, 1890

(Bahasa Malaysia/Indonesia: Ular Senduk, Ular Senduk Sembur. Iban: Ular Belalang)

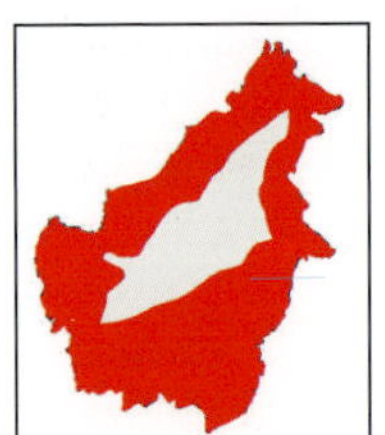

SIZE 1.5m

IDENTIFICATION FEATURES Body robust; head large, distinct from neck; snout rounded; hood rounded in adults, more elongated in juveniles; eyes moderate in size with rounded pupils; tail short; dorsals smooth; midbody scale rows 15–19; ventrals 179–206; subcaudals 40–57; basal ones typically entire; cloacal scute entire.

COLOUR Top of body bluish-black with plain hood; juveniles have pale, narrow cross-bars on dorsum, and pale throat with lateral spots and, often, median spot; juveniles have about 12 pale, narrow cross-bars; pale throat; bright yellow anterior ventrals; belly dark brown or brownish-grey.

HABITAT AND BEHAVIOUR Inhabits lightly forested areas in lowlands and mid-hills, in addition to towns and cities. Nocturnal and may shelter in holes, ditches and sewers in cities. Diet includes rats and other small vertebrates. Reproductive habits unstudied.

VENOM Postsynaptic neurotoxins; possibility of severe and lethal envenoming. No antivenom raised.

DISTRIBUTION Coasts off Peninsular Thailand, Peninsular Malaysia, Singapore, Borneo, Sumatra, Pulau Bangka, Pulau Belitung, Riau Archipelago, Palawan and the Calamianes Archipelago. Bornean localities include Bario in the Kelabit Highlands, Bintulu, Busau, Kuching, Kota Samarahan, Niah National Park, Baleh National Park, Gunung Santubong National Park, Lambir Hills National Park, Oya, Paku, Pangkalan Ampat at foothills of Gunung Penrissen, Bako National Park, Kampung Sebako at foothills of Gunung Pueh, Sungei Mengiong, Sungei Rajang in Sarawak; Labuan; Bandar Seri Begawan in Brunei Darussalam; Bongon, Kota Kinabalu (including Tuaran), Sandakan, Kudat, Gunung Kinabalu (Bundu Tuhan, Kiau and Ranau), Sungei Malutut, Ulu Dusun, Tawau Hills Park and Sungei Padas in Sabah, and Balikpapan, Long Iram, Montrado in Bengkayang Regency, Samarinda, Singkawang and Sungei Duri in Kalimantan.

IUCN THREAT STATUS Least Concern.

Adult

Juvenile

Ophiophagus King Cobras

Characteristics of this genus include paired occipital scales, dorsal banding (at least in juveniles), reduction of midbody scale rows to 15; subcaudal imbrication pattern typically comprises alternate series of fused and split scales; 4/7 renal arteries; paired air-sacs on dorsal side of body; series of tracheal air-sacs; lack of intrapulmonary bronchus; grooved palatine and dentary teeth; reduced number of pterygoid teeth; relatively narrow brain case; proportionately longer and more deeply bifurcate hemipenes.

Sunda King Cobra *Ophiophagus bungarus* (Schlegel, 1837)

(Bahasa Malaysia: Ular Tedung Selar. Bahasa Indonesia: King Kobra. Iban: Ular Belalang)

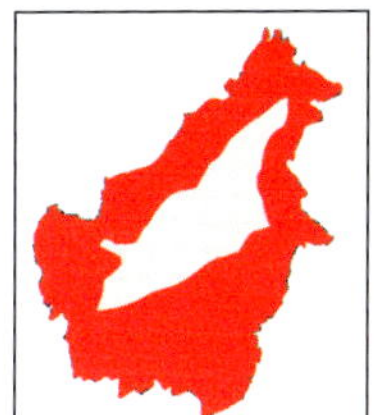

SIZE 5.6m

IDENTIFICATION FEATURES Body robust in adults, more slender in juveniles; head large in adults, distinct from neck; paired occipital scales present; neck dilated as a narrow hood; eyes large; dorsals smooth; midbody scale rows 15; ventrals 215–270; subcaudals 74–125, paired or single, or a combination of the two; cloacal scute entire.

COLOUR Top of body brownish-black, nearly unpatterned to with indistinct yellow bands; scales on posterior and tail lighter in middle; chin and throat yellow, belly dark grey. In juveniles, top of body dark brown or black with 57–87 pale yellow or orange bands; belly anteriorly yellow, turning grey beyond throat region.

HABITAT AND BEHAVIOUR Inhabits lowland and hill dipterocarp forests and mangrove swamps. Terrestrial; juveniles more arboreal. Diet includes other snakes and monitor lizards. Oviparous, eggs produced on mound nests of fallen leaves.

VENOM Postsynaptic neurotoxins; possibility of severe and lethal envenoming. King Cobra Antivenin (Thai Red Cross Society).

DISTRIBUTION Southern Thailand, Peninsular Malaysia, Singapore, Borneo, Sumatra, Borneo, Pulau Simeulue, Pulau Galang, Pulau Bangka, Pulau Belitung, Java, Bali and the southern

Philippines. The distribution above refers to the lineage from Sundaland. The species complex has a distribution from eastern and northern Pakistan and India, across south and east Asia, to eastern China. Bornean localities are from Bako National Park, Kapit, Kota Samarahan, Kubah National Park, Gunung Santubong National Park, Subis Forest Reserve, Niah National Park, Baleh National Park, Lio Matoh, Paku, Pangkalan Ampat at the foot of Gunung Penrissen, foothills of Gunung Pueh, Sibu, Sungei Baram, Sungei Ja'ong, Sungei Labang, Sungei Mengiong, Sungei Nyabau and Sungei Tangap in Sarawak (with historical records from Kuching); Brunei Darussalam; Danum Valley, Madai Caves, Sandakan, Sungei Serudong in Tawau and Sungei Rompon in Sabah, and Balikpapan, Sebulu and Sungei Bulangan in Kalimantan.

IUCN THREAT STATUS Vulnerable.

REMARKS In the recent past, all four species of the genus *Ophiophagus* were referred to as a single species, *O. hannah* (Cantor, 1836).

Dorsal view of forehead

Lateral view of head

Adult

Subfamily Hydrophiinae Sea Snakes

Members of this subfamily have exclusively marine representatives, with a few exceptions.

Key to Bornean species of Hydrophiinae

1a. Midbody scale rows > 73 .... *H. annandalei* (p. 307)
1b. Midbody scale rows < 73 .... **2**

2a. Rostral divided into 4–5 parts .... *H. anomalus* (p. 308)
2b. Rostral entire **3**

3a. Ventrals broadest anteriorly .... *H. viperinus* (p. 325)
3b. Ventrals, if distinct, not broadest anteriorly **4**

4a. Chin shield elongated .... *H. schistosus* (p. 321)
4b. Chin shield not elongated **5**

5a. Head elongated with bill-like snout; colouration includes black dorsum and yellow venter .... *H. platurus* (p. 320)
5b. Head not elongated and snout not bill-like; colouration of dorsum not black in combination with yellow venter **6**

6a. Ventrals reduced; lower middorsal scales enlarged .... *H. curtus* (p. 312)
6b. Ventrals not reduced; lower middorsal scales not enlarged **7**

7a. Ventrals enlarged, at least x 3 as adjacent scales .... *Aipysurus eydouxii* (p. 306)
7b. Ventrals not enlarged **8**

8a. Middorsal scale rows < 24 .... *H. jerdonii* (p. 316)
8b. Middorsal scale rows at least 24 .... **9**

9a. Ventrals divided by longitudinal groove posteriorly .... *H. gracilis* (p. 315)
9b. Ventrals lack longitudinal groove posteriorly **10**

10a. Maxillary teeth behind fangs at least 14 .... *H. caerulescens* (p. 311)
10b. Maxillary teeth behind fangs < 14 .... **11**

11a. Body bands parallel behind head and tail, with 6–11 grey bands .... *H. ornatus* (p. 319)
11b. Body bands not parallel behind head and/or tail, with < 6 bands **12**

12a. Midbody scale rows > 47 .... *H. fasciatus* (p. 314)
12b. Midbody scale rows ≤ 47 .... **13**

13a. Head small (≤ 8 mm **14**
13b. Head large (> 8 mm **16**

14a. Forehead without pale marking **15**
14b. Forehead with curved yellow marking .... *H. brookii* (p. 310)

15a. Forehead unicoloured black .... *H. atriceps* (p. 309)
15b. Forehead with horseshoe-shaped mark .... *H. klossi* (p. 317)

16a. Dark bands on body broader than their interspaces **17**
16b. Dark bands on body not broader than their interspaces **18**

17a. Forehead has yellow horseshoe-shaped mark .... *H. torquatus* (p. 324)
17b. Forehead has yellow dots .... *H. sibauensis* (p. 322)

18a. Bands on dorsum pale bluish-grey; interspaces narrower anteriorly .... *H. melanosoma* (p. 318)
18b. Bands on dorsum not pale bluish-grey; interspaces broader anteriorly **19**

19a. Interspaces 2–4 times broader than bands posteriorly .... *H. spiralis* (p. 323)
19b. Interspaces narrower than bands posteriorly .... *H. cyanocinctus* (p. 313)

Aipysurus Egg-eating Sea Snakes

Large, slender marine species with subcylindrical bodies; body scales imbricate; rostral single; nasals in contact; nostrils dorsally located; ventrals wide, about a third to half of body width; tail short, flattened.

Beaded Sea Snake *Aipysurus eydouxii* (Gray, 1849)

(No vernacular names recorded)

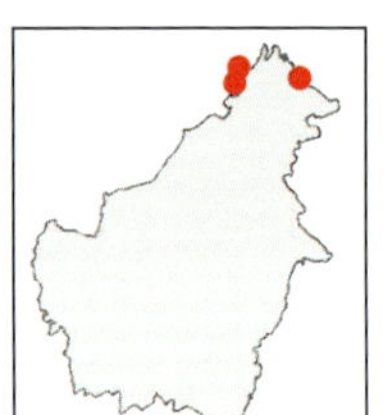

SIZE 1.5m

IDENTIFICATION FEATURES Body slender; head shields symmetrical, some with spines on posterior edges; supralabials not divided horizontally; pupils rounded; tail flattened; dorsals smooth; midbody scale rows 17; ventrals distinct throughout, 124–155; subcaudals 27–30; cloacal scute divided.

COLOUR Top of body brownish-olive, with 44–55 tan or yellowish-cream, dark-edged bands, sometimes broken up on vertebral region; bands widen on flanks; forehead dark brown or olive, nearly black in juveniles.

HABITAT AND BEHAVIOUR Inhabits coastal waters. Diet comprises benthic fish eggs. Reproductive biology unstudied.

VENOM Only member of the family to have lost its fangs and has reduced venom gland.

DISTRIBUTION Coasts off Thailand, Vietnam, Peninsular Malaysia, Singapore, Borneo, Sumatra, Java, the Philippines, New Guinea, Australia and New Caledonia. Bornean records are from Kimanis Bay, off Kota Kinabalu, and Pulau Tiga.

IUCN THREAT STATUS Least Concern.

Hydrophis Typical Sea Snakes

Large, robust or slender marine species with subcylindrical bodies; body scales imbricate; rostral single; nasals separated by internasals; nostrils dorsally located; ventrals narrow or absent, under quarter of body width; tail short and flattened.

Annandale's Sea Snake *Hydrophis annandalei* (Laidlaw, 1901)

(No vernacular names recorded)

SIZE 520mm

IDENTIFICATION FEATURES Body short and robust; head small, narrower than widest part of body; forehead scales fragmented; eyes small with rounded pupils; tail flattened; dorsals keeled; midbody scale rows 70–100; ventrals 320–368; cloacal scutes enlarged, bordered with 1–2 small scales.

COLOUR Top of body grey-blue with 35–46 dark bands, broader than pale interspaces; head olivaceous; belly pale yellow or cream.

HABITAT AND BEHAVIOUR Inhabits sea coasts, in shallow waters with sandy bottoms; also fresh waters. Diet comprises fish. Reproductive habits unstudied.

VENOM Postsynaptic neurotoxins; clinical effects uncertain, and may result in major envenoming. Sea Snake Antivenom (CSL Limited, Victoria).

DISTRIBUTION Coasts off Thailand, Cambodia, Vietnam, Peninsular Malaysia, Singapore, Borneo and possibly Sumatra. The sole Bornean record is from the Muara coast of Brunei Darussalam.

IUCN THREAT STATUS Data Deficient.

ANOMALOUS SEA SNAKE *Hydrophis anomalus* (Schmidt, 1852)

(No vernacular names recorded)

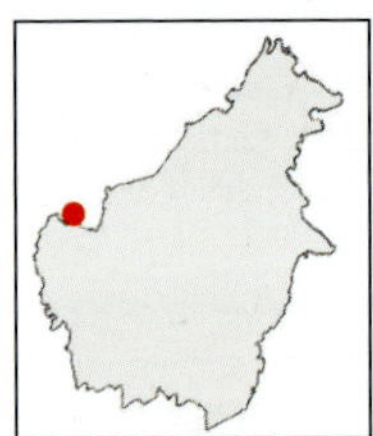

SIZE 810mm

IDENTIFICATION FEATURES Body robust; head short, depressed; rostral fragmented into 4–5 parts; single preocular; tail flattened; scales hexagonal with strong keels; ventrals subequal to dorsals; midbody scale rows 33 (rarely 35), scales hexagonal with strong keels; ventrals 210–256.

COLOUR Top of body pale grey with 30–36 dark cross-bars, including 3–4 on tail; cross-bars broader than interspaces, narrowing to point on flanks; belly unpatterned white.

HABITAT AND BEHAVIOUR Inhabits river mouths, presumably in sandy-bottom seas. Diet and reproductive habits unstudied.

VENOM Presumably with postsynaptic neurotoxins; possibility of severe and lethal envenoming. Sea Snake Antivenom (CSL Limited, Victoria).

DISTRIBUTION Coasts off Thailand, Cambodia, Vietnam, Peninsular Malaysia, Singapore, Borneo, Sumatra, Java, the Sulu Sea (the Philippines), Flores Sea and waters off Maluku (Indonesia). Bornean records are from the Santubong coast in Sarawak, and one generally mentioned as 'North Borneo', presumably the northern coast, off Sabah.

IUCN THREAT STATUS Data Deficient.

Black-headed Sea Snake *Hydrophis atriceps* Günther, 1864

(No vernacular names recorded)

SIZE 1.2m

IDENTIFICATION FEATURES Body robust posteriorly; head small; slender neck and forebody; dorsals have central keel; scales juxtaposed or slightly imbricate; single preocular and postoculars; tail flattened; midbody scale rows 39–49; ventrals 320–455.

COLOUR Top of body, including forehead and neck, dark olive to black with pale, oval yellow spots on sides, sometimes connected to form 50–75 cross-bars; posteriorly grey; yellow spot between nostril and eye or behind eye; belly cream; dark rhomboidal spots sometime extend down flanks to form annuli in juveniles.

HABITAT AND BEHAVIOUR Inhabits river mouths. Diet includes eels and marine invertebrates. Ovoviviparous, producing 2–3 neonates.

VENOM Presumably with postsynaptic neurotoxins; possibility of severe and lethal envenoming. Sea Snake Antivenom (CSL Limited, Victoria).

DISTRIBUTION Coasts off Thailand, Cambodia, Vietnam, Singapore, Borneo, the Philippines, New Guinea and northern Australia. Known from Borneo, without a specific locality.

IUCN THREAT STATUS Least Concern.

Rajah Brook's Sea Snake *Hydrophis brookii* Günther, 1872

(No vernacular names recorded)

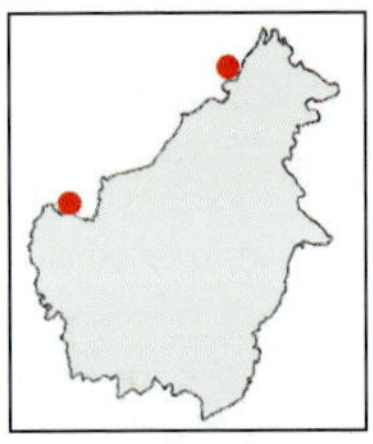

SIZE 1.03m

IDENTIFICATION FEATURES Body robust, elongated; head small; single preocular; tail flattened; dorsals keeled; midbody scale rows 37–45; ventrals 328–453; precloacal scutes enlarged.

COLOUR Top of body blackish-grey, with 60–80 dark grey cross-bars or bands, two times broader than pale interspaces; bands encircle anterior body; head black with curved yellow mark across snout along sides of head; belly yellowish-white.

HABITAT AND BEHAVIOUR Inhabits shallow coastal waters. Diet includes anguiliform eels, and also gobies. Ovoviviparous, producing seven neonates.

VENOM Presumably with postsynaptic neurotoxins; possibility of severe and lethal envenoming. Sea Snake Antivenom (CSL Limited, Victoria).

DISTRIBUTION Coasts off Thailand, Vietnam, Peninsular Malaysia, Singapore, Borneo, Sumatra and Java. Bornean records are from the Santubong coast of Sarawak, and Pulau Tiga off Sabah.

IUCN THREAT STATUS Least Concern.

BLUE-GREY SEA SNAKE *Hydrophis caerulescens* (Shaw, 1802)

(No vernacular names recorded)

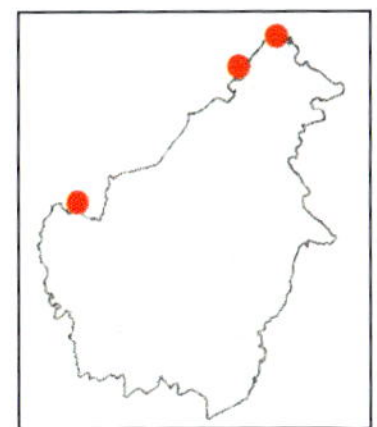

SIZE 1.09m

IDENTIFICATION FEATURES Body moderately robust; head small, with projecting upper jaw; eyes small with rounded pupils; scales quadrangular or hexagonal, weakly imbricate or juxtaposed; tail flattened; dorsals keeled; midbody scale rows 38–54; ventrals 253–334, less than twice as large as adjacent scales; cloacal scute divided.

COLOUR Top of body bluish-white or bluish-grey, with 40–60 dark bands, narrowing on lower flanks; forehead dark (juveniles with yellow or cream, 'U'-shaped mark); belly cream, with bands twice as wide as interspaces, tapering ventrally.

HABITAT AND BEHAVIOUR Inhabits shallow seas and estuaries. Diet comprises burrowing gobies. Ovoviviparous, producing 2–6 neonates.

VENOM Postsynaptic neurotoxins; clinical effects uncertain and may result in major envenoming. Sea Snake Antivenom (CSL Limited, Victoria).

DISTRIBUTION Coasts off Pakistan to northern Australia. In South-east Asia, known from Myanmar, Thailand, Peninsular Malaysia, Singapore, Borneo and Sumatra. Bornean records are from the Santubong coast of Sarawak and Kimanis Bay and Marudu Bay in Sabah.

IUCN THREAT STATUS Least Concern.

Short Sea Snake *Hydrophis curtus* (Shaw, 1802)

(Bahasa Indonesia: Ular Lempe)

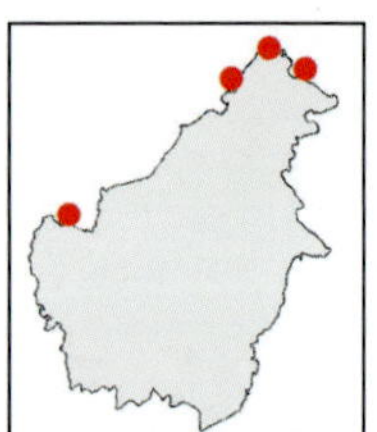

SIZE 972mm

IDENTIFICATION FEATURES Body robust and short; head broad and short; head shields entire; nostrils dorsally situated; eyes small with rounded pupils; tail flattened; scales squarish or hexagonal, with lowermost rows (especially in males) with short keel; midbody scale rows 25–43; ventrals 114–274; cloacal scutes weakly enlarged.

COLOUR Top of body of adults plain olive to dark grey; in juveniles, top of body brownish-grey to olive, with 35–55 olive to dark grey bands that taper to a point on flanks; belly plain yellow or cream, or with narrow, dark ventral stripe or irregular band.

HABITAT AND BEHAVIOUR Inhabits coasts with muddy bottoms; also coral reefs (at depths of 6–15m). Diet includes eels, gobies, catfish, squid and other marine invertebrates. Ovoviviparous, producing 1–6 neonates, each 250–373mm.

VENOM Postsynaptic neurotoxins; possibility of severe and lethal envenoming. Sea Snake Antivenom (CSL Limited, Victoria).

DISTRIBUTION Persian Gulf to Australia. In South-east Asia, known from coasts off Myanmar, Thailand, Peninsular Malaysia, Singapore, Borneo and Sumatra. Bornean records are from the Santubong coast of Sarawak, and Kimanis Bay, Marudu Bay and Kampung Mumiang in Sandakan, in Sabah.

IUCN THREAT STATUS Least Concern.

ANNULATED SEA SNAKE *Hydrophis cyanocinctus* Daudin, 1803

(Bahasa Indonesia: Ular Laut Besar Berdagu Kehitaman)

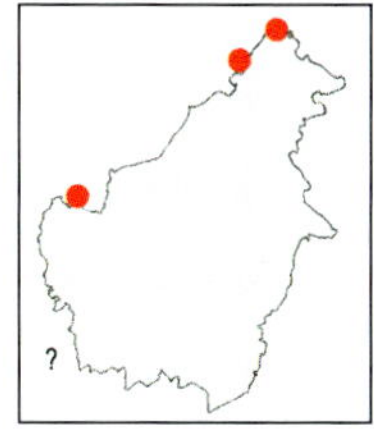

SIZE 1.88m

IDENTIFICATION FEATURES Body elongated and moderately robust anteriorly, thickening posteriorly; head small, almost indistinct from neck; eyes small with rounded pupils; tail flattened; dorsals strongly keeled; midbody scale rows 37–47; ventrals 290–390; cloacal scutes enlarged.

COLOUR Top of body typically olive or yellow, with numerous bluish-black transverse bands that may encircle body; head yellowish-green; belly yellowish-cream.

HABITAT AND BEHAVIOUR Inhabits shallow coastal waters. Diet comprises gobies, eels and marine invertebrates. Ovoviviparous, producing 3–16 neonates, each 381mm.

VENOM Postsynaptic neurotoxins; potential for severe and lethal envenoming. Sea Snake Antivenom (CSL Limited, Victoria).

DISTRIBUTION Extends from Persian Gulf to eastern China, Vietnam, the Philippines and Japan. In the region, also known from Thailand, Peninsular Malaysia, Singapore and Borneo. Bornean records are from the mouth of Sungei Sarawak in Sarawak; Kimanis Bay, Marudu Bay and Kampung Mumpiang in Sandakan in Sabah, and an unspecified locality along the southwestern coast of Kalimantan.

IUCN THREAT STATUS Least Concern.

Close-up of head and forebody

Banded Sea Snake *Hydrophis fasciatus* (Schneider, 1799)

(Bahasa Indonesia: Ular Lait. Bahasa Malaysia: Ular Selimpat. Iban: Ular Birang)

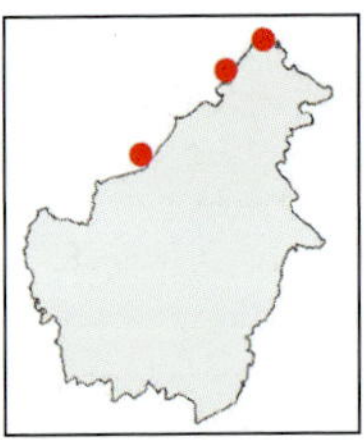

SIZE 1.1m

IDENTIFICATION FEATURES Body slender anteriorly, thickening posteriorly; head small; eyes small with rounded pupils; tail flattened; scales juxtaposed or slightly imbricate; midbody scale rows 47–58; ventrals 410–514; cloacal scutes divided.

COLOUR Top of body and forehead dark olive to black, with pale yellow oval spots on flanks sometimes fused as cross-bars; posterior grey; belly cream; dark rhomboidal spots along flanks, sometimes forming annuli in juveniles.

HABITAT AND BEHAVIOUR Inhabits shallow coastal waters. Diet comprises anguiliform eels. Reproductive habits unstudied.

VENOM Postsynaptic neurotoxins; potential for severe and lethal envenoming. Sea Snake Antivenom (CSL Limited, Victoria).

DISTRIBUTION Coasts off Pakistan, India, Myanmar, Thailand, Vietnam, Peninsular Malaysia, Singapore, Borneo, Sumatra and Java. Bornean records are from Kimanis Bay and Marudu Bay in Sabah; Bintulu in Sarawak, and one mentioned as from 'South China Sea near to Sarawak'.

IUCN THREAT STATUS Least Concern.

Narrow-headed Sea Snake *Hydrophis gracilis* (Shaw, 1802)

(No vernacular names recorded)

SIZE 950mm

IDENTIFICATION FEATURES Body slender anteriorly, thickening posteriorly; head small; projecting upper jaws; eyes small with rounded pupils; neck slender; juxtaposed scales on thickest part of body; tail flattened; dorsals keeled; midbody scale rows 29–43; ventrals 212–298, divided by longitudinal fissure; cloacal scutes weakly enlarged.

COLOUR Top of body yellow with 40–60 black bands or lateral blotches; forehead black; supralabials cream to grey; neck has dark bands with narrow, light grey interspaces; belly cream.

HABITAT AND BEHAVIOUR Inhabits deep, turbid coastal waters; possibly a bottom dweller. Diet includes anguiliform eels. Ovoviviparous, producing up to six neonates.

VENOM Postsynaptic neurotoxins; potential for severe and lethal envenoming. Sea Snake Antivenom (CSL Limited, Victoria).

DISTRIBUTION Persian Gulf to Australia and Melanesia. In the region, known from Myanmar, Thailand, Vietnam, Peninsular Malaysia, Singapore, Borneo, Sumatra and Java. Bornean records are from Sabah and off the Sarawak coast.

IUCN THREAT STATUS Least Concern.

Jerdon's Sea Snake *Hydrophis jerdonii* (Gray, 1849)

(No vernacular names recorded)

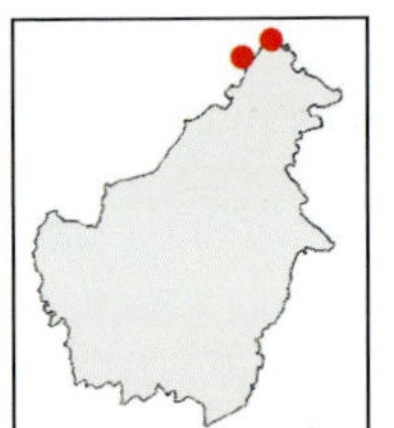

SIZE 1m

IDENTIFICATION FEATURES Body robust, subcylindrical and nearly uniform in diameter; head short, nearly indistinct from neck; single preocular and postocular; prefrontals usually not in contact with supralabials; tail flattened; dorsals keeled and imbricate; midbody scale rows 21–23 (rarely 19); ventrals 200–278, small, distinct; precloacal divided.

COLOUR Top of body yellowish-olive, with black dorsal spots or rhomboidal marks, forming bands, especially in juveniles, adults show less contrast; forehead dark grey; black band across neck; belly yellow or cream; tail black, some scales light centred.

HABITAT AND BEHAVIOUR Inhabits shallow coastal waters. Diet comprises eels. Ovoviviparous, producing four young.

VENOM Presumably with postsynaptic neurotoxins; possibility of severe and lethal envenoming. Sea Snake Antivenom (CSL Limited, Victoria).

DISTRIBUTION West coast of India to eastern China. In the region, recorded from Myanmar, Thailand, Vietnam, Peninsular Malaysia, Singapore, Borneo and Sumatra. The only Bornean reports are from Marudu Bay and Likas Bay in Sabah.

IUCN THREAT STATUS Not Evaluated.

Kloss's Sea Snake *Hydrophis klossi* Boulenger, 1812

(No vernacular names recorded)

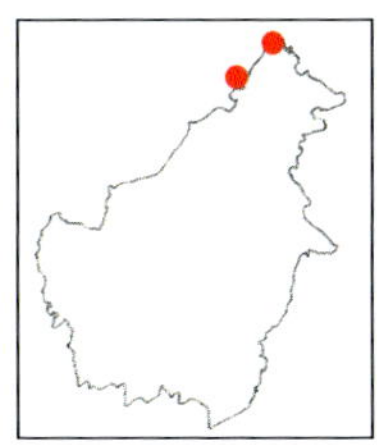

SIZE 1.19m

IDENTIFICATION FEATURES Body robust, elongated; head small; snout projecting; tail flattened; single preocular and postocular; dorsals smooth or weakly keeled; midbody scale rows 31–39; ventrals 360–430, with two tubercles; precloacals enlarged.

COLOUR Top of body blackish-grey to olivaceus-yellow, with 50–75 dark cross-bars; bars broadest dorsally and broader than interspaces; black vertebral stripe sometimes present; forehead black or olivaceous, sometimes with indistinct horseshoe-shaped mark.

HABITAT AND BEHAVIOUR Inhabits shallow coastal waters. Diet and reproductive habits unstudied.

VENOM Presumably with postsynaptic neurotoxins; possibility of severe and lethal envenoming. Sea Snake Antivenom (CSL Limited, Victoria).

DISTRIBUTION Coasts off Thailand, Cambodia, southern Vietnam, Peninsular Malaysia, Singapore, Borneo and Sumatra. The sole Bornean record is from Marudu Bay in Sabah.

IUCN THREAT STATUS Data Deficient.

Black Sea Snake *Hydrophis melanosoma* Günther, 1864

(No vernacular names recorded)

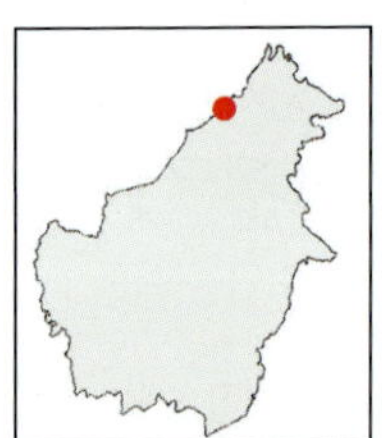

SIZE 1.39m

IDENTIFICATION FEATURES Body robust; single preocular and postocular; dorsals strongly keeled; tail flattened; midbody scale rows 35–45; ventrals 266–368, with two tubercles; cloacal scute divided.

COLOUR Top of body blackish-grey with 50–70 wide white bands, subequal to interspaces; head and neck black with yellow subovoid marks; forehead has yellow speckles; belly cream or yellow.

HABITAT AND BEHAVIOUR Inhabits sea coasts off the mainland and large islands, and rivers. Diet includes anguiliform eels. Reproductive habits unstudied.

VENOM Presumably with postsynaptic neurotoxins; possibility of severe and lethal envenoming. Sea Snake Antivenom (CSL Limited, Victoria).

DISTRIBUTION Coasts off southern Peninsular Malaysia, Singapore, Borneo, Sumatra, eastern Indonesia, New Guinea to northern Australia. The sole Bornean record is from Telok Brunei, off the coast of Sabah.

IUCN THREAT STATUS Data Deficient.

ORNATE SEA SNAKE *Hydrophis ornatus* (Gray, 1842)

(Bahasa Indonesia: Ular Laut Berlurik)

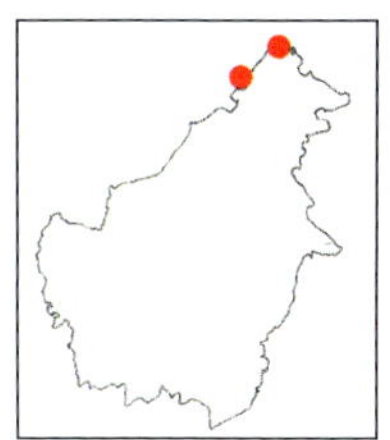

SIZE 1.15m

IDENTIFICATION FEATURES Body robust; head large; eyes small with rounded pupils; scales hexagonal, feebly imbricate or juxtaposed; tail flattened; dorsals tuberculate or with short keel; midbody scale rows 42–54; ventrals 235–298, anteriorly ventrals twice as large as adjacent scales; subcaudals 38–50; cloacal scutes weakly enlarged.

COLOUR Top of body pale grey or greyish-olive to plain cream, with 30–60 broad dark bars or rhomboidal spots separated by narrow interspaces on body, and 6–11 on tail; belly yellow or cream.

HABITAT AND BEHAVIOUR Inhabits shallow waters with coral reefs, river mouths and estuaries. Diet includes marine catfish. Ovoviviparous, producing 1–4 neonates, each c. 34mm.

VENOM Postsynaptic neurotoxins; potential for severe and lethal envenoming. Sea Snake Antivenom (CSL Limited, Victoria).

DISTRIBUTION Persian Gulf, through Vietnam, the Philippines, New Guinea to Australia. In the region, known from Myanmar, Thailand, Peninsular Malaysia, Singapore, Borneo and Sumatra. The two Bornean records are from Kimanis Bay and Marudu Bay in Sabah.

IUCN THREAT STATUS Least Concern.

Yellow-bellied Sea Snake *Hydrophis platurus* (Linnaeus, 1766)

(No vernacular names recorded)

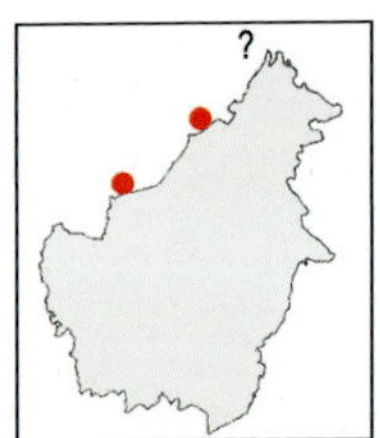

SIZE 1m

IDENTIFICATION FEATURES Body slender, compressed; head elongated, bill-like, slightly flattened and distinct from neck; eyes small with rounded pupils; tail flattened; dorsals keeled; midbody scale rows 49–67; ventrals 264–406, irregular in shape and indistinct after anterior of body; cloacal scutes enlarged.

COLOUR Top of body black or dark brown; belly light brown or yellow, sometimes with series of black spots or bars; dorsal and ventral colours sharply separated; tail has bright yellow diamond-shaped pattern.

HABITAT AND BEHAVIOUR An open sea species, known mostly from coastal stranding. Diet comprises surface-feeding fish. Ovoviviparous, producing clutches of 2–6 neonates, each 220–260mm.

VENOM Postsynaptic neurotoxins; possibility of severe and lethal envenoming. Sea Snake Antivenom (CSL Limited, Victoria).

DISTRIBUTION Pacific and Atlantic Oceans. In the region, known from Myanmar, Thailand, Peninsular Malaysia, Singapore, Borneo, Sumatra and Java. On Borneo, has been recorded in Oya in Sarawak; off the coast of Kuala Belait in Brunei Darussalam, besides one noted as 'northwestern Borneo', presumably off the coast of Sabah.

IUCN THREAT STATUS Least Concern.

Beaked Sea Snake *Hydrophis schistosus* Daudin, 1803

(Bahasa Malaysia: Ular Selimpat. Bahasa Indonesia: Ular Laut Berparuh)

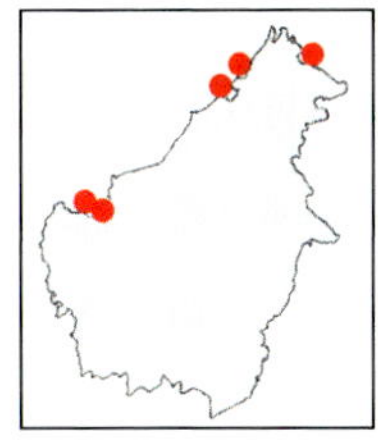

SIZE 1.58m

IDENTIFICATION FEATURES Body robust; head narrow, nearly indistinct from neck; tail flattened; eyes small with rounded pupils; rostral extends ventrally; dorsals keeled; midbody scale rows 49–66; ventrals small, numbering 239–322; cloacal scutes slightly enlarged.

COLOUR Top of body greyish-olive or silvery-grey, with darker forehead; indistinct darker markings, sometimes forming 40–60 dark transverse bands on body; suborbital stripe absent; belly cream anteriorly, darkening to greenish-yellow towards tail.

HABITAT AND BEHAVIOUR Inhabits shallow (under 5m) sea coasts and mangroves; known to occur in freshwater habitats. Diet comprises marine catfish. Ovoviviparous, producing 4–33 neonates, each 150–280mm.

VENOM Postsynaptic neurotoxins; possibility of severe and lethal envenoming. Sea Snake Antivenom (CSL Limited, Victoria).

DISTRIBUTION Extends from Persian Gulf, east to New Guinea and northern Australia. In the region, recorded in Myanmar, Thailand, Peninsular Malaysia, Singapore, Borneo and Sumatra. Bornean records are from the mouth of Sungei Baram, Buntal, Santubong and Sungei Trusan in Sarawak, and Beluran in Sandakan, Kimanis Bay and Kampung Mumiang in Sabah.

IUCN THREAT STATUS Least Concern.

River Sea Snake *Hydrophis sibauensis* Rasmussen, Auliya & Böhme, 2001

(No vernacular names recorded)

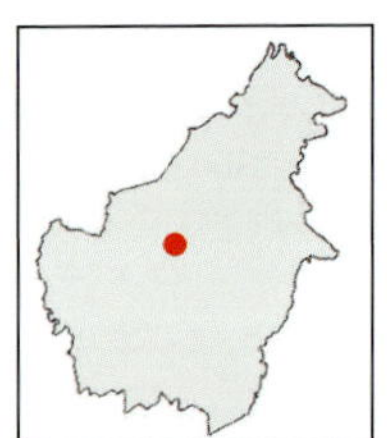

SIZE 735mm

IDENTIFICATION FEATURES Body slender; head narrower than body; nostrils on top of head; forehead shields large and regular; eyes small with rounded pupils; tail flattened; dorsals have median keel; midbody scale rows 35–37; ventrals 257–264, small, distinct throughout; cloacal scutes enlarged.

COLOUR Top of body grey-brown, darkening posteriorly; 49–58 incomplete yellow to light orange bands; forehead black with small yellow spots and arrow-shaped markings; belly and throat black anteriorly and greyish-yellow from mid-body to posterior.

HABITAT AND BEHAVIOUR Known from a freshwater section of Sungei Sibau, a tributary of Sungei Kapuas, more than 1,000km upriver. Diet unstudied. Ovoviviparous, producing seven neonates.

VENOM Presumably with postsynaptic neurotoxins; possibility of severe and lethal envenoming. Sea Snake Antivenom (CSL Limited, Victoria).

DISTRIBUTION Endemic to Borneo. The sole record is from Sungei Sibau in Kalimantan.

IUCN THREAT STATUS Data Deficient.

Spiral Sea Snake *Hydrophis spiralis* (Shaw, 1802)

(No vernacular names recorded)

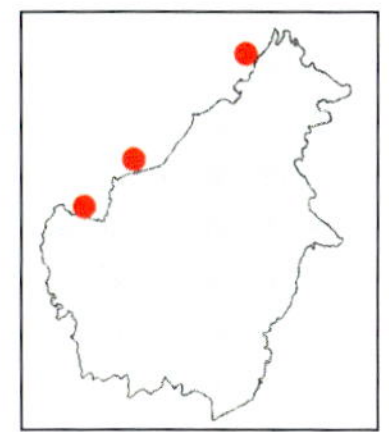

SIZE 2.75m

IDENTIFICATION FEATURES Body elongated and moderately robust anteriorly, thickening posteriorly; head and neck slender; eyes small with rounded pupils; tail flattened; dorsals smooth or keeled; midbody scale rows 33–38; ventrals 295–362, distinct throughout, about twice as broad as adjacent body scales; cloacal scutes enlarged.

COLOUR Top of body and forehead of adults olive-yellow to olive-brown, with 35–50 encircling dark bands, narrower than interspaces, sometimes with black dorsal spot between each; in juveniles, forehead nearly black with yellow horseshoe-shaped mark; flanks yellowish-cream; belly yellow or cream; tip of tail dark brown or black.

HABITAT AND BEHAVIOUR Inhabits deep-water habitats (more than 10m). Diet unstudied. Ovoviviparous, producing 5–14 neonates, each 200–405mm.

VENOM Postsynaptic neurotoxins; possibility of severe and lethal envenoming. Sea Snake Antivenom (CSL Limited, Victoria).

DISTRIBUTION Extends from Persian Gulf to Sulawesi and the Philippines. In the region, known from Myanmar, Thailand, Peninsular Malaysia, Singapore, Borneo and Sumatra. Bornean records are from Muka and Santubong in Sarawak, and Kimanis Bay in Sabah.

IUCN THREAT STATUS Least Concern.

Garlanded Sea Snake *Hydrophis torquatus* Günther, 1864

(No vernacular names recorded)

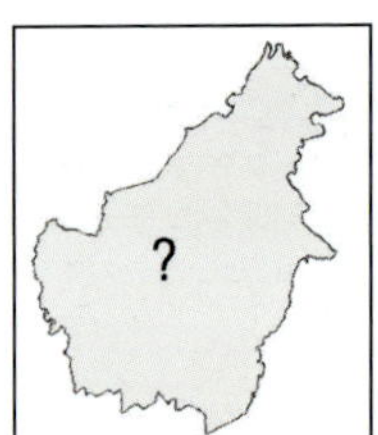

SIZE 895mm

IDENTIFICATION FEATURES Body elongated and moderately robust; head moderate; single preocular; tail flattened; dorsals keeled; midbody scale rows 35–42; ventrals 242–343; precloacal scutes enlarged.

COLOUR Top of body greyish- or greenish-grey, with more than 50 pale bands on body and tail; forehead black or dark olive with yellow horseshoe-shaped mark; venter yellowish-cream.

HABITAT AND BEHAVIOUR Inhabits brackish waters off river mouths. Diet and reproductive biology unstudied.

VENOM Presumably with postsynaptic neurotoxins; possibility of severe and lethal envenoming. Sea Snake Antivenom (CSL Limited, Victoria).

DISTRIBUTION Coasts off Thailand, Cambodia, Vietnam, Peninsular Malaysia, Singapore, Borneo and Sumatra. The only Bornean record lacks a precise location.

IUCN THREAT STATUS Data Deficient.

Grey Sea Snake *Hydrophis viperinus* (Schmidt, 1852)

(Bahasa Indonesia: Ular Laut Abu-abu)

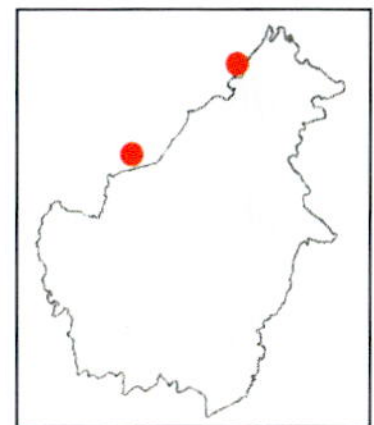

SIZE 925mm

IDENTIFICATION FEATURES Body robust; head short, depressed; forehead scales entire; nostrils on dorsal surface; eyes small with rounded pupils; scales on body hexagonal and keeled; tail flattened; dorsals keeled; midbody scale rows 37–50; scales on body hexagonal and keeled; ventrals as broad as dorsal scales in anterior of body; ventrals 226–274, at anterior covering half body width, narrowing at posterior to twice width of adjacent scales or less; four cloacal scutes, outer two largest.

COLOUR Top of body uniform grey, or with lighter mottling or 25–35 dark cross-bars or spots; forehead grey or black, sometimes with dark mottling; belly cream or white; juveniles have enlarged dark vertebral pattern.

HABITAT AND BEHAVIOUR Inhabits seas up to 32km from the coast; also river mouths. Diet consists of marine invertebrates and eels. Ovoviviparous, producing 3–5 neonates.

VENOM Postsynaptic neurotoxins; possibility of severe and lethal envenoming. Sea Snake Antivenom (CSL Limited, Victoria).

DISTRIBUTION Extends from Persian Gulf to eastern China. In the region, known from Myanmar, Thailand, Peninsular Malaysia, Singapore and Borneo. The only Bornean records are from Oya in Sarawak, and Pulau Tiga off Sabah.

IUCN THREAT STATUS Least Concern.

Subfamily Laticaudinae Sea Kraits

Members of this subfamily are marine, like those of the aforementioned subfamily, but come ashore to lay eggs on offshore islands.

Laticauda Sea Kraits

Large, robust marine species with subcylindrical bodies; body scales imbricate; rostral single; nasals separated by internasals; nostrils laterally located; ventrals wide, about a third to half of body width; tail short, flattened.

Key to Bornean species of *Laticauda*

1a. Supralabials yellow .... *L. colubrina* (below)
1b. Supralabials dark brown or black .... *L. laticaudata* (opposite)

Yellow-lipped Sea Krait *Laticauda colubrina* (Schneider, 1799)

(Bahasa Indonesia: Krait Laut Berbibir Kuning)

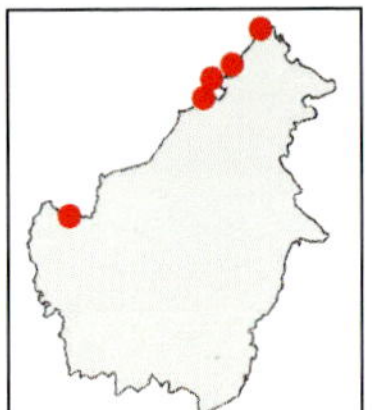

SIZE 1.71m

IDENTIFICATION FEATURES Body robust (especially in adult females), subcylindrical; nostrils lateral; rostral undivided; nasals separated by internasals; eyes small with rounded pupils; tail flattened; dorsals smooth; midbody scale rows 21–25; ventrals 213–245, broad; subcaudals 29–47; cloacal scute divided.

COLOUR Top of body blue-grey with 24–64 black annuli; supralabials yellow; belly cream.

HABITAT AND BEHAVIOUR Inhabits shallow seas, particularly around rocky outcrops, to depths of c. 60m. Comes on land to lay eggs, bask, rest and digest food. Diet includes anguiliform eels. Oviparous, producing 3–13 eggs, each 44.6–92.2 × 20.3–31.1mm.

VENOM Postsynaptic neurotoxins; possibility of severe and lethal envenoming. Sea Snake Antivenom (CSL Limited, Victoria).

DISTRIBUTION Widespread in the Indian and Pacific Oceans, east to Polynesia. In the region, recorded from Myanmar, Thailand, Peninsular Malaysia, Singapore, Borneo, Sumatra, Pulau Belitung, Java and Bali. Bornean records are from offshore, from Bako National Park in Sarawak; Pulau Punjet in Brunei Darussalam; Pulau Layang-Layang and Pulau Kalampunian Damit, Pulau Keraman in Labuan, and Pulau Layang-Layang, Papar in Sabah.

IUCN THREAT STATUS Least Concern.

Close-up of head

Blue Sea Krait *Laticauda laticaudata* (Linnaeus, 1758)

(No vernacular names recorded)

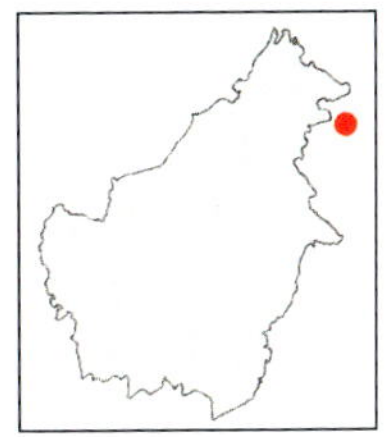

SIZE 1.1m

IDENTIFICATION FEATURES Body robust (especially in adult females), subcylindrical; azygous prefrontal scute missing; rostral undivided; nostrils lateral; nasals separated by internasals; tail flattened; dorsals smooth; midbody scale rows 19; no azygous prefrontal shield; rostral undivided; ventrals 225–243, ventrals large, one-third to over half body width; subcaudals 30–47.

COLOUR Top of body bright blue with 20–70 black bands; supralabials dark brown or black; belly cream.

HABITAT AND BEHAVIOUR Inhabits shallow seas and may rest on small, rocky islands; occasionally found near fresh water. Active by day and night. Diet includes eels. Oviparous, producing 1–7 eggs.

VENOM Postsynaptic neurotoxins; possibility of severe and lethal envenoming. Sea Snake Antivenom (CSL Limited, Victoria).

DISTRIBUTION Widespread in the Indo-Pacific, extending eastwards to Polynesia. In the region, known from Myanmar, Thailand, Peninsular Malaysia, Borneo, Sumatra, Java and Bali. The sole Bornean record is from Pulau Si Amil, off Pulau Mabul in Sabah.

IUCN THREAT STATUS Least Concern.

Family Homalopsidae Puff-faced Water Snakes

This family includes water snakes of the Asia-Pacific region that have reduced eyes and valvular nostrils, features reflecting a life in muddy and/or turbid habitats such as fresh and brackish waters, including flooded agricultural fields, marshes and ponds. They feed on fish, frogs and crustaceans, and are ovoviviparous, producing live young underwater. Homalopsidae species range from the Indian subcontinent in the west, to South-east Asia, northern Australia and Micronesia.

Key to Bornean species of Homalopsidae

1a. Nasals in contact..**2**
1b. Nasals separated..**8**

2a. Dorsals smooth on anterior of body..**3**
2b. Dorsals keeled..**19**

3a. Midbody scale rows 19..**4**
3b. Midbody scale rows > 19..**5**

4a. Internasal single; fails to contact loreal..*Hypsiscopus plumbea* (p. 336)
4b. Internasal divided; in contact with loreal..*Miralia alternans* (p. 337)

5a. Midbody scale rows 21–23..*Enhydris enhydris* (p. 330)
5b. Midbody scale rows 25–33..**6**

6a. Suboculars absent; supralabials contact orbit..*Phytolopsis punctata* (p. 338)
6b. Suboculars present; supralabials fail to contact orbit..**7**

7a. Midbody scale rows 25–27; ventrals 155–159..*Homalophis gyii* (p. 334)
7b. Midbody scale rows 29–33; ventrals 139–152..*Homalophis doriae* (p. 333)

8a. Loreal present..**9**
8b. Loreal absent..*Fordonia leucobalia* (p. 331)

9a. Midbody scale rows 17..*Gerarda prevostiana* (p. 332)
9b. Midbody scale rows > 17..**10**

10a. Parietals fragmented; midbody scale rows 23–25..*Cerberus schneiderii* (opposite)
10b. Parietals entire; midbody scale rows 33–46..*Homalopsis buccata* (p. 335)

Cerberus Dog-faced Water Snakes

Medium-sized, stout aquatic snakes with maxillary teeth range 12–17; parietals fragmented into small scales; midbody scale rows 21–29; nostrils crescentic and dorsally located; eyes small and beady, with rounded pupils and somewhat dorsal in position; tail moderate or short. Includes species of mangroves and other coastal habitats.

Schneider's Dog-faced Water Snake *Cerberus schneiderii*

(Schlegel, 1837)

(Bahasa Malaysia: Ular Air Kadut, Ular Birang. Bahasa Indonesia: Ular Tambak)

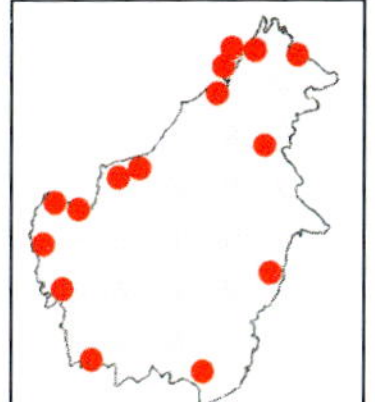

SIZE 886mm

IDENTIFICATION FEATURES Body moderately robust; head long, distinct from neck; scales on crown slightly imbricate, not rounded, parietal scales not plate-like; keels on crown scales anterior to level of angle of jaw; forehead scales flat, imbricate, keels anterior to angle of jaw missing; last supralabial horizontally divided; tail short and slender; dorsals strongly keeled; midbody scale rows 23 (rarely 25); ventrals 122–160; subcaudals 49–72, paired; cloacal scute divided.

COLOUR Top of body dark grey or greyish-green with faint dark blotches; dark postocular stripe; pale stripe on scale rows 1–3; belly cream or yellowish-cream, mottled with dark grey or unpatterned.

HABITAT AND BEHAVIOUR Low-lying coastal areas, including mangrove mudflats. Diet comprises mudskippers, gobies and crabs. Ovoviviparous, producing 3–45 neonates.

DISTRIBUTION Thailand, Cambodia, Vietnam, Peninsular Malaysia, Singapore, Sumatra, Borneo, Mentawai and Natuna archipelagos, Java, Bali, Sulawesi and the Philippines archipelago. On Borneo, widespread along coastal and mangrove regions, with records from the Santubong coast including Buntal, besides Bako National Park, Tanjung Datu National Park, Saratok and Sibu in Sarawak; the Muara coast of Brunei Darussalam; Sandakan, Dewhurst Bay and Pulau Tiga in Sabah; Labuan, and Banjarmasin, Samarinda, Sinkawang, Pontianak and Sungei Bulangan in Kalimantan.

IUCN THREAT STATUS Not Evaluated.

Enhydris Common Water Snakes

Medium-sized aquatic snakes; midbody scale rows 21; reduced to 17 or 19 posteriorly; prefrontal and internasal typically contact loreal, posterior chin shields larger than gulars; hemipenes extend to subcaudals VII–VIII.

Rainbow Water Snake *Enhydris enhydris* (Schneider, 1799)

(Bahasa Malaysia/Indonesia: Ular Air Biasa)

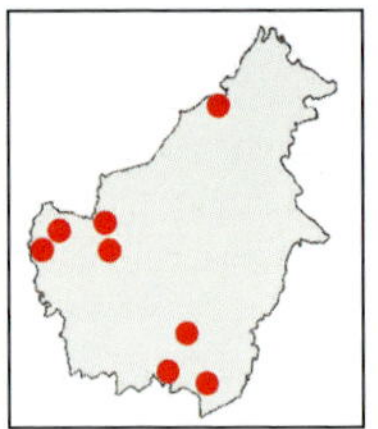

SIZE 882mm

IDENTIFICATION FEATURES Body robust, subcylindrical; head small, depressed, slightly distinct from neck; snout rounded; nostrils situated on upper surface of head; eyes small with vertical pupils; tail short and tapering; dorsals smooth; midbody scale rows 21 or 23 (rarely 19); ventrals 134–177; subcaudals 46–83, paired; cloacal scute divided.

COLOUR Top of body greyish-brown or olive-green, with dark vertebral and two pale lateral stripes from upper surface of head to tail; belly yellowish-cream, ventrals have dark spot on edges, creating dark line on flanks.

HABITAT AND BEHAVIOUR Inhabits fresh and brackish-water habitats, including swamps, sluggish rivers, oxbow lakes and wet rice fields. Diet comprises fish, frogs, tadpoles and lizards. Ovoviviparous, producing 4–20 neonates, each 158–206mm.

DISTRIBUTION Eastern India, Nepal, Bangladesh, Bhutan, eastern China, Myanmar, Thailand, Vietnam, Peninsular Malaysia, Singapore, Sumatra, Borneo and Java. Isolated records from Borneo, including Anna Rais at foothills of Gunung Penrissen, Saribas and Simanggang in Sarawak; Brunei Darussalam (without a precise locality), and Banjarmasin, Monterado in Bengkayang, Singkawang and Sungei Kapuas in Kalimantan.

IUCN THREAT STATUS Least Concern.

Fordonia Crab-eating Mangrove Snakes

Medium-sized, robust aquatic snakes; head short and wide; supralabials 5–6; supralabial III contacts orbit; loreal typically absent; plate-like scales form ocular ring; frontal shorter than parietal; rear fangs grooved.

Crab-eating Mangrove Snake *Fordonia leucobalia* (Schlegel, 1837)

(Bahasa Malaysia: Ular Air Ketam)

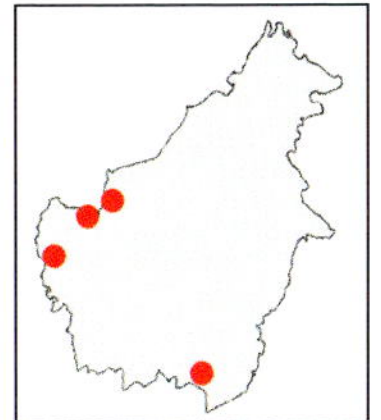

SIZE 950mm

IDENTIFICATION FEATURES Body robust and cylindrical; head short, wide and indistinct from neck; forehead scales large and distinct; snout rounded; lower jaw short; eyes small with rounded pupils; tail short, with acute tip; dorsals smooth; midbody scales 25–29; ventrals 137–159; subcaudals 27–43, paired; cloacal scute divided.

COLOUR Top of body dark grey or brown with light spots, or light grey, yellow or orange with dark or light spots; labials yellowish-cream; belly pale cream, sometimes with small dark spots.

HABITAT AND BEHAVIOUR Associated with tidal rivers and mangrove forests. Aquatic, sheltering in crab burrows. Diet comprises crabs, fish and mud lobsters. Ovoviviparous, with 2–17 neonates, each 176–196mm.

DISTRIBUTION The Sunderbans of eastern India and presumably Bangladesh, Andaman and Nicobar islands, Myanmar, Thailand, Vietnam, Peninsular Malaysia, Singapore, Sumatra, Borneo, Java, Lesser Sundas, the Philippines, New Guinea and northern Australia. Bornean records are from near Kampung Bako, Santubong, Niah, Saratok and Sungei Nyabau (presumably at its lower reaches) in Sarawak, and Banjarmasin and Pontianak in Kalimantan.

IUCN THREAT STATUS Least Concern.

Gerarda Marsh Snakes

Medium-sized, slender aquatic snakes; head somewhat distinct from neck; nasals separated by internasal; middorsal rows 17, smooth; loreal present.

Glossy Marsh Snake *Gerarda prevostiana* (Eydoux & Gervais, 1822)

(No vernacular names recorded)

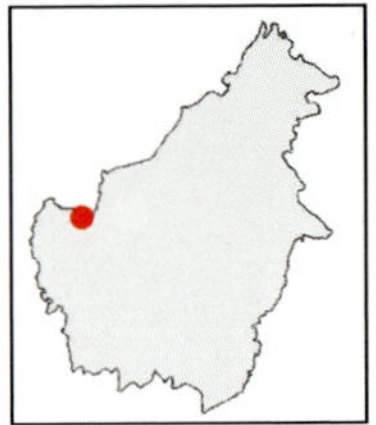

SIZE 530mm

IDENTIFICATION FEATURES Body slender and cylindrical; head slightly distinct from neck; eyes small and situated dorsally, with vertical pupils; tail short with acute tip; dorsals smooth; midbody scale rows 17; ventrals 144–157; subcaudals 29–36, paired; cloacal scute divided.

COLOUR Top of body grey, greyish-green or brown; cream or yellow stripe from labials to tail-tip; labials edged with dark grey or olive; belly grey or brownish-cream with white edges, or white with grey edges; subcaudals dark grey.

HABITAT AND BEHAVIOUR Inhabits mangrove swamps and estuaries. Diet includes soft-shelled crabs, fish and shrimps. Ovoviviparous.

DISTRIBUTION India, Sri Lanka, Myanmar, Thailand, Peninsular Malaysia, Singapore and Borneo. Bornean records are from Kampung Bako and adjacent regions.

IUCN THREAT STATUS Least Concern.

Close-up of head and neck

Homalophis Bornean Water Snakes

Large, robust aquatic snakes; head weakly distinct from neck; middorsal scale rows 25–33, smooth; nasals in contact; subocular scales, divided internasal; first labial does not contact loreal; 2–4 pairs of chin shields, supralabials 13–15, those posterior to eye horizontally divided into three; infralabials 15–20.

Key to Bornean species of *Homalophis*

1a. Midbody scale rows 29–33; ventrals 139–152 .... *H. doriae* (below)
1b. Midbody scale rows 25–27; ventrals 155–159 .... *H. gyii* (p. 334)

Marquis Doria's Water Snake *Homalophis doriae* Peters, 1871

(No vernacular names recorded)

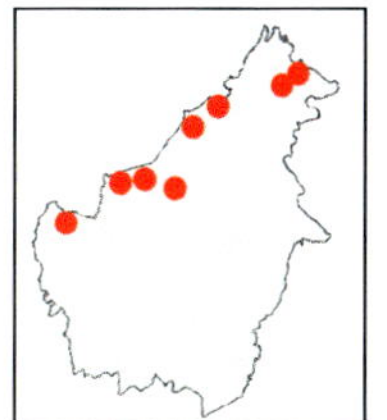

SIZE 796mm

IDENTIFICATION FEATURES Body robust, subcylindrical; head small, depressed, slightly distinct from neck; forehead scales fragmented; eyes small with rounded or slightly oval pupils; tail short; dorsals smooth; midbody scale rows 29–31; ventrals 137–152; subcaudals 39–60, paired; cloacal scute divided.

COLOUR Top of body reddish-brown to greyish-brown; labials blotched with grey or cream; belly yellow, orange, cream or red, sometimes with dark blotches, darkening posteriorly.

HABITAT AND BEHAVIOUR Inhabits small freshwater streams and swamps, including peat lakes at < 500m asl. Aquatic. Diet comprises fish. Reproductive habits unstudied.

DISTRIBUTION Endemic to Borneo. Localities include base of Gunung Pueh and upper or middle reaches of Sungei Baram and Sungei Mengion in Sarawak (with historical records from Kuching); Sandakan Bay and Kampung Kayu Madang in Sabah, and Sungei Kapuas in Kalimantan.

IUCN THREAT STATUS Least Concern.

Gyi's Water Snake *Homalophis gyii* (Murphy, Voris & Auliya, 2005)

(No vernacular names recorded)

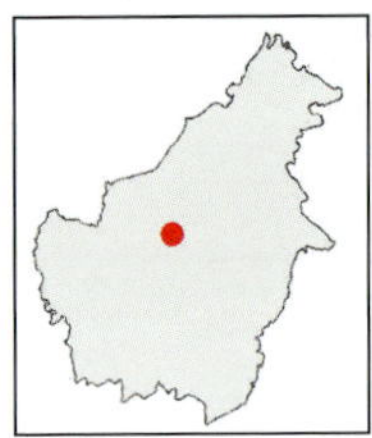

SIZE 762mm

IDENTIFICATION FEATURES Body robust and subcylindrical; head distinct from neck; eyes small with vertical pupils; dorsals smooth; midbody scale rows 25 or 27; ventrals 155–159; subcaudals 44–46, paired; cloacal scute divided.

COLOUR Top of body iridescent greyish-brown, except for scale rows 1–4, which are reddish-brown, each dorsal red spotted (colour changeable to nearly white when kept in the dark); supralabials from anterior to orbit greyish-black; posterior supralabials reddish-brown; black stripe extends from nape to angle of jaws; belly reddish-brown.

HABITAT AND BEHAVIOUR Inhabits swamp forests at c. 50m asl. Diet and reproductive biology unstudied.

DISTRIBUTION Endemic to Borneo. Known from Putussibau, near Sungai Kapuas in Kalimantan.

IUCN THREAT STATUS Data Deficient.

Homalopsis Puff-faced Water Snakes

Medium-sized, stout aquatic snakes; head distinct from neck; nasals in contact, midbody scale rows 33–49; some supralabials posterior to eyes horizontally divided into three tiers; pattern on body dorsum comprises dark brown or black transverse bands.

Puff-faced Water Snake *Homalopsis buccata* (Linnaeus, 1758)

(Bahasa Malaysia: Ular Air Tembam, Ular Kadut. Bahasa Indonesia: Ular Air Belalang. Cantonese: Sui Seh)

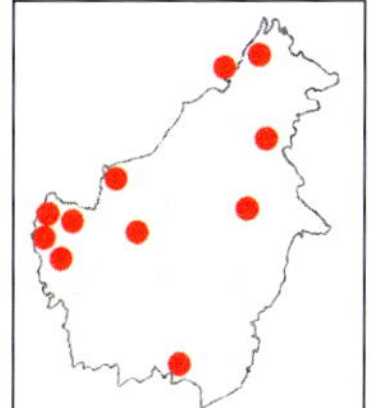

SIZE 1.4m

IDENTIFICATION FEATURES Body robust and flattened dorsoventrally; head large, distinct from neck; snout squarish; eyes and nostrils directed upwards; eyes small with vertically oval pupils; tail short with acute tip; dorsals keeled; midbody scale rows 33–49; ventrals 155–180; subcaudals 68–101, paired; cloacal scute divided.

COLOUR Top of body greyish, dark brown or black with 19–51 narrow, black-edged, yellow cross-bars; belly cream with black spots.

HABITAT AND BEHAVIOUR Inhabits slow-moving and stagnant waterways, including peat swamps, ponds, marshes, rice fields and coastal areas. Diet comprises fish, crustaceans and frogs. Ovoviviparous, producing 2–37 neonates, each 192–260mm.

DISTRIBUTION Southern Thailand, Peninsular Malaysia, Singapore, Borneo, Sumatra, Pulau Belitung and Java. Bornean records are from Buntal, Kuching, Kota Samarahan, Gunung Santubong National Park, and Sungei Rajang in Sarawak; Labuan, and Banjarmasin, Monterado in Bengkayang, Pontianak, Singkawan, Sebruang Valley and Sungei Bulangan, Sungei Landak and Sungei Mahakam in Kalimantan.

IUCN THREAT STATUS Least Concern.

Hypsiscopus Mud Snakes

Short, plump aquatic snakes; midbody scale rows 19–21; internasals not in contact with loreal, supralabials II–III or II–IV contact loreal; infralabials I–V typically contact anterior chin shields; tail relatively short.

Plumbeous Water Snake *Hypsiscopus plumbea* (Boie, 1827)

(Bahasa Malaysia: Ular Air Sawah. Bahasa Indonesia: Ular Lumpur)

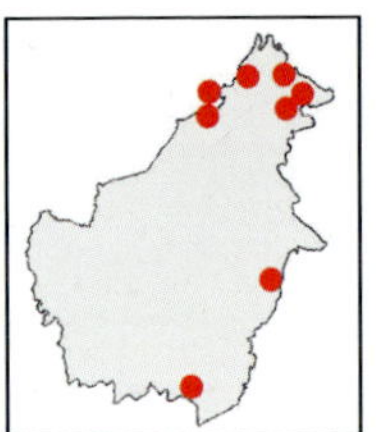

SIZE 713mm

IDENTIFICATION FEATURES Body robust, subcylindrical; head short, rounded and indistinct from neck; eyes small with rounded pupils; tail short, compressed; dorsals smooth; midbody scale rows 19; ventrals 112–139; subcaudals 22–46, paired; cloacal scute divided.

COLOUR Top of body grey or greyish-olive, scales edged with dark brown or black; supralabials and belly cream or yellow, the latter black spotted; cloacal scute and subcaudals have dark grey median line.

HABITAT AND BEHAVIOUR Inhabits swamps, marshes, streams, ditches and paddy fields; occasionally reported from brackish waters. Diet comprises crustaceans, frogs and their eggs, tadpoles and fish. Ovoviviparous, producing 2–30 neonates.

DISTRIBUTION Southern China, Nicobar Islands, Myanmar, Thailand, Laos, Cambodia, Vietnam, Peninsular Malaysia, Sumatra, Borneo, Pulau Belitung, Java, Bali and Sulawesi. Bornean records are from Ulu Temburong in Brunei Darussalam; Labuan; Kota Kinabalu, Penampang, Ranau, Tuaran, Kampung Kayu Madang, Batu Putih and Menggatal in Sabah, and Banjarmasin and Samarinda in Kalimantan.

IUCN THREAT STATUS Least Concern.

Miralia Banded Snakes

Freshwater snakes with robust build; nasals in contact; middorsal scale rows 19, smooth; reduced preocular; lower postocular fails to extend below eye; supralabials eight; supralabials IV–V contact orbit; nasal cleft contacts supralabial II.

RED-BANDED WATER SNAKE *Miralia alternans* (Reuss, 1833)

(No vernacular names recorded)

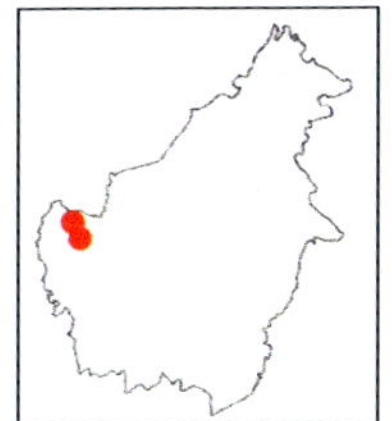

SIZE 610mm

IDENTIFICATION FEATURES Body moderate, subcylindrical; nostrils dorsal; loreal present; single preocular; infralabials contact anterior chin shields; paired internasals; two posterior teeth enlarged; eyes small; tail short; dorsals smooth; midbody scale rows 19; ventrals 125–164; subcaudals 23–36, paired; cloacal scute divided.

COLOUR Top of body purplish-brown, with reddish-orange cross-bars, numbering 40–45 on body and 8–9 on tail, sometimes broken up into spots; belly purplish-brown, with 39–61 yellow transverse bars; subcaudals have yellowish-brown and yellow variegation.

HABITAT AND BEHAVIOUR Inhabits lowland forest swamps and their edges. Diet includes eels. Reproductive habits unstudied.

DISTRIBUTION Borneo, Sumatra, Pulau Bangka, Pulau Belitung and Java. Bornean localities include Bau, Lundu and the Matang Range.

IUCN THREAT STATUS Data Deficient.

Phytolopsis Spotted Water Snakes

Large, robust aquatic species; middorsal scale rows 25–27 rows; supralabials 11–13; supralabials posterior to eyes horizontally divided; infralabials 12–15; suboculars absent; two pairs of chin shields, the first pair widened.

SPOTTED WATER SNAKE *Phytolopsis punctata* Gray, 1849

(No vernacular names recorded)

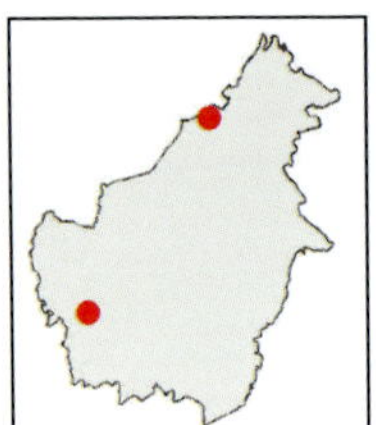

SIZE 730mm

IDENTIFICATION FEATURES Body robust, subcylindrical; head short, rounded, depressed; indistinct from neck; single preocular; two postoculars; single supraocular; suboculars absent; supralabials 10–14; supralabials VI–VII contact orbit; infralabials 14–16; 3–4 infralabials contact anterior chin shields; anterior chin shields larger than posterior; tail short and compressed; dorsals smooth; midbody scale rows 25 (rarely, 23 or 27); ventrals 137–160; subcaudals 27–48, paired; cloacal divided.

COLOUR Dorsum brown; yellow cross-bar, one scale wide, on occipital region; transverse rows of yellow spots, of which 6–7 anterior rows form transverse bands; caudal scales have yellow spot; venter, along with adjacent three rows of dorsals, yellow; subcaudals yellow edged with brown.

HABITAT AND BEHAVIOUR Inhabits lowlands. Aquatic and associated with rivers. Diet and reproductive habits unstudied, and presumed to be piscivorous and ovoviviparous.

DISTRIBUTION Peninsular Malaysia (Gunung Pulai, Johor State), Sumatra, Borneo, Pulau Bangka and Pulau Belitung. The Bornean record is from Sinkawang in Kalimantan.

IUCN THREAT STATUS Data Deficient.

Family Pareidae Slug-eating Snakes

Members of this family are specialized forest dwellers. They have short skulls, large eyes, chin shields without a midline groove, and a low mid-body scale count of 13–15 rows. Slug-eating snakes are nocturnal and arboreal, with a specialist diet of snails and slugs, and are oviparous. They are restricted to South-east Asia.

Key to Bornean species of Pareidae

1a. Subcaudal scales in single row; 13 midbody scale rows .... *Aplopeltura boa* (p. 340)
1a. Subcaudal scales paired; 15 midbody scale rows **2**

2a. Preoculars and suboculars present; loreal and preocular between nasal and eye; supralabials do not touch eye............. *Pareas nuchalis* (p. 346)
2b Preoculars and suboculars absent; loreal between nasal and eye; one or more supralabials in contact with eye............ **3**

3a. First or third pair of infralabials in contact **4**
3b. First and third pairs of infralabials in contact .... *Asthenodipsas ingeri* (p. 342)

4a. Infralabials III (rarely II) pair in contact.. **5**
4b. Infralabial I in contact.. *Asthenodipsas vertebralis* (p. 345)

5a. Supralabials 5–6.. **6**
5b. Supralabials 7–8 .... *Asthenodipsas borneensis* (p. 341)

6a. Dorsal scale rows 15/15/15; sharp vertebral keel........................ **7**
6b. Dorsal scale rows 15/15/13; vertebral keel not sharp .... *Asthenodipsas laevis* (p. 343)

7a. Subcaudals 35–47; ventrals in males 165.. *Asthenodipsas stuebingi* (p. 344)
7b. Subcaudals 52–53; ventrals in males 173–175............................... *Asthenodipsas jamilinaisi* (p. 342)

Aplopeltura Common Slug-eating Snakes

Small, slender arboreal snakes, common in areas of limestone karst and identifiable due to having head distinct from neck; blunt snout; middorsal scale rows 13; subcaudals entire and tail long.

Blunt-headed Slug-eating Snake *Aplopeltura boa* (Boie, 1828)

(No vernacular names recorded)

SIZE 850mm

IDENTIFICATION FEATURES Body slender, laterally compressed; head short, rounded, distinct from neck; snout short; eyes large with vertical pupils; tail third of body length; dorsals smooth; midbody scale rows 13; ventrals 148–191; subcaudals 88–131; cloacal scute entire.

COLOUR Top of body brown to greyish-brown, with dark-edged, saddle-like markings; flanks have large white spots; forehead dark brown; labials cream; cream patch with dark subtriangular area under eye; belly brown to dark grey.

HABITAT AND BEHAVIOUR Inhabits lowland and submontane forests at < 1,500m asl. Associated with low vegetation of bushes and undergrowth. Diet includes slugs, snails and even lizards. Oviparous, laying 4–8 eggs, each 18–23 × 10–13.5mm.

DISTRIBUTION Myanmar, Thailand, Peninsular Malaysia, Borneo, Sumatra, Pulau Nias, Natuna Archipelago, Java and the Philippines. Bornean records are from Saribas, Sungei Seran, Sungei Baram and Sungei Segaham regions, Gunung Penrissen, Gunung Niah National Park and Gunung Mulu National Park in Sarawak (with historical records from the Kuching region); Ulu Temburong in Brunei Darussalam, and Danum Valley, Maliau Basin, Sandakan Bay, Sepagaya Forest Reserve, Ulu Dusun, Tawau Hills Park and Sungei Rurulon and Togopi in Sabah.

IUCN THREAT STATUS Least Concern.

Asthenodipsas Slug-eating Snakes

Small, slender arboreal snakes, identifiable by having head distinct from neck; blunt snout; middorsal scale rows 15, smooth; preoculars and suboculars absent, supralabials in contact with orbit; two pairs of posterior inframaxillaries; a premental placed anterior to the chin shields; subcaudals divided; long tail.

Bornean Slug-eating Snake *Asthenodipsas borneensis* Quah, Grismer, Lim, Anuar & Chan, 2020

(No vernacular names recorded)

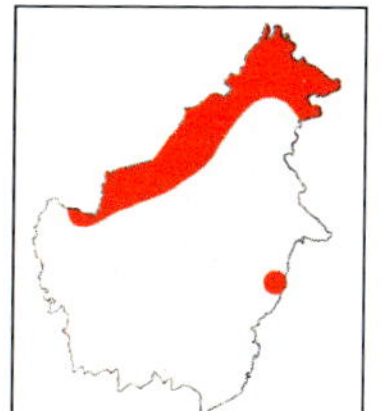

SIZE 441mm

IDENTIFICATION FEATURES Body robust, laterally compressed; head short, rounded, distinct from neck; snout short; eyes small with vertical pupils; tail short; vertebral scale row enlarged; vertebrals have distinct keel; dorsals weakly keeled; midbody scale rows 15; ventrals 166–179; subcaudals 35–48, paired; cloacal scute entire.

COLOUR Top of body beige to orange-brown; dark brown to black patch on nuchal region, followed by 23–40 rhomboidal dark brown bands along body and tail and on flanks, 2–4 scales wide; pale vertebral stripe from nape or after dark patch; forehead dark grey or cream with dark speckling; venter beige or yellow, mottled with dark spots on belly

HABITAT AND BEHAVIOUR Inhabits lowland dipterocarp forests at < 1,000m asl. Semi-arboreal and terrestrial; active on low vegetation and forest floor. Diet comprises snails and slugs. Reproductive habits unstudied.

DISTRIBUTION Endemic to Borneo. Has been encountered at Pa Brayong, Kubah National Park and Wind and Fairy Caves National Park in Sarawak, and Sipitang, Tenom, Tawau Hills Park, as well as the Kinabalu region of Porting and Sayap of Sabah.

IUCN THREAT STATUS Not Evaluated.

INGER'S SLUG-EATING SNAKE *Asthenodipsas ingeri* Quah, Lim & Grismer, 2021

(No vernacular names recorded)

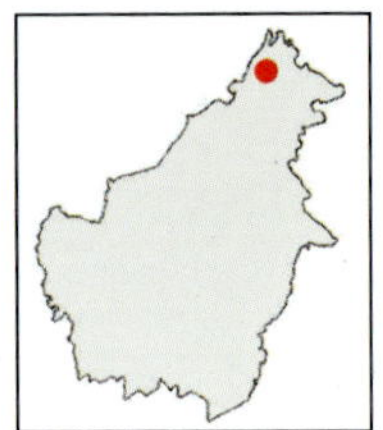

SIZE 715mm

IDENTIFICATION FEATURES Body slender, laterally compressed; head short, rounded and distinct from neck; snout short; eyes large with vertical pupils; tail short; vertebrals slightly enlarged with distinct keel; preoculars and suboculars absent; infralabials I and III in contact; supralabials III–V (sometimes III–IV) in contact with orbit; middorsal scale rows 15; vertebral keel sharp; midbody scale rows 15, smooth; ventrals 201; subcaudals 53, paired; cloacal scute entire.

COLOUR Top of body brown to grey, with 51 rhomboidal dark bands from nape to tail; narrow cream vertebral stripe; head, body and tail dark speckled; venter white to cream, with dark speckling along edges.

HABITAT AND BEHAVIOUR Associated with montane forests at elevations of 1,000–2,000m asl. Arboreal; active on low vegetation and nocturnal. Diet assumed to comprise snails. Reproductive habits unstudied.

DISTRIBUTION Endemic to Borneo. The sole locality is Gunung Kinabalu in Sabah.

IUCN THREAT STATUS Not Evaluated.

JAMILI'S SLUG-EATING SNAKE *Asthenodipsas jamilinaisi* Quah, Grismer, Lim, Anuar & Imbun, 2019

(No vernacular names recorded)

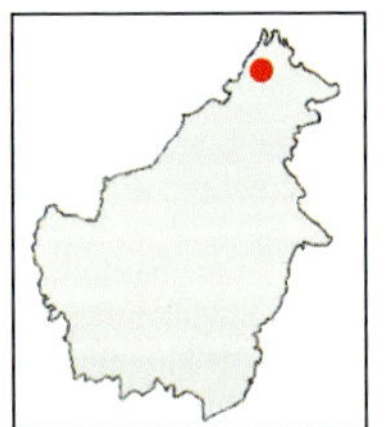

SIZE 378mm

IDENTIFICATION FEATURES Body slender, laterally compressed; head short, rounded and distinct from neck; snout short; eyes large with vertical pupils; tail short; vertebrals slightly enlarged with distinct keel; midbody scale rows 15, smooth; ventrals 173–175; subcaudals 52–53, paired; cloacal scute entire.

COLOUR Top of body dark brown, with indistinct bands, and ground colour paler on flanks; 44 dark bands on body and tail; belly cream to pale yellow, with dark lateral blotches; chin and throat dark brown.

HABITAT AND BEHAVIOUR Associated with montane forests at 1,500–2,500m asl. Arboreal and nocturnal. Diet likely to comprise snails and slugs. Reproductive habits unstudied.

DISTRIBUTION Endemic to Borneo. Known localities include Gunung Trus Made and Marai-Parai in Gunung Kinabalu in Sabah.

IUCN THREAT STATUS Not Evaluated.

SMOOTH SLUG-EATING SNAKE *Asthenodipsas laevis* (Boie, 1827)

(Bahasa Malaysia: Ular Kapak Rimau)

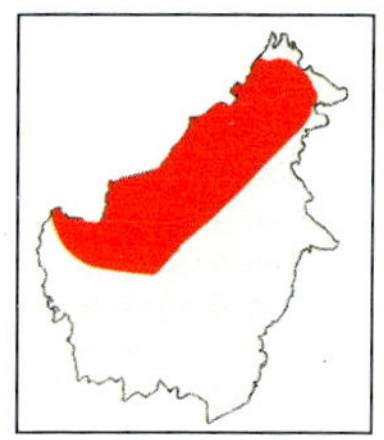

SIZE 600mm

IDENTIFICATION FEATURES Body slender, laterally compressed; head short, rounded and distinct from neck; snout short; eyes large with vertical pupils; tail short; vertebrals slightly enlarged but not sharp; midbody scale rows 15, smooth; ventrals 148–173; subcaudals 34–51, paired; cloacal scute entire.

COLOUR Top of body medium brown to dark brown, with dark vertical bars extending to belly; forehead darker than dorsum and lacks lines; throat brown; belly cream or pale yellow; edges of each scale dark spotted.

HABITAT AND BEHAVIOUR Inhabits lowland forests at < 1,150m asl. Semi-arboreal and terrestrial; active on low vegetation and on the forest floor. Diet comprises slugs and snails. Reproductive biology unstudied.

DISTRIBUTION Thailand, Peninsular Malaysia, Borneo, Sumatra, Pulau Bangka, Natuna and Mentawai archipelagos and Java. Bornean records are from Bukit Brayong, Bidi near Kuching (currently, Deded Krian National Park), Lawas, Gunung Santubong National Park, Pangkalan Ampat at the foothills of Gunung Penrissen, Sungei Baram, Sungei Labang and Sungei Mengiong in Sarawak; Bongon, the Gunung Kinabalu region, and Sungei Mendolong and Sungei Purulon in Sabah, and Banjarmasin in Kalimantan.

IUCN THREAT STATUS Least Concern.

Stuebing's Slug-eating Snake *Asthenodipsas stuebingi* Quah, Grismer, Lim, Anuar & Imbun, 2019

(No vernacular names recorded)

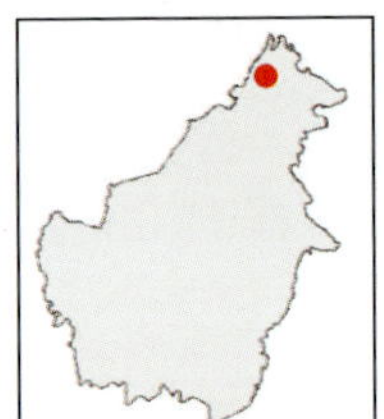

SIZE 619mm

IDENTIFICATION FEATURES Body robust; head short, rounded and distinct from neck; snout short; eyes large with vertical pupils; tail short; vertebrals slightly enlarged with distinct keel; midbody scale rows 15, smooth; ventrals 165–175; subcaudals 35–47, paired; subcaudals paired; cloacal scute entire.

COLOUR Top of head, body and tail pale brown with dark speckling, the pattern paler on flanks; dark nuchal patch from back of jaw to about a fourth of body, followed by 42 dark bands along body and tail; narrow, pale yellow vertebral stripe; belly yellow with dark lateral blotches.

HABITAT AND BEHAVIOUR Inhabits submontane and montane forests at 900–2,000m asl. Arboreal and associated with shrubs, with some terrestrial activity. Diet presumably snails and slugs. Reproductive habits unknown.

DISTRIBUTION Endemic to Borneo. Records are from the Crocker Range Park and Gunung Kinabalu Park and its vicinity (including Kundasang, Kamborangoh and near Kiau) in Sabah.

IUCN THREAT STATUS Not Evaluated.

DARK-SPINED SLUG-EATING SNAKE *Asthenodipsas vertebralis*

(Boulenger, 1900)

(No vernacular names recorded)

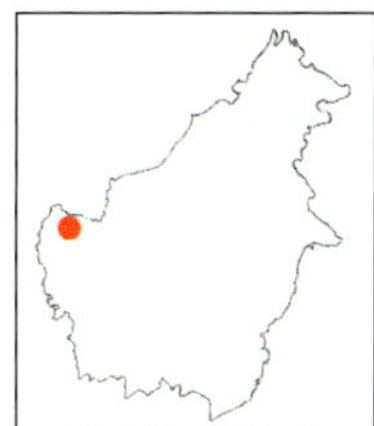

SIZE 771mm

IDENTIFICATION FEATURES Body slender and laterally compressed; head short, rounded, distinct from neck; snout short; eyes large with vertical pupils; infralabial I in contact; tail short; midbody scale rows 15, smooth; ventrals 190–202; subcaudals 68–78; cloacal scute entire.

COLOUR Top of body reddish-brown or dark brown, with small, dark brown spots and indistinct dark cross-bars; interrupted yellow vertebral stripe; belly yellow with brown spots laterally.

HABITAT AND BEHAVIOUR Inhabits submontane and montane forests at 1,585–2,012m asl. Arboreal; active on low vegetation. Diet comprises snails and slugs. Reproductive biology unstudied.

DISTRIBUTION Borneo and Peninsular Malaysia. Bornean record is from Gunung Berembput in Sarawak.

IUCN THREAT STATUS Least Concern.

Pareas Slug-eating Snakes

Small, slender arboreal snakes, identifiable in having head distinct from neck; blunt snout; middorsal scale rows 15; subcaudals divided; no prementals; three pairs of chin shields and long tail.

Barred Slug-eating Snake *Pareas nuchalis* (Boulenger, 1900)

(No vernacular names recorded)

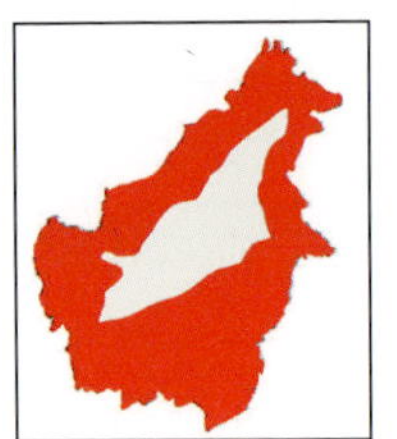

SIZE 715mm

IDENTIFICATION FEATURES Body slender and laterally compressed; head short, rounded and distinct from neck; snout short; orbit separated from supralabials by small scales; eyes large with vertical pupils; tail short; vertebrals enlarged; dorsals weakly keeled; midbody scale rows 15; ventrals 195–218; subcaudals 105–118, paired; cloacal scute entire.

COLOUR Top of body mid-brown to tan; anterior of body has narrow bands, fused vertebrally, and separated by 2–5 scales; dark postocular stripe to angle of jaws, rising to nuchal region, fusing with a dark, irregular 'V'-shaped nuchal patch; belly yellowish-cream.

HABITAT AND BEHAVIOUR Inhabits lowland forests at up to 850m asl. Nocturnal and arboreal; associated with shrubs and other low vegetation up to 2m above substratum, including within bamboo internodes. Diet comprises snails and slugs. Reproductive habits unstudied.

DISTRIBUTION Endemic to Borneo. Localities include Kubah National Park, Niah National Park, Gunung Penrissen, Saribas, Tanjung Datu National Park, Pangkalan Lobang in Balingean, and Sungei Segaham, Sungei Trusan and Sungei Rajang in Sarawak, and Gunung Kinabalu (Sungei Langanan) and Sungei Purulon in Sabah.

IUCN THREAT STATUS Least Concern.

Family Viperidae Vipers & Pit Vipers

Members of this familiar family include vipers and pit vipers, a group with haemotoxic venom. They are identifiable by their broad, triangular head, hinged fang that folds back against the palate of the mouth when not in use, and an elongated ectopterygoid bone that functions as a lever to move the fang. Additionally, their dorsal scales tend to be heavily keeled. In pit vipers, there are specialized pit organs between the nostrils and eyes; these contain infrared receptors that permit the detection of prey through thermal cues. Members of the Viperidae are encountered in habitats as diverse as scrubland, mangrove swamps, tropical rainforests and temperate forests. Their diet comprises primarily birds and mammals, although insects and other large invertebrates may also be eaten. They are mainly ovoviviparous.

Key to Bornean species of Viperidae

1a. Cephalic scales large, symmetrical .... *Calloselasma rhodostoma* (p. 348)
1b. Cephalic scales small, fragmented **2**

2a. Gular and intersupraoculars keeled .... *Tropidolaemus subannulatus* (p. 354)
2b. Gular and intersupraoculars smooth **3**

3a. Subcaudals < 35 .... *Garthius chaseni* (p. 350)
3b. Subcaudals > 40 .... **4**

4a. Snout strongly projecting .... *Craspedocephalus borneensis* (p. 349)
4b. Snout not strongly projecting **5**

5a. Midbody scale rows 19 .... *Trimeresurus malcolmi* (p. 351)
5b. Midbody scale rows 21 .... **6**

6a. Ventrals 148–157 .... *Trimeresurus sabahi* (p. 352)
6b. Ventrals 182–191 .... *Trimeresurus sumatranus* (p. 353)

Calloselasma Malayan Pit Vipers

Mid-sized terrestrial species; body stout in adults, with a weak vertebral ridge; head distinct from neck; snout pointed; cephalic scales large, symmetrical; hemipenes long, slender, deeply forked; enlarged spines at base of lobes.

Malayan Pit Viper *Calloselasma rhodostoma* (Kuhl, 1824)

(Bahasa Malaysia: Ular Kapak Daun. Bahasa Indonesia: Ular Donda)

SIZE 1450mm

IDENTIFICATION FEATURES Body robust; head triangular and distinct from neck; snout acute and slightly upturned; nostril between two nasals; tail thin and short; vertebral ridge distinct; eyes with vertical pupil; dorsals 21, smooth; supralabials 7–8; infralabials 11–12; ventrals 142–163; subcaudals 34–55; cloacal scute entire.

COLOUR Top of body reddish-brown or purplish-brown; flanks paler, speckled with dark brown; a series of 19–31 dark brown subtriangular marks on each side, their tips meeting or alternating; dark postocular stripe, scalloped ventrally, extends to angle of jaws; belly pinkish-cream, mottled with brown.

HABITAT AND BEHAVIOUR Inhabits lowland forests and plantations, from sea level, reaching submontane limits (<1,524m asl). Terrestrial, associated with dense undergrowth and rocky landscapes. Diet comprises small rodents and birds. Oviparous, producing 13–40 eggs (diameter 20–32mm), taking 28–49 days to hatch and producing young, measuring 148–200mm in TL.

VENOM Fibrinogenases and haemorrhagins; linked to severe envenoming that is potentially lethal. Malayan Pit Viper Antivenin (Thai Red Cross Society); Antivenin Polyvalent (Equine) (P.T. Bio Farma, Persero, Bandung); Malayan Pit Viper Antivenom (Thai Government Pharmaceutical Organisation).

DISTRIBUTION Southern Myanmar, Thailand, Laos, Cambodia, Vietnam, Peninsular Malaysia, Java, Karimun Jawa and Kangean archipelagos, Laos, Cambodia, Vietnam and Borneo. Bornean records are from West Kalimantan.

IUCN THREAT STATUS Least Concern.

Craspedocephalus Palm Pit Vipers

Small terrestrial species; body stout; head distinct from neck; snout strongly projecting, spatulate; supraoculars 1–3, flattened or button shaped; gular and intersupraocular scales smooth; subcaudals 45–58; tail short.

Bornean Palm Pit Viper *Craspedocephalus borneensis* (Peters, 1872)

(No vernacular names recorded)

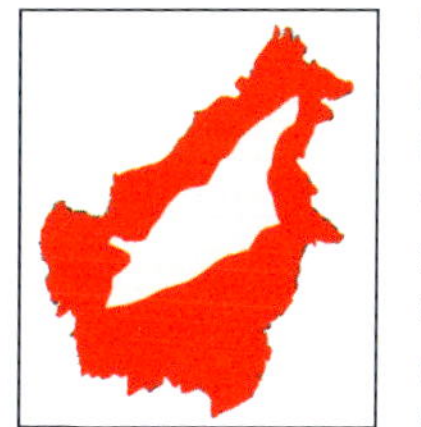

SIZE 830mm

IDENTIFICATION FEATURES Body robust; head triangular and distinct from neck; snout projecting; forehead scales smooth; eyes small with vertical pupils; tail prehensile; dorsals smooth or weakly keeled; midbody scale rows 19 or 21; ventrals 149–166; subcaudals 41–67; cloacal scute entire.

COLOUR Top of body mottled light brown or mid-brown with dark brown, saddle-like pattern with 20–30 blotches or cross-bars, or even bright yellow with darker mottling; oblique, pale postocular stripe extends to neck; belly paler.

HABITAT AND BEHAVIOUR Inhabits lowland swamps and mid-hill forests at < c. 1,130m asl. Terrestrial as juveniles, semi-arboreal as adults. Diet comprises mammals. Oviparous, producing 7–14 eggs.

VENOM Procoagulants, haemorrhagins and necrotoxins possibly present. Bites poorly documented but has potential for serious envenoming. Green Pit Viper Antivenin (Thai Red Cross Society).

DISTRIBUTION Endemic to Borneo. Records are from Kubah National Park, Kidi District, Lundu, Gunung Dulit, Gunung Mulu National Park, Baleh National Park, Pangkalan Ampat near Gunung Penrissen, Tinbarap Palm Oil Estate and middle or upper reaches of Sungei Baram, Sungei Labang and Sungei Mengion in Sarawak; Ulu Temburong in Brunei Darussalam; Gunung Kinabalu (Poring), Mendolong, Paitan, Sandakan Bay, and Maliau Basin, Sungei Kallang, Sungei Malutut and Sungei Purulon in Sabah, and Samarinda in Kalimantan.

IUCN THREAT STATUS Least Concern.

Garthius Brown Pit Vipers

Small terrestrial species; body stout; head flat and distinct from neck; eyes small; forehead scales relatively large and subimbricate; supraocular scales large; gular and intersupraocular scales smooth; subcaudals under 35; tail short.

Kinabalu Brown Pit Viper *Garthius chaseni* (Smith, 1931)

(No vernacular names recorded)

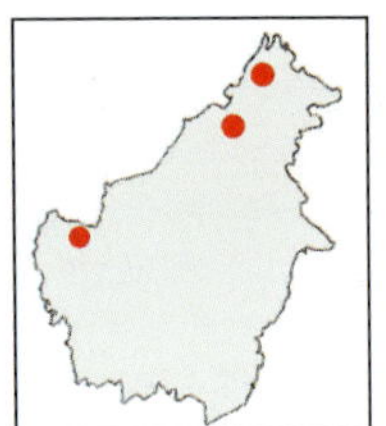

SIZE 690mm

IDENTIFICATION FEATURES Body robust; head triangular, broad, flattened and distinct from neck; snout blunt; eyes small with vertical pupils; tail not prehensile; dorsals strongly keeled anteriorly, weakly keeled posteriorly; midbody scale rows 17 or 19; ventrals 131–143; subcaudals 20–30, paired; cloacal scute entire.

COLOUR Top of body dark brown or reddish-brown, marked with irregular dark brown blotches in paired rows anteriorly, fused to form bands towards posterior of body; smaller dark blotches on lower flanks; dark postocular stripe, edged ventrally by line of yellow scales, extends to neck; belly yellow with large grey areas.

HABITAT AND BEHAVIOUR Inhabits submontane forests at 915–1,550m asl. Encountered on leaf litter. Diet unknown. Ovoviviparous (neonate numbers and size unknown).

VENOM Mixture of procoagulants and probably anticoagulants, and venom haemorrhagins. Bites with potentially lethal envenoming. No antivenom raised.

DISTRIBUTION Endemic to Borneo. Restricted to upper-middle elevations of Gunung Penrissen, Gunung Mulu National Park and the Bario Highlands of Sarawak, and Gunung Kinabalu Park (Kiau and Lumu-Lumu) and Crocker Range Park in Sabah. One isolated lowland record from Maliau Basin in Sabah.

IUCN THREAT STATUS Least Concern.

Close-up of head

Trimeresurus Green Pit Vipers

Medium-sized arboreal species; body slender to moderately robust; head distinct from neck; gular and intersupraocular scales smooth; subcaudals > 40; tail short.

Kinabalu Green Pit Viper *Trimeresurus malcolmi* Loveridge, 1938

(No vernacular names recorded)

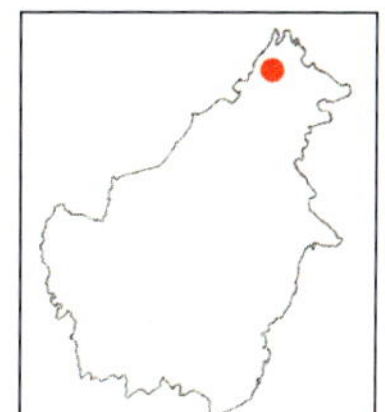

SIZE 1.33m

IDENTIFICATION FEATURES Body robust; head narrow and distinct from neck; forehead scales smooth; eyes small with vertical pupils; tail prehensile; dorsals weakly keeled; midbody scale rows 19; ventrals 168–174; subcaudals 61–81; cloacal scute entire.

COLOUR Top of body black, each scale with subtriangular green pattern, creating indistinct bands; forehead black with green-centred scales; juveniles have white stripe along flanks, absent in adults; belly pale green, with green ventrals that are dark posterior edged, and with pale green anterior edge; tail has parallel red dots within dark green scales; subcaudals light green edged with black.

HABITAT AND BEHAVIOUR Inhabits submontane and montane forests at 1,000–1,700m asl, particularly in oak forests; also secondary vegetation. Diet unknown. Reproductive habits unstudied.

VENOM Procoagulants, anticoagulants and haemorrhagins possibly present. Bites unstudied and likely to cause severe envenoming. No antivenom raised.

DISTRIBUTION Endemic to Borneo. Restricted to upper-middle elevations of Gunung Kinabalu Park (Kiau, Bundu Tuhan, Lumu-Lumu and Sungei Silau-Silau), and outside park in Kundasang, in Sabah.

IUCN THREAT STATUS Near Threatened.

Sabah Green Pit Viper *Trimeresurus sabahi* Regenass & Kramer, 1981

(No vernacular names recorded)

SIZE 810mm

IDENTIFICATION FEATURES Body slender in males, robust in females; head triangular, distinct from neck; occipitals and temporals smooth or weakly keeled; eyes small with vertical pupils; dorsals keeled; midbody scale rows 21, ventrals 147–157; subcaudals 59–76; cloacal scute entire.

COLOUR Top of body bright green, without cross-bars or a postocular streak; ventrolateral stripe red or rusty-red in males, and white or yellow in females; iris red or orange in adults, orange or yellowish-green in juveniles.

HABITAT AND BEHAVIOUR Inhabits submontane forests at 1,000– 1,150m asl. Arboreal, on low vegetation of shrubs and branches. Diet and reproductive biology unstudied.

VENOM Venom characteristics and bite unstudied. Green Pit Viper Antivenin (Thai Red Cross Society).

DISTRIBUTION Endemic to Borneo. Localities include Gunung Dulit, Gunung Gading National Park and Pangkalan Ampat at the mid-hills of Gunung Penrissen in Sarawak; Brunei Darussalam (without a precise locality), and Gunung Kinabalu Park (Sayap, Kiau, Kenokok and Bundu Tuhan), Crocker Range Park and Gunung Lumaku in Sabah.

IUCN THREAT STATUS Least Concern.

SUMATRAN PIT VIPER *Trimeresurus sumatranus* (Raffles, 1822)

(Bahasa Malaysia/Indonesia: Ular Kapak. Iban: Ular Beliung, Ular Engkaradau)

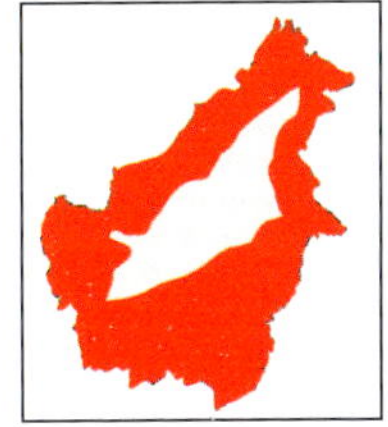

SIZE 1.35m

IDENTIFICATION FEATURES Body robust; head distinct from neck; forehead scales smooth; internasals present; eyes small with vertical pupils; tail prehensile; dorsals weakly keeled; midbody scales 19 or 21 (rarely 23); ventrals 182–191; subcaudals 55–66; cloacal scute entire.

COLOUR Top of body green with black cross-bars, in adults; forehead scales edged with black; supralabials light blue; tail reddish-brown, brighter in juveniles; belly yellowish-green.

HABITAT AND BEHAVIOUR Inhabits lowlands and mid-hill forests at < 1,000m asl. Arboreal, on low vegetation. Diet comprises mammals, birds and frogs. Reproductive habits unstudied.

VENOM Anticoagulants and haemorrhagins present; procoagulants possibly present. Bites unstudied and likely to cause severe envenoming. Green Pit Viper Antivenin (Thai Red Cross Society).

DISTRIBUTION Thailand, Peninsular Malaysia, Singapore, Borneo, Sumatra, Mentawai Archipelago, Pulau Belitung and possibly Palawan. Bornean localities include the base of Gunung Penrissen, Kubah National Park, Gunung Dulit, Gunung Mulu National Park, Baleh National Park, Nanga Segerak along Sungei Engkari, Nanga Serembuang along Sungei Skrang and the upper reaches of Sungei Baram in Sarawak; Ulu Temburong in Brunei Darussalam; Ranau, Mendolong, Kamborangoh, Lumu-Lumu, Gunung Kinabalu Park, including Bundu Tuhan and Marak-Parak, Danum Valley, Pulau Gaya, and Sungei Purulon and Sungei Malutut in Sabah, and Banjarmasin in Kalimantan.

IUCN THREAT STATUS Least Concern.

Tropidolaemus Keeled Green Pit Vipers

Medium-sized to large arboreal snakes; body slender to moderately robust, particularly in adult females; head distinct from neck; nasal pores absent; upper surfaces of snout and forehead have small, keeled scales; gular scales strongly keeled; supralabial II not bordering anterior margin of loreal pit; tail short.

Bornean Keeled Green Pit Viper *Tropidolaemus subannulatus* (Gray, 1842)

(Bahasa Malaysia/Indonesia: Ular Kapak. Iban: Ular Engkaradau)

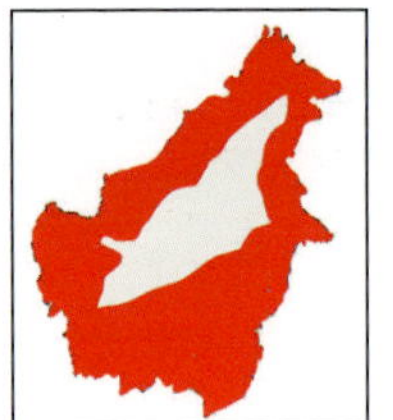

SIZE 963mm

IDENTIFICATION FEATURES Body slender in juveniles, relatively thick in adults; head distinct from neck; eyes small with vertical pupils; tail prehensile; occipitals distinctly keeled in males; midbody scale rows 21–23 in males, 21–29 in females; ventrals 127–148; subcaudals 40–54, paired; cloacal scute divided.

COLOUR Top of body green or greenish-blue, with dark cross-bars that are white or red in males and juveniles, and pale blue in adult females; postocular stripe cream or yellow in adult females, and white and red in juveniles and males; belly uniform pale green, or blotched or spotted with blue or red.

HABITAT AND BEHAVIOUR Inhabits lowland forests and strictly arboreal. Encountered on both low vegetation and tall trees in riparian and dipterocarp forests. Diet comprises birds, rodents and possibly bats. Ovoviviparous (numbers and size of neonates unknown).

VENOM Mixture of procoagulants; anticoagulants, haemorrhagins and necrotoxins possibly present. Envenoming may be lethal. Green Pit Viper Antivenin (Thai Red Cross Society).

DISTRIBUTION Borneo, Pulau Belitung, Buton, Sangihe Archipelago and Sulawesi in Indonesia, as well as Balabac, Basilan, Bohol, Dinagat, Jolo, Leyte, Luzon, Mindanao, Negros, Palawan, Panay, Samar, Sibutu and Tumindao in the Philippines. The best Bornean site to see it is Bako National Park in western Sarawak. Other localities include Kubah National Park, Santubong National Park, Gunung Mulu National Park, Lambir Hills National Park, Gunung Gading National Park, Gunung Dulit, Gunung Penrissen, Baleh National Park, Samunsam Wildlife Sanctuary, Ranchan Pool in Serian, Nanga Segerak along Sungei Engkari, Busau, Tinbarap Palm Oil Estate and Oya in Sarawak; Ulu Temburong in Brunei Darussalam; Labuan; Bukit Padang in Kota Kinabalu, Kota Belud, Pulau Gaya, Pulau Sulug, Penampang, Sandakan, Tawau Hills Park, Gunung Kinabalu Park and Sungei Purulon in Sabah, and Landak, Muara Jawa in Kutai, Sebruang Valley and the upper reaches of Sungei Kapuas in Kalimantan.

IUCN THREAT STATUS Least Concern.

Female (left); male (right)

Male

Family Xenodermatidae Strange-skinned Snakes

Species with scales nearly completely fused to underlying skin (in most snakes, only one scale edge is embedded in underlying dermis). They are restricted to South-east Asia.

Key to Bornean species of Xenodermatidae

1a. Enlarged head scales present..*Paraxenodermus borneensis* (p. 356)
1b. Enlarged head scales (apart from rostral) absent....................*Xenodermus javanicus* (p. 357)

Paraxenodermus Bornean Stream Snakes

Long, slender, laterally compressed terrestrial and semi-arboreal species from montane regions, diagnosable in having the following characteristics: head distinct from neck; posterior third of forehead and posterior temporal region covered with small scales; numerous small scales between parietal and supralabials behind eye; nostril in a large concave nasal; ventrals large and tail long.

Bornean Stream Snake (p. 356)

BORNEAN STREAM SNAKE *Paraxenodermus borneensis* (Boulenger, 1899)

(No vernacular names recorded)

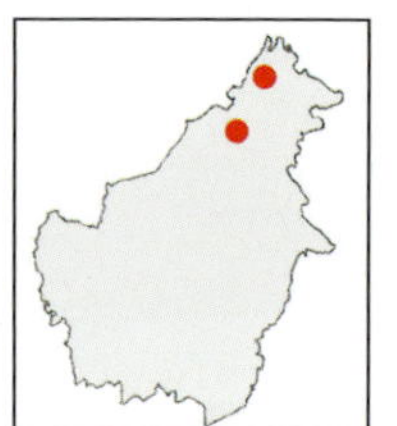

SIZE 750mm

IDENTIFICATION FEATURES Body slender and laterally compressed, with sharp ridge on vertebral region; head large and distinct from neck; paired rows of small scales in front of eyes; nostrils large, flaring; eyes small and beady with vertical pupils; tail long and slender; dorsals strongly keeled; vertebrals enlarged; midbody scale rows 30–35; ventrals 205–210; subcaudals 117–128, single; cloacal scute entire.

COLOUR Top of body bluish-brown or mid-brown, with dark, squarish marks, as broad as or broader than interspaces; several dark bars behind forehead; belly plain brown.

HABITAT AND BEHAVIOUR Inhabits submontane and montane forests at 800–1,800m. Nocturnal; active on stream banks, as well as low vegetation. Diet and reproductive habits unstudied.

DISTRIBUTION Endemic to Borneo, with records from Gunung Kinabalu, Trus Madi and Crocker Range in Sabah, and Gunung Murud in Sarawak.

IUCN THREAT STATUS Least Concern.

REMARKS In the recent past, this species was allocated to the north-east Indian genus *Stoliczkia*.

Xenodermus Rough-backed Snakes

Small, slender snakes; head distinct from neck; eyes moderate with rounded pupils; forehead scales, apart from rostral, not enlarged; nostrils located within large concave scute; three longitudinal series of tubercles along body dorsum; slender tail.

Rough-backed Litter Snake *Xenodermus javanicus* Reinhardt, 1836

(No vernacular names recorded)

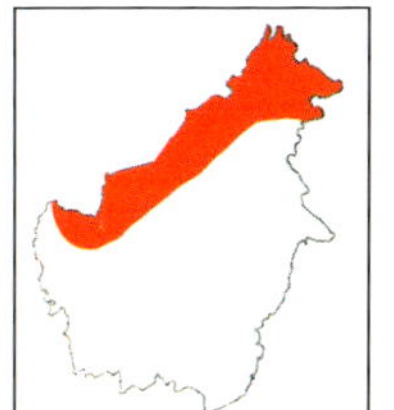

SIZE 650mm

IDENTIFICATION FEATURES Body slender and compressed; head large and distinct from neck; three rows of large keeled scales on dorsum; nostrils flaring and pointed forwards; eyes large with rounded pupils; tail long; midbody scale rows 40–51; ventrals 165–186; subcaudals 129–165, single; caudal scale entire.

COLOUR Top of body nearly unpatterned grey; ridges on scales cream; snout slightly paler than rest of forehead; labials and throat cream; belly cream with black areas.

HABITAT AND BEHAVIOUR Inhabits lowland dipterocarp forests and agricultural areas at 500–1,100m asl. Terrestrial and subfossorial. Diet comprises frogs. Oviparous, with clutches comprising 2–4 eggs, each 23–28 × 9–11mm.

DISTRIBUTION Myanmar, Thailand, Peninsular Malaysia, Borneo, Sumatra and Java. Records from Borneo are from Kubah National Park, Baleh National Park, Gunung Mulu National Park, Sungei Labang, Sungei Mengion and Sungei Pesu in Sarawak, and Danum Valley and foothills of Mt Lucia at Tawau Hills Park in Sabah.

IUCN THREAT STATUS Least Concern.

Close-up of head and forebody

Family Typhlopidae Blind Snakes

Blind-snake species have blunt heads, short tails, small eyes concealed under scales, distinct with pupil or indistinct dark spot, four supralabials, ventrals not enlarged, upper jaw with 3–4 teeth, and lower jaw toothless. They are nocturnal and subfossorial, sometimes active on the surface, after rains, and some display arboreality. Their diet comprises small arthropods such as termites, ants and their eggs, larvae and pupae. They are oviparous, with parthenogenesis known in a species, and cosmopolitan in tropical and subtropical regions.

Key to Bornean species of Typhlopidae

1a. Postocular 1 .... *Indotyphlops braminus* (p. 360)
1b. Postoculars 2–3 .... **2**

2a. Left lung present .... *Argyrophis muelleri* (opposite)
2b. Left lung absent **3**

3a. Hemipenis lacking retrocloacal sacs .... *Malayotyphlops koekkoeki* (p. 361)
3b. Hemipenis with retrocloacal sacs **4**

4a. Preocular absent .... *Ramphotyphlops lineatus* (p. 362)
4b. Preocular present **5**

5a. Rostral over half head width .... *Ramphotyphlops olivaceus* (p. 364)
5b. Rostral under half head width .... *Ramphotyphlops lorenzi* (p. 363)

Argyrophis Giant Blind Snakes

Small, and sometimes large, stout or slender fossorial snakes, identifiable by having head that is rounded laterally and in profile; rostral moderate; postoculars 2–3; eyes moderate with distinct pupil; presence of left lung; midbody scale rows 20–30; apical spine present; tail short to moderate.

Müller's Blind Snake *Argyrophis muelleri* (Schlegel, 1839)

(No vernacular names recorded)

SIZE 540mm

IDENTIFICATION FEATURES Body moderately built; head indistinct from neck; snout rounded and strongly projecting; upper rostral narrow, widening ventrally; eyes distinct; forehead scales lack depressions and with glands along sutures; parietals twice as wide as deep; body scales smooth; midbody scale rows 24–26 (rarely 22, 28 or 30); total length 25–45 times midbody diameter; caudal spine present.

COLOUR Top of body dark brown, purple or black; belly cream, yellow or gold, clearly demarcated from top colour.

HABITAT AND BEHAVIOUR Inhabits forested lowlands and mid-hills, and areas of wet agriculture at < 1,676m. Subfossorial, in waterlogged areas, under rocks and logs. Diet includes larvae of ants and termites. Ovoviviparous, producing 2–22 neonates.

DISTRIBUTION Myanmar, Thailand, Laos, Cambodia, Vietnam, Peninsular Malaysia, Singapore, Borneo, Sumatra, Pulau Bangka, Pulau Nias, Pulau Weh, Pulau Nias and New Guinea. The Bornean record lacks a precise locality.

IUCN THREAT STATUS Least Concern.

Indotyphlops Common Blind Snakes

Small, slender fossorial snakes, identifiable by head that is rounded dorsally and in profile; rostral narrow to moderate; postoculars single; eyes small; left lung absent; midbody scale rows 18–22; apical spine present; tail short or long.

Brahminy Blind Snake *Indotyphlops braminus* (Daudin, 1803)

(Bahasa Malaysia: Ular Buta Biasa; Ular Hitam. Bahasa Indonesia: Ular Buta Brahminy)

SIZE 197mm

IDENTIFICATION FEATURES Body moderately built; head indistinct from neck; snout rounded; nostrils situated laterally; eyes distinct; total length 30–45 times midbody diameter; midbody scale rows 20; middorsal scales 261–368; subcaudals 8–15; caudal spine present.

COLOUR Top of body pale brown to black; pale ventrally; lower snout, chin, cloaca and tail-tip white or cream; in paler individuals, a dark triangular apical spot on each scale.

HABITAT AND BEHAVIOUR Inhabits lightly forested areas and frequently encountered in human habitation at < 2,000m asl. Subfossorial. Diet includes termites and ants, and their larvae. Parthenogenetic, producing 1–8 eggs, each 11–20mm. Hatchlings 61mm.

DISTRIBUTION Widespread in tropical, subtropical and temperate regions worldwide, and in South-east Asia known from Myanmar, Thailand, Cambodia, Laos, Vietnam, Peninsular Malaysia, Singapore, Borneo, Sumatra, Java and Bali. Records for Borneo include Kuching, Gunung Santubong National Park, Kota Samarahan and Lambir Hills National Park in Sarawak; Labuan; Bandar Seri Begawan in Brunei Darussalam; Kota Kinabalu and Madai Cave in Sabah, and Buntok, Samarinda and Sinkawang in Kalimantan.

IUCN THREAT STATUS Least Concern.

Close-up of head and forebody

Malayotyphlops Malay Blind Snakes

Small to medium-sized fossorial snakes with moderate to stout bodies, identifiable by head that is rounded dorsally and in profile; postoculars 2–4; eyes small with distinct pupils; left lung absent; eversible hemipenis lacking retrocloacal sacs; midbody scale rows 20–30; apical spine present; tail short to moderate.

Koekkoek's Blind Snake *Malayotyphlops koekkoeki* (Brongersma, 1934)

(No vernacular names recorded)

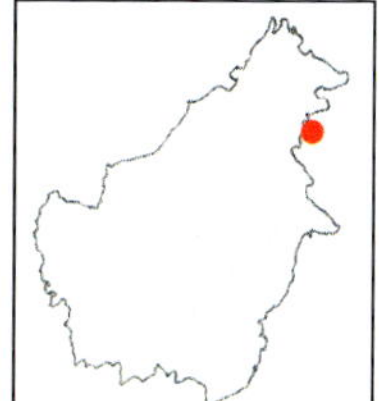

SIZE 445mm

IDENTIFICATION FEATURES Body slender; head indistinct from neck; snout rounded and projecting; rostral broad, being 1.5 times head width; preocular present; ocular in contact with supralabials III–IV; eye indistinct; midbody scale rows 26; middorsal scales 285; caudal spine present; tail short.

COLOUR Top of body greyish-brown in preservative; live colouration unknown.

HABITAT AND BEHAVIOUR Habitat and behaviour unknown, although assumed to be a lowland species that is fossorial. Diet unstudied; and a clutch size of 14 has been reported.

DISTRIBUTION Endemic to Borneo, with records from Sarawak and Pulau Bunyu, a 114 km² island off the east coast in Kalimantan Timur Province. A third probable record is from Gunung Kinabalu, in Sabah.

IUCN THREAT STATUS Data Deficient.

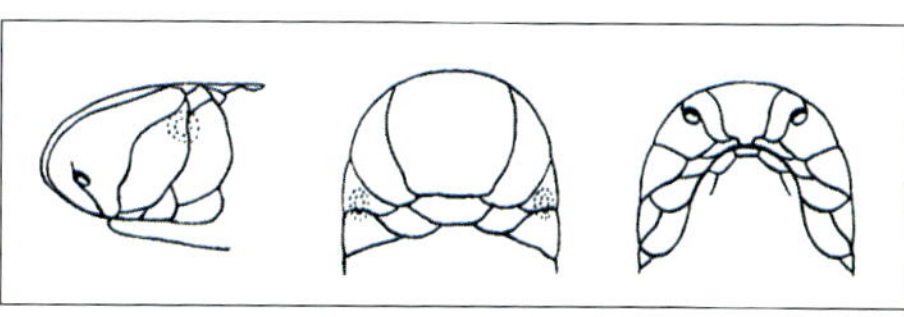

After Brongersma (1834).

Ramphotyphlops Common Blind Snakes

Small to large fossorial snakes with slender to stout bodies, identifiable by rounded head; rostral narrow to moderate; postoculars 2–3; eye moderately large with distinct pupil that may be reduced to a spot; prefrontals paired; hemipenis with retrocloacal sacs; midbody scale rows 18–30; apical spine present; tail short to long.

Lined Blind Snake *Ramphotyphlops lineatus* (Schlegel, 1839)

(Bahasa Malaysia: Ulat)

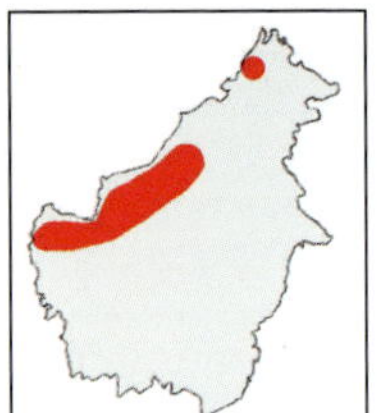

SIZE 480mm

IDENTIFICATION FEATURES Body moderate; head indistinct from neck; snout rounded, projecting; rostral enlarged, about half head width; nasal incompletely divided; single large ocular; preoculars and suboculars absent; eye indistinct; midbody scale rows 22–23; subcaudals nine; caudal spine present.

COLOUR Top of body cream or yellow, with 10 brown longitudinal stripes from head to tail-tip; venter cream or yellow.

HABITAT AND BEHAVIOUR Inhabits tropical forests in the lowlands, up to submontane limits, at 0–1,420m asl. Fossorial. Diet and reproductive biology unstudied.

DISTRIBUTION Thailand, Laos, Cambodia, Vietnam, Peninsular Malaysia, Singapore, Borneo, Sumatra, Pulau Bangka, Pulau Nias and Java. On Borneo, records are from Sungei Pesu, Gunung Pueh and Gunung Penrissen in Sarawak; Penampang in Sabah, and Singkawang in Kalimantan.

IUCN THREAT STATUS Least Concern.

Preserved specimen from Sarawak.

LORENZ'S BLIND SNAKE *Ramphotyphlops lorenzi* (Werner, 1909)

(No vernacular names recorded)

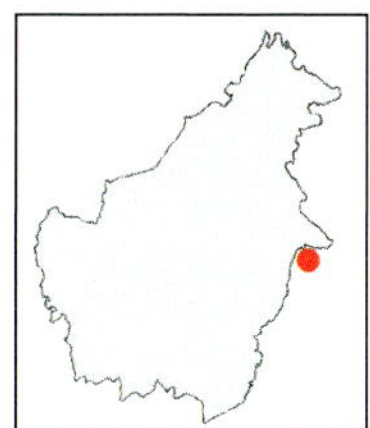

SIZE 337mm

IDENTIFICATION FEATURES Body slender; head indistinct from neck; snout has projecting sharp, horizontal edge; rostral less than half head width; nasal incompletely divided; preocular present; eye distinct; tail longer than broad; midbody scale rows 22; caudal spine present.

COLOUR Top of body light greyish-brown or greyish-green; rostral brown edged with yellow; venter pale olive-green.

HABITAT AND BEHAVIOUR Apparently inhabits lowland forests. Diet and reproductive habits unstudied. Presumably subfossorial, associated with litter, and consumes small arthropods.

DISTRIBUTION Endemic to Pulau Miang Besar in Kalimantan, off the east coast of Borneo.

IUCN THREAT STATUS Data Deficient.

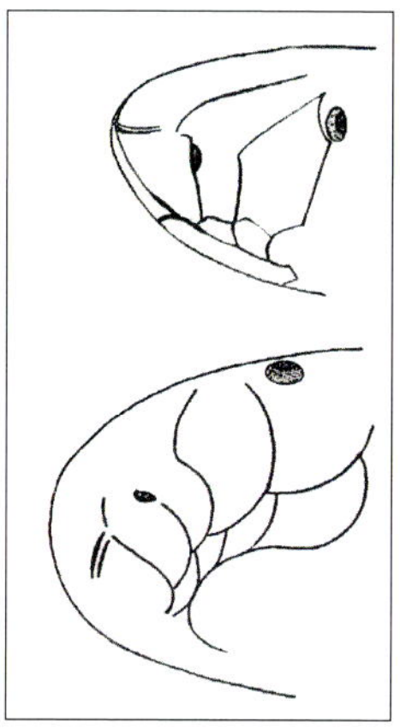

After de Rooij (1917)

Olive Blind Snake *Ramphotyphlops olivaceus* (Gray, 1845)

(No vernacular names recorded)

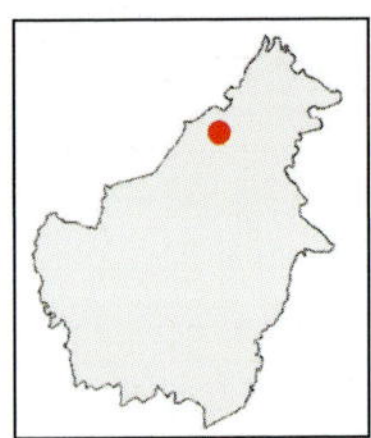

SIZE 430mm

IDENTIFICATION FEATURES Body slender; head indistinct from neck; snout projecting, with narrow transverse edge; rostral scute large, over half head width; eyes distinct; midbody scale rows 20–22; caudal spine present.

COLOUR Top of body pale olive-brown, each scale paler basally; belly paler; scales on lips, throat and vent pale; occasionally, indistinct dark longitudinal stripes present between scale rows.

HABITAT AND BEHAVIOUR Inhabits lowland rainforests. Fossorial, living in leaf litter. Diet unknown. Oviparous; clutch size three.

DISTRIBUTION Borneo, the Sangihe Archipelago, Seram and Mysool in Indonesia, and Samar and Babuan in the Philippines. Sole Bornean record is from Sungei Baram basin in Sarawak.

IUCN THREAT STATUS Least Concern.

Family Xenophidiidae Spine-jawed Snakes

The two known members of this family are characterized by compressed bodies, short tails, rounded pupils, enlarged prefrontals, the absence of a loreal scute, undivided nasals and a single cloacal scute. The right lung is not vascularized, a pelvis is absent and there is an ectopterygoid process on the maxilla.

Xenophidion Spiny-jawed Snakes

Small terrestrial or subfossorial snakes; body laterally compressed; head indistinct from neck; nasal undivided; prefrontals enlarged; premaxilla lacking teeth; dorsal scales keeled; subcaudals and cloacal scutes undivided; tail short.

Bornean Spiny-jawed Snake *Xenophidion acanthognathus*

Günther & Manthey, 1995

(No vernacular names recorded)

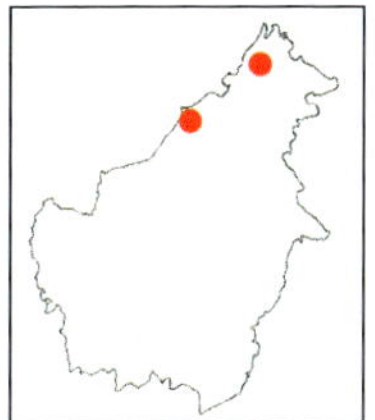

SIZE 335mm

IDENTIFICATION FEATURES Body slender, strongly compressed; head slightly flattened and indistinct from neck; loreal absent; scales keeled (except two lowest rows on body); midbody scale rows 23, keeled (except two lowest rows); ventrals 181; subcaudals 51.

COLOUR Top of body brown with zigzag longitudinal pattern; forehead mid-brown with pale areas; broad white postocular stripe extends to sides of neck; zigzag tan stripes on flanks; belly black with small, squarish yellow spots.

HABITAT AND BEHAVIOUR Inhabits forested mid-hills at about 600m asl. Possibly subfossorial. Diet includes skinks. Oviparous; clutch size unknown.

DISTRIBUTION Endemic to Borneo. Only known from the Kinabalu Massif of Sabah and Lambir Hills National Park in Sarawak.

IUCN THREAT STATUS Data Deficient.

Close-up of head and forebody

Checklist of the Reptiles of Borneo

Reptiles of Borneo (current: 30 October 2024), and their occurrence in the political divisions within the island. References: * (preceding scientific name) = invasive species established on Borneo. Unchecked boxes within distributional ranges imply a Bornean record without a specific locality.

No.	Current Name	Occurrence			
		Sarawak	Sabah	Brunei	Kalimantan
	Order CROCODYLIA				
	Family CROCODYLIDAE True Crocodiles				
1	*Crocodylus porosus* Schneider, 1801	•	•	•	•
2	*Crocodylus raninus* Müller & Schlegel, 1844	•			
3	*Crocodylus siamensis* Schneider, 1801				•
	Family GAVIALIIDAE Gharials				
4	*Tomistoma schlegelii* (Müller, 1838)	•	•	•	•
	Order CHELONIA				
	Family GEOEMYDIDAE Asian Hard-shelled Turtles				
5	*Batagur borneoensis* (Schlegel & Müller, 1845)	•	•	•	•
6	*Cuora couro* (Leschenault de la Tour in Schweigger, 1812)	•	•	•	•
7	*Cyclemys dentata* (Gray, 1831)	•	•	•	•
8	*Cyclemys enigmatica* Fritz, Guicking, Auer, Sommer, Wink & Hundsdörfer, 2008	•			
9	*Heosemys spinosa* (Gray, 1831)	•	•	•	•
10	*Notochelys platynota* (Gray, 1834)	•	•	•	•
11	*Orlitia borneensis* Gray, 1873	•			•
12	*Siebenrockiella crassicollis* (Gray, 1831)	•	•		•
	Family CHELONIIDAE Sea Turtles				
13	*Caretta caretta* (Linnaeus, 1758)				
14	*Chelonia mydas* (Linnaeus, 1758)	•	•	•	•
15	*Eretmochelys imbricata* (Linnaeus, 1766)	•	•	•	•
16	*Lepidochelys olivacea* (Eschscholtz, 1829)	•	•	•	
	Family DERMOCHELYIDAE Leatherback Sea Turtles				•
17	*Dermochelys coriacea* (Vandelli, 1761)	•		•	•
	Family EMYDIDAE New World Hard-shelled Turtles				
18	**Trachemys scripta* (Thunberg, 1792)	•		•	
	Family TESTUDINIDAE Land Tortoises				
19	*Manouria emys* (Schlegel & Müller, 1840)	•	•		•
	Family TRIONYCHIDAE Softshell Turtles				
20	*Amyda cartilaginea* (Boddaert, 1770)	•	•	•	•
21	*Chitra chitra* Nutaphand, 1986				•
22	*Dogania subplana* (Geoffroy Saint-Hilaire, 1809)	•	•	•	•
23	*Pelochelys cantorii* Gray, 1864	•	•		•
24	**Pelodiscus sinensis* Wiegmann, 1835	•	•		•
	Order SQUAMATA – SAURIA				
	Family AGAMIDAE Dragon Lizards				
25	*Aphaniotis acutirostris* Modigliani, 1889	•			•
26	*Aphaniotis fusca* (Peters, 1864)	•			•
27	*Aphaniotis ornata* (van Lidth de Jeude, 1893)	•	•	•	•
28	*Bronchocela cristatella* (Kuhl, 1820)	•	•	•	•
29	*Bronchocela jubata* Duméril & Bibron, 1837				•
30	**Calotes versicolor* (Daudin, 1802)	•	•	•	•
31	*Draco cornutus* Günther, 1864	•	•	•	•
32	*Draco cristatellus* Günther, 1872	•	•		•
33	*Draco fimbriatus* Kuhl, 1820	•			
34	*Draco haematopogon* Boie, Gray, 1831	•	•		•
35	*Draco maximus* Boulenger, 1893	•	•		•

No.	Current Name	Occurrence			
		Sarawak	Sabah	Brunei	Kalimantan
36	*Draco melanopogon* Boulenger, 1887	•	•	•	•
37	*Draco obscurus* Boulenger, 1887	•	•		•
38	*Draco punctatus* Boulenger, 1900	•	•		•
39	*Draco quinquefasciatus* Hardwicke & Gray, 1827	•	•	•	•
40	*Draco sumatranus* Schlegel, 1844	•	•	•	•
41	*Gonocephalus bornensis* (Schlegel, 1848)	•	•	•	•
42	*Gonocephalus doriae* (Peters, 1871)	•	•		•
43	*Gonocephalus grandis* (Gray, 1845)	•	•	•	•
44	*Gonocephalus liogaster* (Günther, 1872)	•	•	•	
45	*Gonocephalus mjobergi* Smith, 1925	•			
46	*Harpesaurus borneensis* (Mertens, 1924)	•			•
47	*Hypsicalotes kinabaluensis* (de Grijs, 1937)		•		
48	*Pelturagonia anolophium* Harvey, Larson, Jacobs, Shaney, Streicher, Hamidy, Kurniawan & Smith, 2019				•
49	*Pelturagonia borneensis* (Inger, 1960)	•	•		•
50	*Pelturagonia cephalum* Mocquard, 1890	•	•		
51	*Pelturagonia nigrilabris* (Peters, 1864)	•			•
52	*Pelturagonia spiniceps* Smith, 1925	•			
53	*Pseudocalotes nigrigularis* (Ota & Hikida, 1991)		•		
54	*Pseudocalotes saravacensis* Inger & Stuebing, 1994	•			
	Family ANGUIDAE Glass Snakes				
55	*Dopasia buettikoferi* (van Lidth de Jeude, 1905)	•	•		•
	Family DIBAMIDAE Worm Lizards				
56	*Dibamus ingeri* Das & Lim, 2003		•		
57	*Dibamus leucurus* (Bleeker, 1860)	•	•		•
58	*Dibamus vorisi* Das & Lim, 2003		•		
	Family EUBLEPHARIDAE Eyelid Geckos				
59	*Aeluroscalabotes felinus* (Günther, 1864)	•	•		•
	Family GEKKONIDAE True Geckos				
60	*Cnemaspis dringi* Das & Bauer, 1998	•			
61	*Cnemaspis kendallii* (Gray, 1845)	•			
62	*Cnemaspis lagang* Nashriq, Davis, Bauer & Das, 2022	•			
63	*Cnemaspis leucura* Kurita, Nishikawa, Matsui & Hikida, 2017	•			
64	*Cnemaspis matahari* Nashriq, Davis, Bauer & Das, 2022	•			
65	*Cnemaspis nigridia* (Smith, 1925)	•			
66	*Cnemaspis paripari* Grismer & Chan, 2009	•			
67	*Cnemaspis sirehensis* Nashriq, Davis, Bauer & Das, 2022	•			
68	*Cyrtodactylus baluensis* (Mocquard, 1890)	•	•		
69	*Cyrtodactylus cavernicolus* Inger & King, 1961	•			
70	*Cyrtodactylus consobrinus* (Peters, 1871)	•			•
71	*Cyrtodactylus hamidyi* Riyanto, Fauzi, Sidik, Mumpuni, Irham, Kurniawan, Ota, Okamoto, Hikida & Grismer, 2021		•		•
72	*Cyrtodactylus hantu* Davis, Bauer, Jackman, Nashriq & Das, 2021	•			
73	*Cyrtodactylus hutan* Davis, Nashriq, Woytek, Wikramanayake, Bauer, Karin, Brennan, Iskandar & Das, 2023	•		•	
74	*Cyrtodactylus ingeri* Hikida, 1990	•		•	
75	*Cyrtodactylus kapitensis* Davis, Nashriq, Woytek, Wikramanayake, Bauer, Karin, Brennan, Iskandar & Das, 2023	•			
76	*Cyrtodactylus limajalur* Davis, Bauer, Jackman, Nashriq & Das, 2019	•			
77	*Cyrtodactylus malayanus* (de Rooij, 1915)	•		•	•
78	*Cyrtodactylus matsuii* Hikida, 1990		•		
79	*Cyrtodactylus miriensis* Miri Bent-toed Gecko	•		•	

No.	Current Name	Occurrence			
		Sarawak	Sabah	Brunei	Kalimantan
80	*Cyrtodactylus muluensis* Davis, Bauer, Jackman, Nashriq & Das, 2019	•			
81	*Cyrtodactylus pubisulcus* Inger, 1957	•			
82	*Cyrtodactylus yoshii* Hikida, 1990		•		
83	*Gehyra mutilata* (Wiegmann, 1834)	•	•	•	•
84	*Gekko* cf. *albofasciolatus* (Günther, 1867)	•	•	•	•
85	*Gekko browni* (Russell, 1979)	•			•
86	*Gekko gecko* Linnaeus, 1758		•		
87	*Gekko horsfieldii* (Gray, 1827)	•	•	•	•
88	*Gekko kuhli* (Stejneger, 1902)	•	•	•	•
89	*Gekko monarchus* (Schlegel in Duméril & Bibron, 1836)	•	•	•	•
90	*Gekko rhacophorus* (Boulenger, 1899)	•	•		
91	*Gekko sorok* (Das, Lakim & Kandaung, 2008)	•	•		
92	*Gekko yasumai* (Ota, Sengoku & Hikida, 1996)				•
93	*Hemidactylus brookii* Gray, 1845	•			
94	*Hemidactylus craspedotus* (Mocquard, 1890)	•	•		
95	*Hemidactylus frenatus* Duméril & Bibron, 1836	•	•	•	•
96	*Hemidactylus garnotii* Duméril & Bibron, 1836		•		•
97	*Hemidactylus platyurus* (Schneider, 1792)	•	•	•	•
98	*Hemiphyllodactylus typus* Bleeker, 1860	•	•	•	•
99	*Lepidodactylus lugubris* (Duméril & Bibron, 1836)	•	•		•
100	*Lepidodactylus ranauensis* Ota & Hikida, 1988		•		
	Family LACERTIDAE Eurasian Lizards				
101	*Takydromus sexlineatus* Daudin, 1802	•			•
	Family LANTHANOTIDAE Earless Monitors				
102	*Lanthanotus borneensis* Steindachner, 1877	•		•	•
	Family SCINCIDAE Skinks				
103	*Brachymeles apus* Hikida, 1982	•	•		
104	*Dasia grisea* (Gray, 1845)	•	•		•
105	*Dasia olivacea* Gray, 1839	•	•		•
106	*Dasia semicincta* (Peters, 1867)	•			
107	*Dasia vittata* (Edeling, 1864)	•	•	•	•
108	*Dasia vyneri* (Shelford, 1905)	•	•	•	•
109	*Emoia atrocostata* (Lesson, 1830)	•	•	•	
110	*Emoia caeruleocauda* (De Vis, 1892)		•		
111	*Emoia cyanura* (Lesson, 1830)				•
112	*Eutropis indeprensa* (Brown & Alcala, 1980)		•		
113	*Eutropis multifasciata* (Kuhl, 1820)	•	•	•	•
114	*Eutropis rudis* (Boulenger, 1887)	•	•	•	•
115	*Eutropis rugifera* (Stoliczka, 1870)	•	•	•	
116	*Larutia kecil* Fukuyama, Hikida, Hossman & Nishikawa, 2019	•			
117	*Larutia puehensis* Grismer, Leong & Yaakob, 2003	•			
118	*Lipinia inexpectata* Das & Austin, 2007	•	•		•
119	*Lipinia miangensis* (Werner, 1910)				•
120	*Lipinia nitens* (Peters, 1871)	•			
121	*Lipinia vittigera* (Boulenger, 1894)	•	•	•	
122	*Lygosoma bampfyldei* Bartlett, 1895	•			
123	*Lygosoma kinabatanganensis* Grismer, Quah, Dzulkafly & Yambun, 2018		•		
124	*Sphenomorphus alfredi* (Boulenger, 1898)		•		
125	*Sphenomorphus buettikoferi* (van Lidth de Jeude, 1905)				•
126	*Sphenomorphus crassus* Inger, Tan, Lakim & Yambun, 2001		•		
127	*Sphenomorphus cyanolaemus* Inger & Hosmer, 1965	•	•	•	
128	*Sphenomorphus haasi* Inger & Hosmer, 1965	•	•		
129	*Sphenomorphus kinabaluensis* (Bartlett, 1895)		•		

No.	Current Name	Occurrence			
		Sarawak	Sabah	Brunei	Kalimantan
130	*Sphenomorphus maculicollus* Bacon, 1967	•	•		
131	*Sphenomorphus multisquamatus* Inger, 1958	•	•		
132	*Sphenomorphus murudensis* Smith, 1925	•			
133	*Sphenomorphus sabanus* Inger, 1958		•		?
134	*Sphenomorphus shelfordi* (Boulenger, 1900)	•			
135	*Sphenomorphus stellatus* (Boulenger, 1900)	•			
136	*Sphenomorphus tanahtinggi* Inger, Tan, Lakim & Yambun, 2001		•		
137	*Sphenomorphus tenuiculus* (Mocquard, 1890)		•		
138	*Subdoluseps bowringii* (Günther, 1864)	•	•	•	•
139	*Subduloseps samajaya* (Karin, Freitas, Shonleben, Grismer, Bauer & Das, 2018)	•			
140	*Tropidophorus beccarii* Peters, 1871	•	•	•	•
141	*Tropidophorus brookei* (Gray, 1845)	•	•	•	•
142	*Tropidophorus iniquus* van Lidth de Jeude, 1905				•
143	*Tropidophorus micropus* van Lidth de Jeude, 1905	•			•
144	*Tropidophorus mocquardii* Boulenger, 1894		•		
145	*Tropidophorus perplexus* Barbour, 1921	•	•		
146	*Tropidophorus sebi* Pui, Karin, Bauer & Das, 2017	•			
147	*Tytthoscincus aesculeticola* (Inger, Tan, Lakim & Yambun, 2001)		•		
148	*Tytthoscincus batupanggah* Karin, Das & Bauer, 2016	•			
149	*Tytthoscincus hallieri* (van Lidth de Jeude, 1905)	•	•	•	•
150	*Tytthoscincus leproauricularis* Karin, Das & Bauer, 2016	•			
	Family VARANIDAE Monitor Lizards				
151	*Varanus dumerilii* (Schlegel, 1839)	•	•	•	•
152	*Varanus rudicollis* Gray, 1845	•	•		•
153	*Varanus salvator* (Laurenti, 1768)	•	•	•	•
	Order SQUAMATA – SERPENTES				
	Family ACROCHORDIDAE Wart Snakes				
154	*Acrochordus granulatus* (Schneider, 1799)	•	•	•	•
155	*Acrochordus javanicus* Hornstedt, 1787	•		•	•
	Family ANOMOCHILIDAE Giant Blind Snakes				
156	*Anomochilus leonardi* Smith, 1940		•		
157	*Anomochilus monticola* Das, Lakim, Lim & Tan, 2008		•		
158	*Anomochilus weberi* (van Lidth de Jeude, 1890)				•
	Family CYLINDROPHIIDAE Pipe Snakes				
159	*Cylindrophis engkariensis* Stuebing, 1994	•			
160	*Cylindrophis lineatus* Dennys, 1880	•			
161	*Cylindrophis ruffus* (Laurenti, 1768)	•		•	•
	Family PYTHONIDAE Pythons				
162	*Malayopython reticulatus* (Schneider, 1801)	•	•	•	•
163	*Python breitensteini* Steindachner, 1881	•	•	•	•
	Family XENOPELTIDAE Sunbeam Snakes				
164	*Xenopeltis unicolor* Reinwardt, 1827	•	•	•	•
	Family COLUBRIDAE 'Typical' Snakes				
165	*Ahaetulla fasciolata* (Fischer, 1885)	•		•	•
166	*Ahaetulla prasina* (Boie, 1827)	•	•	•	•
167	*Boiga cynodon* (Boie, 1827)	•	•	•	•
168	*Boiga dendrophila* (Boie, 1827)	•	•	•	•
169	*Boiga drapiezii* (Boie, 1827)	•	•	•	
170	*Boiga jaspidea* (Duméril, Bibron & Duméril, 1854)	•	•	•	•
171	*Boiga nigriceps* (Günther, 1863)	•	•	•	
172	*Calamaria battersbyi* Inger & Marx, 1965				•
173	*Calamaria bicolor* Duméril, Bibron & Duméril, 1854	•	•		•

No.	Current Name	Occurrence			
		Sarawak	Sabah	Brunei	Kalimantan
174	*Calamaria borneensis* Bleeker, 1860	•	•		•
175	*Calamaria everetti* Boulenger, 1893	•	•		•
176	*Calamaria grabowskyi* Fischer, 1885	•	•	•	•
177	*Calamaria gracillima* (Günther, 1872)	•			
178	*Calamaria griswoldi* Loveridge, 1938		•	•	
179	*Calamaria hilleniusi* Inger & Marx, 1965	•	•		•
180	*Calamaria lateralis* Mocquard, 1890		•		
181	*Calamaria leucogaster* Bleeker, 1860	•	•		•
182	*Calamaria lovii* Boulenger, 1887	•	•	•	•
183	*Calamaria lumbricoidea* Boie, 1827	•	•	•	•
184	*Calamaria lumholtzi* Andersson, 1923				•
185	*Calamaria melanota* Jan, 1862	•			•
186	*Calamaria modesta* Duméril, Bibron & Duméril, 1854		•		
187	*Calamaria prakkei* van Lidth de Jeude, 1893		•		
188	*Calamaria rebentischi* Bleeker, 1860				•
189	*Calamaria schlegeli* Duméril, Bibron & Duméril, 1854	•	•		
190	*Calamaria schmidti* Marx & Inger, 1955		•		
191	*Calamaria suluensis* Taylor, 1922	•	•		
192	*Calamaria virgulata* Boie, 1827	•	•		•
193	*Chrysopelea paradisi* H. Boie, in F. Boie, 1827	•	•	•	•
194	*Chrysopelea pelias* (Linnaeus, 1758)	•	•	•	
195	*Coelognathus erythrurus* (Duméril, Bibron & Duméril, 1854)		•		
196	*Coelognathus flavolineatus* (Schlegel, 1837)	•	•	•	•
197	*Coelognathus radiatus* (Boie, 1827)			•	•
198	*Dendrelaphis caudolineatus* (Gray, 1834)	•	•	•	•
199	*Dendrelaphis formosus* (Boie, 1827)	•	•		•
200	*Dendrelaphis haasi* van Rooijen & Vogel, 2008			•	•
201	*Dendrelaphis kopsteini* Vogel & van Rooijen, 2007	•		•	
202	*Dendrelaphis pictus* (Gmelin, 1789)	•	•	•	•
203	*Dendrelaphis striatus* (Cohn, 1905)				•
204	*Dryophiops rubescens* (Gray, 1835)	•	•	•	
205	*Gongylosoma baliodeirum* Boie, 1827	•	•	•	•
206	*Gongylosoma longicauda* (Peters, 1871)	•	•		
207	*Gonyosoma margaritatum* Peters, 1871	•	•	•	
208	*Gonyosoma oxycephalum* (Boie, 1827)	•	•	•	•
209	*Liopeltis tricolor* (Schlegel, 1827)	•	•	•	
210	*Lycodon albofuscus* (Duméril, Bibron & Duméril, 1854)	•	•		
211	**Lycodon capucinus* (Boie in F. Boie, 1827)	•	•		
212	*Lycodon effraenis* Cantor, 1847	•		•	•
213	*Lycodon sealei* Leviton, 1955	•	•	•	•
214	*Lycodon subannulatus* (Duméril, Bibron & Duméril, 1854)	•	•	•	
215	*Lycodon tristrigatus* (Günther, 1858)	•	•	•	
216	*Oligodon annulifer* (Boulenger, 1893)		•	•	
217	*Oligodon cinereus* (Günther, 1864)		•		
218	*Oligodon everetti* Boulenger, 1893		•		•
219	*Oligodon meyerinkii* (Steindachner, 1891)				
220	*Oligodon octolineatus* (Schneider, 1801)	•	•	•	•
221	*Oligodon purpurascens* (Schlegel, 1837)	•	•	•	•
222	*Oligodon signatus* (Günther, 1864)	•	•		
223	*Oligodon vertebralis* Günther, 1865	•	•		•
224	*Oreocalamus hanitschi* Boulenger, 1899	•	•		
225	*Orthriophis taeniurus* (Cope, 1961)	•	•	•	•
226	*Pseudorabdion albonuchalis* (Günther, 1896)	•	•	•	
227	*Pseudorabdion collaris* (Mocquard, 1892)	•	•		

No.	Current Name	Occurrence			
		Sarawak	Sabah	Brunei	Kalimantan
228	*Pseudorabdion longiceps* (Cantor, 1847)	•		•	•
229	*Pseudorabdion saravacense* (Shelford, 1901)	•			
230	*Ptyas carinata* (Günther, 1858)	•	•	•	•
231	*Ptyas fusca* (Günther, 1858)	•	•	•	•
232	*Ptyas korros* (Schlegel, 1837)		•		•
233	*Sibynophis geminatus* (Boie, 1826)	•			
234	*Sibynophis melanocephalus* (Gray, 1834)	•	•	•	
235	*Stegonotus borneensis* Inger, 1967	•			
236	*Stegonotus caligocephalus* Kaiser, Lapin, O'Shea & Kaiser, 2020		•		
237	*Xenelaphis ellipsifer* Boulenger, 1900	•	•		
238	*Xenelaphis hexagonotus* (Cantor, 1847)	•	•		•
	Family NATRICIDAE Water Snakes				
239	*Elapoidis fusca* Boie, 1826		•		
240	*Hebius arquus* (David & Vogel, 2010)				
241	*Hebius flavifrons* (Boulenger, 1887)	•	•	•	
242	*Hebius frenatus* (Dunn, 1923)	•			
243	*Hebius petersii* (Boulenger, 1893)	•	•		
244	*Hebius saravacensis* (Günther, 1872)	•	•		•
245	*Hydrablabes periops* (Günther, 1872)	•	•	•	•
246	*Hydrablabes praefrontalis* (Mocquard, 1890)		•		
247	*Opisthotropis typica* (Mocquard, 1890)	•	•	•	
248	*Psammodynastes pictus* Günther, 1858	•	•	•	•
249	*Psammodynastes pulverulentus* (Boie, 1827)	•	•	•	•
250	*Rhabdophis chrysargos* (Schlegel, 1837)	•	•	•	•
251	*Rhabdophis conspicillatus* (Günther, 1872)	•	•		•
252	*Rhabdophis flaviceps* (Duméril, Bibron & Duméril, 1854)	•			•
253	*Rhabdophis murudensis* (Smith, 1925)	•	•		
254	*Rhabdophis rhodomelas* (Boie, 1827)	•	•		
255	*Xenochrophis maculatus* (Edeling, 1864)	•	•	•	•
256	*Xenochrophis trianguligerus* (Boie, 1827)	•	•	•	•
	Family PSEUDOXENODONTIDAE False Cobras				
257	*Pseudoxenodon baramensis* (Smith, 1921)	•			
	Family ELAPIDAE Cobras, Kraits, Coral Snakes & Sea Snakes				
	Subfamily ELAPINAE Cobras, Kraits & Coral Snakes				
258	*Bungarus fasciatus* (Schneider, 1801)	•	•	•	•
259	*Bungarus flaviceps* Reinhardt, 1843	•	•	•	
260	*Calliophis bivirgatus* (Boie, 1827)	•	•	•	•
261	*Calliophis intestinalis* (Laurenti, 1768)	•	•	•	•
262	*Calliophis nigrotaeniatus* (Peters, 1863)	•			
263	**Naja kaouthia* Lesson, 1831	•			
264	*Naja sumatrana* Müller, 1890	•	•	•	•
265	*Ophiophagus bungarus* (Schlegel, 1837)	•	•	•	•
	Subfamily HYDROPHIINAE Sea Snakes				
266	*Aipysurus eydouxii* (Gray, 1849)		•		
267	*Hydrophis annandalei* (Laidlaw, 1901)			•	
268	*Hydrophis anomalus* (Schmidt, 1852)	•	•		
269	*Hydrophis atriceps* Günther, 1864				
270	*Hydrophis brookii* Günther, 1872	•	•		
271	*Hydrophis caerulescens* (Shaw, 1802)	•	•		
272	*Hydrophis curtus* (Shaw, 1802)	•	•	•	
273	*Hydrophis cyanocinctus* Daudin, 1803	•	•	•	•
274	*Hydrophis fasciatus* (Schneider, 1799)	•	•		
275	*Hydrophis gracilis* (Shaw, 1802)		•		
276	*Hydrophis jerdonii* (Gray, 1849)		•	•	

No.	Current Name	Occurrence			
		Sarawak	Sabah	Brunei	Kalimantan
277	*Hydrophis klossi* Boulenger, 1812		•		
278	*Hydrophis melanosoma* Günther, 1864		•		
279	*Hydrophis ornatus* (Gray, 1842)		•		
280	*Hydrophis platurus* (Linnaeus, 1766)	•	•	•	
281	*Hydrophis schistosus* Daudin, 1803	•	•		
282	*Hydrophis sibauensis* Rasmussen, Auliya & Böhme, 2001				•
283	*Hydrophis spiralis* (Shaw, 1802)	•	•		
284	*Hydrophis torquatus* Günther, 1864				
285	*Hydrophis viperinus* (Schmidt, 1852)	•	•		
	Subfamily LATICAUDINAE Sea Kraits				
286	*Laticauda colubrina* (Schneider, 1799)	•	•	•	•
287	*Laticauda laticaudata* (Linnaeus, 1758)		•		
	Family HOMALOPSIDAE Puff-faced Water Snakes				
288	*Cerberus schneiderii* (Schlegel, 1837)	•	•	•	•
289	*Enhydris enhydris* (Schneider, 1799)	•		•	•
290	*Fordonia leucobalia* (Schlegel, 1837)	•			•
291	*Gerarda prevostiana* (Eydoux & Gervais, 1822)	•			
292	*Homalophis doriae* Peters, 1871	•	•	•	•
293	*Homalophis gyii* (Murphy, Voris & Auliya, 2005)				•
294	*Homalopsis buccata* (Linnaeus, 1758)	•	•		•
295	*Hypsiscopus plumbea* (Boie, 1827)	•	•	•	•
296	*Miralia alternans* (Reuss, 1833)	•			
297	*Phytolopsis punctata* Gray, 1849			•	•
	Family PAREIDAE Slug-eating Snakes				
298	*Aplopeltura boa* (Boie, 1828)	•	•	•	•
299	*Asthenodipsas borneensis* Quah, Grismer, Lim, Anuar & Chan, 2020	•	•	•	•
300	*Asthenodipsas ingeri* Quah, Lim & Grismer, 2021		•		
301	*Asthenodipsas jamilinaisi* Quah, Grismer, Lim, Anuar & Imbun, 2019	•	•		
302	*Asthenodipsas laevis* (Boie, 1827)	•	•		•
303	*Asthenodipsas stuebingi* Quah, Grismer, Lim, Anuar & Imbun, 2019		•		
304	*Asthenodipsas vertebralis* (Boulenger, 1900)	•			
305	*Pareas nuchalis* (Boulenger, 1900)	•	•		•
	Family VIPERIDAE Vipers & Pit Vipers				
306	*Calloselasma rhodostoma* (Kuhl, 1824)				•
307	*Craspedocephalus borneensis* (Peters, 1872)	•	•	•	•
308	*Garthius chaseni* (Smith, 1931)	•	•		
309	*Trimeresurus malcolmi* Loveridge, 1938		•		
310	*Trimeresurus sabahi* Regenass & Kramer, 1981	•	•	•	
311	*Trimeresurus sumatranus* (Raffles, 1822)	•	•	•	•
312	*Tropidolaemus subannulatus* (Gray, 1842)	•	•	•	•
	Family XENODERMATIDAE Strange-skinned Snakes				
313	*Paraxenodermus borneensis* (Boulenger, 1899)	•	•		
314	*Xenodermus javanicus* Reinhardt, 1836	•	•		•
	Family TYPHLOPIDAE Blind Snakes				
315	*Argyrophis muelleri* (Schlegel, 1839)	?	?	?	?
316	*Indotyphlops braminus* (Daudin, 1803)	•	•	•	•
317	*Malayotyphlops koekkoeki* (Brongersma, 1934)	•	?		•
318	*Ramphotyphlops lineatus* (Schlegel, 1839)	•	•		•
319	*Ramphotyphlops lorenzi* (Werner, 1909)		•		
320	*Ramphotyphlops olivaceus* (Gray, 1845)	•			
	Family XENOPHIDIIDAE Spine-jawed Snakes				
321	*Xenophidion acanthognathus* Günther & Manthey, 1995	•	•		

Further Reading

Auliya, M. 2006 *Taxonomy, Life History and Conservation of Giant Reptiles in West Kalimantan* (Indonesian Borneo). Natur und Tier Verlag GmbH.

Das, I. 2004 *Lizards of Borneo*. Natural History Publications (Borneo) Sdn Bhd.

Das, I. 2010. *A Field Guide to the Reptiles of South-east Asia*. New Holland Publishers (UK) Ltd.

Das, I. 2021. *A Naturalist's Guide to the Snakes of South-east Asia* (3rd edition). John Beaufoy Publishing.

David, P. & Vogel, G. 1996. *The Snakes of Sumatra. An Annotated Checklist and Key with Natural History Notes*. Edition Chimaira.

de Lang, R. & Vogel, G. 2005. *The Snakes of Sulawesi. A Field Guide to the Land Snakes of Sulawesi with Identification Keys*. Edition Chimaira.

de Rooij, N. 1915. *The Reptiles of the Indo-Australian Archipelago. Vol. I. Lacertilia, Chelonia, Emydosauria*. E. J. Brill,

de Rooij, N. (1917). *The Reptiles of the Indo-Australian Archipelago. II. Ophidia*. E. J. Brill.

Gumprecht, A., Tillack, F., Orlov, N. L., Captain, A. & Ryabov, S. 2004. *Asian Pitvipers*. GeitjeBooks.

Ismail, A. K., Teo, E. W., Das, I., Vasaruchapong, T. & Weinstein, S. A. 2022. *Land Snakes of Medical Significance in Malaysia* (3rd edition). Malaysian Society for Toxinology/Ministry of Natural Resources and Environment.

Lim, B.-L. & Das, I. 1999. *Turtles of Borneo and Peninsular Malaysia*. Natural History Publications (Borneo), Sdn. Bhd.

McKay, L. 2006. *A Field Guide to the Reptiles and Amphibians of Bali*. Krieger Publishing.

Malkmus, R., Manthey, U., Vogel, G., Hoffmann, P. & Kosuch, J. 2002. *Amphibians and Reptiles of Mount Kinabalu (North Borneo)*. Koeltz Scientific Books.

Manthey, U. & Grossmann, W. 1997. *Amphibien und Reptilien Südostasiens*. Natur und Tier GmbH.

O'Shea, M. 2007. *Boas and Pythons of the World*. New Holland Publishers (UK) Ltd.

Schulz, K.-D. 1996. *A Monograph of the Colubrid Snakes of the Genus* Elaphe *Fitzinger*. Koeltz Scientific Books.

Somaweera, R. 2017. *A Naturalist's Guide to the Reptiles and Amphibians of Bali*. John Beaufoy Publishing.

Stuebing, R. B., Inger, R. F. & Lardner, B. 2014. *A Field Guide to the Snakes of Borneo* (2nd edition). Natural History Publications (Borneo) Sdn Bhd.

Tweedie, M. W. F. 1983. *The Snakes of Malaya*. (3rd edition). Singapore National Printers.

Vogel, G. 2006. *Venomous Snakes of Asia/ Giftschlangen Asiens. Terralog 15*. Edition Chimaira.

World Health Organization. 2016. *WHO Guidelines for the Production, Control and Regulation of Snake Antivenom Immunoglobulins*. WHO Press.

Acknowledgements

I thank John Beaufoy and Rosemary Wilkinson at John Beaufoy Publishing for pushing me to complete the volume.

For information/translations/companionship during field trips, I am grateful to Kraig Adler, Natalia Ananjeva, Christopher Austin, Aaron M. Bauer, Rafe M. Brown, Chan Chew Lun, Joseph K. Charles, Patrick David, Maximilian Dehling, Uwe Fritz, Ibuki Fukuyama/Magnolia Press, David Gower, Allen Greer, Wolfgang Grossmann, Andreas Gumprecht, Alexander Haas, Jacob Hallermann, Stefan Hertwig, Tsutomu Hikida, Marinus Hoogmoed, Awang Khairul Ikhwan, the late Robert Inger, Djoko Iskandar, John Iverson, Karen Jensen, David Jones, Awang Ikhwan Khairul, the late Vladimir Kharin, Arnold Kluge, Gunther Köhler, Thanisha Kumar, Veronica Leah, Alan Leviton, Harvey Lillywhite, the late Lim Boo Liat, Kelvin Kok Peng Lim, Colin McCarthy, Jimmy McGuire, Ulrich Manthey, Veronica Martin, G. Mumpuni, John Murphy, the late Jonathan Murray, the late Jarujin Nabhitabhata, Izneil Nashriq, Ngo Van Tri, Peter Kee Lin Ng, Samhan bin Nyawa, Nikolai Orlov, Mark O'Shea, Hidetoshi Ota, the late Peter C. H. Pritchard, Yong Min Pui, Arne Rasmussen, Anders Rhodin, Herbert Rösler, Klaus-Dieter Schulz, the late Joseph Slowinski, Robert B. Stuebing, Tan Heok Hui, Frank Tillack, Peter Paul van Dijk, Gernot Vogel, Harold Voris, Van Wallach, Wolfgang Wüster, Norsham Yaakob, the late Er-Mi Zhao, Thomas Ziegler and George Zug.

I am grateful to Mohd-Azlan Jayasilan, Director, Institute of Biodiversity and Environmental Conservation, Universiti Malaysia Sarawak, and my other colleagues at the Institute. Finally, I thank Genevieve V. A. Gee for looking after the household, while permitting me to write this book.

Index

Common Names

Scientific Names